FOREST POLITICS

David Humphreys is Research Fellow in Global Environmental Change at the Open University. He travelled extensively, overland and at sea, before commencing his studies in the UK. He holds degrees from the University of Kent at Canterbury and City University, London. He is the author of several book chapters, journal articles and occasional papers on global forest politics.

FOR MY MOTHER AND FATHER
WITH LOVE AND THANKS

So you cut all the tall trees down
You poisoned the sky and the sea
You've taken what's good from the ground
But you left precious little for me

(Midnight Oil, 'River Runs Red',
from the album *Blue Sky Mining*, 1990)

FOREST POLITICS

The Evolution of International Cooperation

David Humphreys

Earthscan Publications Ltd, London

First published in 1996 by
Earthscan Publications Limited
120 Pentonville Road, London N1 9JN
email: earthinfo@earthscan.co.uk

A catalogue record for this book is available from the British Library

ISBN: 1 85383 378 9 (Paperback)
ISBN: 1 85383 379 7 (Hardback)

Copy-edited and typeset by Selro Publishing Services, Oxford
Printed and bound by Biddles Ltd, Guildford and King's Lynn

Front cover photo © Nick Robinson / Panos Pictures

Earthscan Publications Limited is an editorially independent subsidiary of
Kogan Page Limited and publishes in association with the International
Institute for Environment and Development and WWF–UK.

Contents

List of Figures, Tables and Boxes   x
Acknowledgements   xiii
Foreword by Stanley Johnson   xv
Introduction   xvii

1 Deforestation as a Global Political Issue   1
■ The political complexity of deforestation   1
■ The causes of deforestation   2
■ The effects of deforestation   15
■ The emergence of forest conservation as an international
   political issue   18
■ The forest conservation problematic   21

2 The Tropical Forestry Action Programme   31
■ The origins of the TFAP   31
■ The organizational structure of the TFAP at the international
   level   34
■ The formulation and evolution of a National Forestry Action
   Programme   36
■ The 1990 legitimacy crisis   42
■ The TFAP restructuring process   46

3 The International Tropical Timber Organization   55
■ International Tropical Timber Agreement, 1983   55
■ The institutional structure of the ITTO   60
■ The financing of the ITTO   62
■ The history of the ITTO   63
■ The international relations of the ITTO   75

4 The Forest Negotiations of the UNCED Process   83
■ Proposals for a global forests instrument   83
■ The draft global forests convention of the FAO   85

■ UNCED: an introduction   88
■ Explaining North–South disagreement during the UNCED
    forest negotiations   89
■ PrepCom 1, Nairobi, 6–31 August 1990   90
■ PrepCom 2, Geneva, 18 March–5 April 1991   91
■ PrepCom 3, Geneva, 12 August–4 September 1991   93
■ Environmental diplomacy between PrepComs 3 and 4:
    Abidjan and Caracas   99
■ PrepCom 4, New York, 2 March–3 April 1992   100
■ Second Ministerial Meeting of Developing Countries, Kuala
    Lumpur, April 1992   101
■ UNCED, Rio de Janeiro, 3–14 June 1992   101
■ Conclusions   102

5 *The Negotiation of the International Tropical
    Timber Agreement, 1994*   **105**
■ The preparatory process   105
■ First part of conference, Geneva, 13–16 April 1993   111
■ Environmental diplomacy between the first and second
    parts of the conference   114
■ Second part of conference, Geneva, 21–25 June 1993   117
■ Environmental diplomacy between the second and third
    parts of the conference   117
■ Third part of conference, Geneva, 4–15 October 1993   119
■ Fourth part of conference, Geneva, 10–26 January 1994   120
■ The North's demand for a global forests convention   122
■ An analysis of the International Tropical Timber
    Agreement, 1994   125
■ Concluding remarks   132

6 *The Global Politics of Forest Conservation Since the UNCED*   **135**
■ Confidence-building initiatives   136
■ The Criteria and Indicators processes   138
■ The third session of the CSD: creation of the Intergovernmental
    Panel on Forests   144
■ The World Commission on Forests and Sustainable
    Development   146
■ Prospects for the future   147

7 *Conclusions*   **155**
■ The continuing demand from the North for a global forests
    convention   155
■ Explaining North–South disagreement in global forest politics   157

Contents   ix

■   The forest conservation problematic   166
■   Conclusions   170

Notes   173
Annex A: Possible main elements of an instrument (convention,
    agreement, protocol, charter) for the conservation and develop-
    ment of the world's forests   203
Annex B: Non-legally binding authoritative statement of principles
    for a global consensus on the management, conservation and
    sustainable development of all types of forests   215
Annex C: International Tropical Timber Agreement, 1994 221
Annex D: Draft text for a Convention for the Conservation and
    Wise Use of Forests   246
Recommended further reading   277
Acronyms   281
Index   285

Figures, Tables and Boxes

FIGURES

2.1 The organization of the TFAP at the international level
4.1 Organization chart: forest groups in the UNCED process

TABLES

1.1 Tabulation of deforestation figures for selected tropical forest countries, 1980–93
2.1 National Lead Institutions for NFAPs as at March 1991
2.2 Steering committees in NFAP host countries as at March 1991
2.3 Bilateral/multilateral donor participation in NFAPs as at March 1992
2.4 Status of NFAPs as at November 1994
2.5 Chronology of key events and meetings during the TFAP restructuring process
3.1 Chronology: May 1977–March 1994, the International Tropical Timber Agreement, 1983 and ITTC and Permanent Committee sessions held during the lifespan of this Agreement
3.2 Project work of the International Tropical Timber Organization as at 31 December 1992
4.1 Nine proposals for a global forests instrument
4.2 The UNCED forests debate and related issues: North and South negotiating positions
5.1 Meetings, conferences and seminars of significance in the negotiation of a successor agreement to the International Tropical Timber Agreement, 1983
5.2 Positions of delegations, environmental NGOs and timber traders at the preparatory committee meetings for the United Nations Conference for a Successor Agreement to the International Tropical Timber Agreement, 1983

6.1 The criteria for sustainable forest management of the Helsinki, Montreal and Amazonian processes

BOXES

2.1 The five action programmes of the Tropical Forestry Action Plan
3.1 Producing and consuming countries to have acceded to the International Tropical Timber Agreement, 1983 by 31 March 1985 and their votes according to the annexes of the Agreement
3.2 Countries to have acceded to the International Tropical Timber Agreement, 1983 by 31 March 1994
5.1 Non-ITTO members that took part in the negotiation of the International Tropical Timber Agreement, 1994
5.2 The five new articles in the International Tropical Timber Agreement, 1994
5.3 The ten substantially-reworded articles in the International Tropical Timber Agreement, 1994
5.4 Principal international forest-related instruments
6.1 Post-UNCED international forest political processes
6.2 Agenda of the Intergovernmental Working Group on Forests
6.3 Programme of work of the Intergovernmental Panel on Forests
6.4 The Forest Stewardship Council's nine principles for forest management

Acknowledgements

One of the greatest pleasures of researching this book has been that it has introduced me to a wide range of interesting people. Those who work on forest conservation are, almost without exception, sincere and helpful people and this book would not have been possible in its present form without the help of the following. Simon Counsell, Chris Elliott, Jean-Paul Jeanrenaud, Tony Juniper, Elaine Morrison and Francis Sullivan were generous with their time and were never too busy to discuss the latest political developments. Stanley Johnson was an entertaining host who shared his many interesting insights. Bernardo Zentilli provided a fascinating insider's view on the UNCED forests negotiations. Thanks are also due to Carlos Marx Ribeiro Carneieo, Jane Clark, Marcus Colchester, the Earl of Cranbrook, Rod Harbinson, Nicholas Hildyard, Terene Hpay, Jean-Paul Lanly, Jag Maini, George Marshall, Martin Mathers, Catherine McCluskey, Jill McIntosh, John Moncrieff, the late Arthur Morrell, Sunita Narain, Rupert Oliver, Helena Paul, Gareth Porter, Kilaparti Ramakrishna, Michael Rands, Bart Romijn (who kindly supplied the software for Annex D), H W O Röbbel, Caroline Sargent, Ian Symons, Koy Thomson, Fiona Watson and Farhana Yamin. All gave their help freely and willingly, in many cases at considerable inconvenience to themselves.

The quote at the start of this book is from a song by Midnight Oil, an Australian rock band whose perceptive lyrics and environmental charity concerts have inspired many; my thanks to Georgia Kambouridis from Melbourne who sent me the tape during the early stages of my research. The research was funded for three years by the Science and Engineering Research Council and the Economic and Social Research Council (Quota Award 9130262X). I am pleased to acknowledge the many contributions of friends in the Transgovernmental Relations Research Group, Department of Systems Science, City University, London: Mandy Bentham and Lewis Clifton generously shared their relevant research material; both supplied useful articles and made constructive comments to early drafts, as did Mira Filipovic, Pete Hough and Chris Parks. Peter Willetts shared his personal archives, made countless suggestions and read the first draft of the manuscript; his encouragement, challenging criticisms and keen eye for detail were instrumental factors in the completion of this study. Chris Brown, Stephen Chan, Kelley Lee, Ian Rowlands and Keith Webb showed an interest in this work at an early stage and their encouragement and advice were very much appreciated. Beate Münstermann and Susanne Staab kindly invested time and effort in obtaining and translating German government policy documents. Jean Claude Clément and Sara Oldfield commented upon earlier drafts of the chapters on the Tropical Forestry Action Programme and International Tropical Timber Organization respectively. Stephen Bass, Ewart Carson, James Mayers and John Vogler read the manuscript (at various stages of completion) and I am grateful to them for their many helpful suggestions. Thanks are due to my colleagues at the Faculty of Social Sciences at the Open University and in the Global Environmental Change Open University (GECOU) programme; they were always supportive during the completion of the manuscript. The draft text for a convention for the conservation and wise use of forests of the Global Legislators Organisation for a Balanced Environment (GLOBE), which forms Annex D of this book, is reproduced with the kind permission of GLOBE–European Union and AIDEnvironment. Thanks also to Simon Ashley, Carl Backland, Graham Clewer, Sally Eaton, Howard Gibbins, John Hunt, Petr Jehlicka, Syed Kamall, Val Kirby, Dave Murley, Steve Rees, Judy Richardson, Pia Schmitzer, Francis Stickland, Paul Waters and Paul Worthy. At Earthscan, Jo O'Driscoll, Selina Cohen and Rowan Davies competently and cheerfully guided the project through the final months before publication. Warmest thanks are due to Colette de Jaeger who continues to be inspirational in many ways. Any remaining errors are my own, although they pale into insignificance when compared to the destructive policies of many governments, business concerns and international agencies in the world's forested regions.

Foreword
by Stanley Johnson

This is an important and timely book. The emergence of forest conservation as an international political issue was never more evident than in the run-up to the United Nations Conference on Environment and Development (UNCED), popularly known as the Earth Summit, held in Rio de Janeiro, Brazil, in June 1992. The preparations for UNCED revealed in the starkest terms the clash between those, on the one hand, who saw the fate and future of the forests as being the common responsibility of mankind — and requiring therefore an international, legally-binding regime to ensure their conservation and wise use — and, on the other hand, those who maintained that every aspect of forests fell fairly and squarely within the purview of national sovereignty.

At Rio, the proponents of an international regime were shot down in flames. Instead of the legally-binding convention vigorously proposed by the United Nations Food and Agriculture Organization (FAO) and supported by the G7 nations and the European Union, UNCED adopted a watered-down, wholly voluntary set of guidelines known as the Statement of Forest Principles, relegating to the future the search for more constraining measures with no clear commitments as to the eventual outcome.

David Humphreys covers Rio's forest fiasco in detail. He sets out the lessons to be learned from that failure. But his searchlight is not confined to the Earth Summit. The work of the Tropical Forestry Action Programme (TFAP) and of the International Tropical Timber Organization is also scrutinized in detail and it does not make for comfortable reading. Though much effort has been put into restructuring the TFAP and renegotiating the International Tropical Timber Agreement (ITTA), those instruments by themselves, even if effectively implemented — and implementation is always an issue — are unlikely to be a sufficient international response to the worldwide problem of forest destruction and degradation.

That is why the discussions on forests now being undertaken within the framework of the UN's Commission on Sustainable Development (CSD), and its Intergovernmental Panel on Forests, are so crucial. In a sense, the CSD itself is on trial here. If it fails in this area where it has been given a substantive responsibility by governments (even though there are other more obviously competent international bodies working in the field), many may question whether the institutional arrangements which resulted from Rio are after all the most appropriate.

Stanley Johnson
8 June 1996

Introduction

Forest Politics had its early origins in my twenties when I travelled through Africa and Asia, journeys that enabled me to witness at first hand some of the devastation that humanity has wreaked on nature in the name of 'progress'. When, a few years later, I enrolled for a degree course in Politics and International Relations at the University of Kent at Canterbury these impressions were still vivid in my mind. Shortly before graduation the academic staff at Kent informed me of a PhD place at City University, London that would enable me to combine the formal training I had gathered as an undergraduate with my concern for the environment. A week later an interview with Peter Willetts, who would become my PhD supervisor, was concluded in a pub in Clerkenwell, and shortly afterwards I received the letter stating that the place was mine. The three years that followed were challenging, stimulating, hard-working and ultimately rewarding. This book draws in large part from the doctoral thesis that I submitted to City University.

The book sets out to provide a critical analysis of global forest politics in the period from 1983, when the negotiations were concluded for the first International Tropical Timber Agreement, to the mid-1990s, when the third session of the UN's Commission on Sustainable Development created the Intergovernmental Panel on Forests. The study is based upon research carried out between 1991 and 1996, principally using primary source documents. This information has been supplemented with interviews carried out with policy-makers, environmental campaigners and international civil servants. The book has been written to appeal to undergraduate students pursuing courses on the environment or on international relations. It should also attract policy-makers, campaigners and concerned members of the public seeking a text that deals comprehensively with the history of the subject.

Chapter 1 traces the emergence of deforestation as an issue on the international political agenda. It begins with an assessment of the causes of deforestation and its environmental and social consequences, and considers the problems facing the international community in dealing with the array of issues involved. Deforestation, it is argued, emerged as a global issue because global élites perceived it to be of global concern, with global effects that required global policy responses. Yet there has never been a coherent widely-accepted formulation of the problem and at present, with rates of deforestation in the tropics increasing, international policies are demonstrably failing. The chapter concludes with the author's conceptualization of the forest conservation problematic which, it is argued, needs to be addressed if international forest conservation initiatives are to stand any chance of succeeding. There are three dimensions to the problematic, namely: the causal dimension (the need to arrest the transnational causes of deforestation), the institutional dimension (the need to design qualitatively new international and national institutions) and the proprietorial dimension (the need to move away from the dominant principle that forests are a sovereign resource of the state towards a new commonly-accepted notion of forest ownership). The case is made that these three dimensions are inter-linked, and that they require a common solution.

Chapters 2 to 6 are case studies dealing with the main forest and forest-related international processes and negotiations to take place since the early 1980s. Chapter 2 provides a detailed history of the Tropical Forestry Action Programme (TFAP), analyses the problems this initiative encountered in the early-1990s and outlines the efforts to 'restructure' the TFAP, including the failed attempt to create an independent consultative group. Chapter 3 provides a history of the International Tropical Timber Organization (ITTO), the first international commodity organization to have a conservation mandate. The chapter analyses the ITTO's role as a standard-setting organization, its relationships with other intergovernmental institutions and the main debates to have arisen during the ITTO's history, in particular the protracted and as yet unresolved debate on incentives for 'sustainable forest management'.

The next two chapters are case studies of intergovernmental negotiations. Chapter 4 takes the reader through the forest negotiations that preceded the 1992 United Nations Conference on Environment and Development (UNCED) when the aspirations of governments from the developed North for a global forests convention were thwarted by governments from the developing South; the reasons for this disagreement are examined. Chapter 5 deals with the first forest-related negotiations to take place in the post-UNCED era, namely the negotiations for a successor agreement to the International Tropical Timber Agreement, 1983. Like the UNCED forest negotiations, this process was dominated by deep North–South divisions, with the South arguing unsuccessfully that the scope of the successor agreement should be broadened to include non-tropical timbers.

Chapter 6 examines the main international processes to have arisen in the post-UNCED era and analyses the significance of and relationships between them. In particular, consideration is given to three intergovernmental processes designed to arrive at criteria and indicators for 'sustainable forest management'. It is argued that a more cooperative spirit between North and South has replaced the intense disagreements that prevailed during the UNCED process and its immediate aftermath. However, and as Chapter 7 details, serious North–South divisions remain, and genuine cooperation is unlikely unless a trade-off can be reached between the environmental concerns of the North and the developmental concerns of the South. This chapter also argues that a decade of international cooperation on forest conservation has failed to deal meaningfully with the core problems; none of the three dimensions of the forest conservation problematic have been properly addressed by the international community. The result is that the prospects for the future remain as grim as the destruction of the past. The book finishes with a plea for local communities to have an equitable say in forest policy-making.

Chapter 1

Deforestation as a Global Political Issue

THE POLITICAL COMPLEXITY OF DEFORESTATION

Deforestation as an issue has assumed a global prominence in the past decade. In Africa, the Amazon and Asia tropical forests have been disappearing at an alarming rate. The problem is by no means confined to the tropics, and deforestation of temperate forests has accelerated in recent years in, to name just two areas, British Colombia and Siberia. This chapter will examine the problem of deforestation as a political issue. Why is deforestation important? Who is it important to? Why does it occur? And who are the political actors that contest the issue? Deforestation can be defined as the conversion of forests to other land uses. It is necessary to distinguish between deforestation and forest degradation. The United Nations has defined *deforestation* as occurring when 'a forest is cleared to give way to another use of the land'. *Forest degradation* refers to changes in forest quality and occurs 'when the species diversity and the biomass are significantly reduced through, for instance, unsustainable forms of forest utilization'.[1] Myers refers to deforestation as 'the complete destruction of forest cover ... [so] that not a tree remains and the land is given over to non-forest purposes'.[2] Tropical deforestation became a source of international concern in the early 1980s, whereas deforestation of temperate and boreal forests received high-level international attention only in the early 1990s. Indeed the destruction of non-tropical forests has only recently assumed the same prominence as tropical deforestation on the international agenda.

Deforestation is an intrinsically complex global issue. This complexity arises from two broad factors. First, deforestation introduces a wide range of political actors, from government and international civil society, with a direct or an indirect stake in forest use. These actors include: government departments; private profit-making companies, including transnational corporations (TNCs); UN programmes, such as the United Nations Environment Programme (UNEP); intergovernmental organizations (IGOs), including UN specialized agencies such as the Food and Agriculture Organization (FAO) and IGOs operating outside of the UN system, such as the parties to the Amazonian Cooperation Treaty; and non-governmental organizations (NGOs), such as conservation groups and research fora, operating at the international, national and local levels. Local government structures, community institutions and traditional authorities may also be key political actors at the local level. In addition, various hybrid and *ad hoc* fora with an interest or a stake in the forests may emerge from time to time. Hence a diverse range of frequently competing actors are involved in forest politics, although the specific actors do, of course, vary from area to area.

Second, the complexity of deforestation arises from its linkage to other issues. As we shall see below, deforestation is both an outcome and a causal factor. As an outcome, deforestation is the end product of an array of political, economic and social dynamics arising at the international and national levels. These dynamics seldom act in isolation; rather they interact in dense and complex ways. As a causal factor, deforestation contributes to other environmental problems, such as global warming, soil erosion and the destruction of biological diversity (biodiversity).

The sections that follow will examine the causes and effects of deforestation, the reasons for the emergence of deforestation as an international political issue, and the defining features of deforestation as a global political problem.

THE CAUSES OF DEFORESTATION

In order to consider the causes of deforestation it is necessary first to distinguish between cause and agency. Agents of deforestation always operate at the local level; to cut down a tree or to burn an area of forest is a highly localized act, often performed by local people. Yet these people are frequently agents acting on behalf of wider economic interests, often outside the country in question, so that the causes of deforestation frequently have their loci in complex global economic processes. To arrest deforestation it is therefore necessary to call into question the functioning of the global economic system.

The causes outlined below are particularly relevant to tropical forests, although they may also apply to other forest types. Ten categories of causal process will be explored. The degree to which these processes operate varies from region to region, country to country, and between forest types. The processes are distinguished here for analytical purposes, and should not be seen as separate in the real world. In practice they interact and sustain each other. It should also be noted that the acceleration of forest removal in the tropics over the last half century has, in almost all instances, been deliberate and viewed by many political and economic élites as a 'progressive' way to advance economic development by, for example, increasing food security or earning hard currency through the exports of cash crops and timber.

Internationally-sponsored development

National development and infrastructure projects have often been funded by international capital channelled through multilateral and bilateral agencies. Many of these projects, such as the Grande Carajás programme in the Brazilian Amazon, have attracted criticism for environmental destruction, including deforestation.[3] The use of international capital to build roads has been widely condemned by environmental NGOs. The most infamous example here is the case of the BR–364 (Polonoroeste) highway through the Brazilian Amazon to Peru, funded in part by the World Bank and the Inter-American Development Bank.[4] Plans to complete the construction of the Pan-American Highway will lead to forest clearance in Colombia and Panama. International capital has also played a role in the construction of dams for hydroelectric power in, for example, the Amazon and in Malaysia.[5]

By 1987, the inhabitants of the Amazon, including indigenous forest peoples and rubber tappers, were organizing themselves to attempt to stop the World Bank and other multilateral agencies from funding infrastructure causing deforestation and the displacement of Amazonian inhabitants. The World Bank, responding to charges that its projects have degraded the environment, created an Environment Department in 1987 and claims that it now integrates environmental criteria in its decision-making procedures.[6] In 1991 the World Bank published a Forest Sector Policy Paper which included the commitment that it 'will not under any circumstances finance commercial logging in primary tropical moist forests'.[7] However, in 1992 Friends of the Earth found that a World Bank project in Gabon — the first major Bank intervention in central Africa after the publication of this policy — would lead to logging of primary forests. They recommended a major project review.[8] The role of international capital, especially when channelled through the World Bank,[9] remains a fundamental concern for environmental campaigners.

External debt and structural adjustment programmes

An oft-cited example of a cause of deforestation that has its locus in the structure of global economic relations is external debt, much of it accrued from loans to finance development projects. The debt crisis erupted in August 1982 when Mexico defaulted. Some governments from the South, including Brazil, have accumulated large debts, but a report by the United Nations Centre on Transnational Corporations concluded that competition among the transnational banks to secure loans led to overlending in the developing countries and was a major contributory factor in the evolution of the debt crisis.[10]

The World Commission on Environment and Development noted a link between deforestation and debt, and concluded that the need for foreign exchange encourages 'many developing countries to cut timber faster than forests can be regenerated'.[11] A study for the International Tropical Timber Organization (ITTO) noted that the large external debts of many African countries lead to 'pressure on the forestry sector not only to generate foreign currency, but also to relinquish forest fees and taxes in principle designated for forest resource management to the general state budget'.[12] In recent years, interest and repayments accruing from debt have exceeded North-to-South aid with the result that there now exists substantial net financial flows running from the South to the North.[13] However, the linkage between debt and deforestation is a contested one, and it is important that correlation is not mistaken for cause. Hurrell notes that in the Brazilian Amazon 'the main policy decisions facilitating large-scale development were taken well before the emergence of the debt crisis'.[14] Hecht and Cockburn argue that the claim that Brazil's indebtedness to the banks led to overexploitation of the Amazon 'is difficult to substantiate'.[15] But such views are not shared by all analysts. Diegues argues that the Brazilian government targeted the Amazon and established export-oriented mining and livestock projects in order to help in the solution of the country's foreign exchange problems.[16] George notes two debt–environment connections: first, the act of borrowing, often to finance projects that are environmentally destructive; and second, paying for them by 'cashing in' natural resources.[17] She emphasizes that the relationship between debt and the environment is one of feedback, and not one of linear connections.[18]

Patricia Adams forwards the view that a large debt burden can help *save* the environment. External debt was accumulated from borrowing for large environmentally destructive industrial development policies. With the debt crisis, many creditors are no longer willing to lend money, so that it is no longer possible to fund such projects.[19] Adams's argument is more of an indictment against the destructive effects of transnational capital rather than an argument in favour of permanently high debts in the South.

The fact that a linkage has been asserted between debt and deforestation has led to the emergence of debt-for-nature swaps. This idea was first proposed by Thomas Lovejoy of WWF–US in 1984.[20] Opinion is divided on the subject of debt-for-nature swaps among NGO activists. Some support the concept as a means of conserving forests.[21] However, others such as Mahony oppose swaps on the grounds that only commercial banks and other financial institutions benefit, that swaps do nothing to improve the conditions of local peoples, and that they deflect attention from crucial issues such as land tenure.[22] Tropical forest governments may be critical on the grounds that such arrangements are an infringement on sovereignty. The WWF continues to favour debt-for-nature swaps; it notes that the 'swaps carried out to date are too small to reduce indebtedness but they can be an effective tool for financing conservation activities'.[23] The UK branch of Friends of the Earth, like George, argues that there is a correlation between debt and deforestation and favours debt cancellation as opposed to debt-for-nature swaps.[24] It is also necessary to note the recent advent of 'debt-for-timber' swaps, whereby debt swaps become a mechanism that destroys, rather than conserves, tropical forests. In 1994 the government of France agreed to cancel approximately half of Cameroon's external debt in exchange for a guarantee that French timber companies be given access to Cameroon's tropical forests.[25]

The debt crisis resulted in intervention in the economies of indebted countries by the imposition of International Monetary Fund (IMF) Structural Adjustment Programmes (SAPs). Formerly called austerity programmes, SAPs are intended to curb inflation, promote economic growth and prune government spending thus restoring a country's balance of payments equilibrium. However, a criticism of SAPs is that the export-oriented growth strategy they aim to promote has damaged environments. Research by Friends of the Earth concludes that the link between SAPs and deforestation is generalized in nature.[26] To Susan George, the model of growth advocated by the IMF and the World Bank 'is a purely extractive one involving more the "mining" than the management — much less the conservation — of resources'.[27] Berthoud sees SAPs as putting economic efficiency above social justice; they are 'the attempts of the IMF and the World Bank to impose liberalism on a worldwide scale'.[28]

Export-led industries and the role of TNCs

According to this view, global consumption imperatives lead to the growth of export-led industries, which in turn leads to deforestation. Brief consideration will be given to four such industries, namely gold, beef, oil and timber. Gold prospecting in Brazil has resulted in forest clearance in the Amazon, especially in Roraima, by *garimpeiros* (independent mineral prospectors). The present Amazonian gold rush began in 1980, with the *garimpeiros* clashing with the interests of the established Brazilian mining companies. Some gold prospecting by *garimpeiros* has taken place on Indian reserves leading to intense conflict with local Indians. The small placer mines in Amazonia, worked by the *garimpeiros*, were generating almost US$ 1 billion a year by 1989.[29]

A second export-led industry that has led to deforestation is cattle ranching. According to the 'hamburger connection' view, forests in Central and South America have been cleared to provide rangelands to breed cattle for the Northern beefburger market.[30] Deforestation for cattle ranching is a cumulative process. Approximately two years after forest clearance the soil is no longer suitable for pasture, with the result that further deforestation occurs. In the Brazilian Amazon, 30 per cent of government-assisted cattle ranches have been abandoned,[31] with the agricultural frontier steadily expanding into the Amazon.

There is controversy as to how valid the 'hamburger connection' is as a cause of deforestation in the Brazilian Amazon. Hecht argues that deforestation for livestock grazing has its origins not in the international demand for beef, but in Brazilian speculators seeking to acquire land and financial benefits such as tax holidays and fiscal incentives.[32] Harrison also notes the importance of state subsidies to cattle ranchers, and agrees with Hecht that there is no international hamburger connection in Brazil because most beef is produced for the home market.[33] However, the hamburger connection is less contested as a cause of deforestation in Central America.

A third export-led industry that impacts on deforestation is oil. A TNC formerly involved in oil production in Ecuador is British Gas which in 1987 signed a contract with Petro Ecuador, the state oil company.[34] British Gas attracted criticism from NGOs as its operations impacted upon land claimed by native forest dwellers.[35] The company eventually pulled out of Ecuador in 1992, although it did not publicly state its reasons for this. WWF questions whether 'it is technically possible to explore and exploit oil without damaging the environment?',[36] while Friends of the Earth notes that environmental degradation has ensued in all forested regions in Ecuador opened to oil companies.

In recent years three initiatives have sought to modify the behaviour of TNCs involved in oil production in forests. The first is a set of guidelines published by the International Union for the Conservation of Nature and Natural Resources (IUCN) in 1991.[37] In the same year the Oil Industry Exploration and Production Forum (E & P Forum) issued a set of guidelines for oil companies operating in tropical rainforests; the purpose of these guidelines is to establish an 'internationally acceptable uniform guideline for environmental conservation of rainforests in conjunction with petroleum operations'.[38] The third initiative involves a set of forestry guidelines for tree plantations. In conjunction with the WWF, Shell published in 1993 a set of guidelines on tree plantations, although it was emphasized that the guidelines are not a policy statement from WWF or Shell.[39]

A fourth export-led industry which impacts upon deforestation is the international tropical timber trade. With respect to tropical timber, TNCs provide the economic linkages between the felling of timber and the international timber market. They have also altered the social relations of forestry by introducing new technology, such as chainsaws and chipping machines, that result in swifter tree felling.[40] TNCs in both consumer and producer countries play a significant role in the trade. In 1991 the government of Honduras signed a 40 year agreement with the Chicago-based TNC, Stone Container Corporation, which gave the latter the sole right to log, for 40 years, an 11,000 square mile area of tropical forest, including UNESCO-designated World Heritage Sites.[41] In Indonesia, a powerful economic actor is Apkindo (Association of Plywood Makers), the world's largest exporter of plywood whose influential chairman, Mohamad Hasan, has represented his country at the ITTO (see Chapter 3).

The practices of Japanese TNCs have attracted growing attention from NGOs. In the case of Papua New Guinea, Japanese TNCs have stood accused by the Barnett Report[42] of participating in illegal practices. The report documented widespread transfer pricing, a mechanism whereby TNCs can transfer profits from a tropical forest country to another country by declaring a sale price for timber purchased that is below the current market value. The sum of money transferred in this way escapes tax in the tropical forest country. Hence, although there is no impact on the gross (pre-tax) profits of the TNC, there is an impact on the net (after-tax) profits. The Barnett Commission concluded that in the Papua New Guinean timber industry transfer pricing was a major activity of most of the companies studied. During 1986/7 the practice caused the Papua New Guinea economy a loss of 'up to US$ 27,500,000 in foreign currency earnings', an average of between $5 and $10 per cubic metre of timber.[43] For a TNC to operate a transfer pricing policy it is first necessary for the parent company to establish a subsidiary in the host country; hence, Japanese TNCs created subsidiary companies in Papua New Guinea. To give just one example, during the period 1986–7 the Stettin Bay Lumber Company made a hidden profit in excess of US$ 3 million for its parent company, the Japanese TNC Nissho Iwai, which at the time had an 83 per cent stake in Stettin Bay.[44]

The Barnett Commission also documented widespread corruption among the political élite of Papua New Guinea and Japanese-owned, or Japanese-dominated, companies. The inquiry reported other illegal practices, such as misdeclaration of timber, whereby high-value species were declared to customs authorities as low-value species.[45] In the 1990s the pattern of foreign domination of Papua New Guinea's timber industry shifted towards Malaysian ownership; local commentators charge that the range of malpractices identified by Barnett continues. The Barnett Commission's report is, perhaps, the most comprehensive account of illegal practices in the tropical timber industry to date, although NGOs have reported on other instances. TRAFFIC International has revealed six types of illegal activity prevalent in the Asia–Pacific region: illegal logging; timber smuggling; transfer pricing; under-grading of timber; misclassification of species; and illegal processing of timber. The statistics it gathered point to massive illegal activity, so that losses in foreign revenue and uncollected forestry-related charges 'have been measured in millions, and in some cases, billions of dollars'.[46]

In the Malaysian state of Sabah allegations of transfer pricing by various companies were made in 1993 by the Malaysian primary industries minister.[47] In the Philippines, Bautista has noted that Filipino figures of timber exports to Japan exceed Japanese figures of timber imports from the Philippines, the difference being largely attributable to unreported and illegally-shipped log exports.[48] Investigations by Friends of the Earth have revealed illegal practices by companies involved in the export of timber from Ghana, as well as the import of illegally-felled Brazilian mahogany into the UK.[49]

Drug cultivation

The international drugs trade also causes deforestation. This is frequently linked with armed groups, usually terrorists, but sometimes members of national armed forces. In Peru, both the Peruvian armed forces and the terrorist group *Sendero Luminoso* (Shining Path) have become deeply involved in the cocaine trade. *Narcotraficante* operations in the Upper Huallaga Valley of Peru have caused widespread deforestation, along with other environmental problems such as polluted waterways.[50] By 1990, seven million hectares had been deforested in Peru due to coca bush cultivation, which, according to the International Narcotics Control Board (INCB), 'threatens to alter the whole ecological balance of vast areas of the country'.[51] In 1991, Peruvian President Alberto Fujimori announced that 140,000 acres of rainforest were being destroyed each year to grow coca.[52] Cocaine is not the only drug to threaten South America's forests. A 1991 report by the commander of Colombia's police narcotics brigade noted that virgin cloud forest had been cut and burnt to clear land for opium poppy plantations.[53] The INCB notes that Colombian poppy cultivation has expanded to virgin forests and estimates that in Colombia 'the area under illicit poppy cultivation has expanded to an estimated 18,000 hectares, thereby equalling the size of the area under illicit coca bush cultivation'.[54] It also reports that 'expansion of coca bush cultivation to remote zones of the Amazon poses further problems, not only for eradication programmes but also for the ecological equilibrium of this vital area. Guerrilla groups continue to provide protection to traffickers and cultivators in exchange for arms and money.'[55] The drug problem also extends to other Latin American countries. In Guatemala, cannabis cultivators have cleared large areas of jungle in Peten province with some corrupt Guatemalan military commanders establishing a working relationship with drug traffickers.[56]

As well as arguing that a relationship exists between debt and deforestation, George also notes a relationship between the supply of cocaine and debt. SAPs do not just lead to pressure to sell natural resources such as timber for hard currency; they also lead to impoverishment as old unprofitable industries are closed, such as the tin mines in Bolivia. The unemployed are then driven into the coca economy.[57] The relationship between drug cultivation and deforestation is thus a complex, deeply-embedded structural one that also involves other factors such as industrial development, SAPs and external debt.

Government development policies

Industrial development policies in forest regions, agricultural policies on the fringe of the forests, energy policies and land settlement policies may all interact with each other and with global dynamics, such as the demand for tropical timber, to cause deforestation. Agricultural credit policies, fiscal incentives for capital investment and investment regulations often increase levels of deforestation by effectively subsidizing conversion of forests to other land uses. National legislation on issues such as tax incentives helps determine the area of land cleared for cattle grazing in Brazil, while in Ecuador the Land Colonization Laws consider virgin forest to be 'unproductive'; landless farmers can lay claim to this land by deforesting it.[58]

National development policies often lead to population movements. Roads act as migratory channels. The cases of the BR–364 and the Pan-American Highway, both financed in part by international capital, have already been noted. A further such project, also partly financed by outside agencies, is the Trans-Sumatra Highway in Indonesia. Highway construction inevitably leads to further deforestation following the construction of feeder roads, new settlements and small-scale agriculture. The construction of the Trans-Sumatra Highway was accompanied by a much-criticized transmigration programme designed to relieve population pressure on Java by moving people to less densely-populated areas such as Sumatra, Kalimantan and West Papua. This programme has been criticized by *The Ecologist* magazine and Survival International, both for its contribution to deforestation and for its role in the eviction of traditional forest peoples from their customary land.[59]

The fact that many population movements in tropical forest countries are enforced, as opposed to voluntary, led Westoby to adopt the term 'shifted cultivators',[60] in distinction to shifting cultivators. The latter are those people, usually indigenous to the forests, engaged in traditional slash-and-burn, or shifting, agriculture. However, shifted cultivators are the landless poor who are displaced as the result of loss of land. While shifted cultivators are agents of deforestation, the causes are those factors that cause displacement. These include dam construction, large-scale industrial agriculture and transmigration programmes. Myers notes that shifted cultivators are involved in more forest destruction 'than all the other agents of deforestation combined'.[61]

The degree to which the dynamics arising from export-led dynamics impact on forests is partially a function of the host government's policy responses. National policies may accentuate the environmental impact of global dynamics, or they may attenuate them. Infrastructure created to service industries, such as roads or oil pipelines, has its own environmental impact. The effects of these causes can be modified, but not fully arrested, and regulated at the national level. Unless a government and all other actors in a national economy were to follow a purely isolationist line, with no international trade, and with no connections to international agencies such as the World Bank and the IMF, it is inevitable that such structural factors will impact on the environment.

Finally in this section it is necessary to note the acid rain phenomenon. Acidic depositions of sulphur dioxide and nitrogen oxides may cause forest (and also crop and building) destruction. Countries that do not contribute to acid rain may suffer from its effects. This is an example of transboundary vulnerability. Heavily industrialized areas are particularly prone to this type of forest destruction, especially in Europe and North America.

Wars and the role of the military

It is well-established in the public domain that military operations caused deforestation and defoliation during the Vietnam war. More recently, linkages have been documented between the civil war in Myanmar (formerly Burma) and the timber trade between Myanmar and Thailand. Harbinson alleges that the Myanmar regime sells timber to the Thais to finance its civil war against the Karen hill tribe,[62] while Hopkinson notes a possible link between the Moung Tai Army, a group engaged against the Myanmar government, and the Thai military.[63] Both writers note the importance of drug cultivation in the region, a factor that places additional strain on the forests. There is also a linkage between the financing of the Khmer Rouge and the illegal timber trade across the Thai/Cambodia border, with Khmer Rouge guerrillas in western Cambodia having entered into partnership with foreign timber merchants.[64]

Utting notes that forest destruction in El Salvador has resulted from war. He also observes the contradictory positions of the US government in Central America in the 1980s: 'While USAID took a leading role in environmental protection initiatives in the region, the United States government also became involved in the internal affairs of certain countries by actively financing, arming and training parties in armed conflict.'[65] In 1994, the civil war between the Hutu and Tutsi tribes led to a refugee exodus from Rwanda to Tanzania and Zaire, leading a United Nations spokesperson to express fears for the area's forests and national parks.

Having established a correlation between war and environmental degradation, it is necessary to note that a reverse linkage may also exist. Weinberg argues that in Central America forest destruction, driven in part by cattle ranching to supply the North American beef market, contributed to landlessness and social unrest, which helped fuel the growth of guerrilla movements in the 1970s and 1980s. This in turn led to increased government military expenditure so that a feedback loop arose, with militarization 'worsening the very crisis to which it is a response — including the ecological roots of that crisis'.[66]

Apart from military involvements in wars, the role of the military in deforestation has been documented in Southeast Asia and South America. Schücking and Anderson note that the international timber industry has been supported by military élites in Indonesia, Myanmar and the Philippines.[67] Hecht and Cockburn argue that the role of the military was of paramount importance in the opening of the Brazilian Amazon, which senior military officers viewed as a frontier to be developed. Hecht and Cockburn focus purely on domestic actors, primarily the military; they consider that the role of international capital in causing deforestation in Brazilian Amazonia has been 'relatively minor'.[68] However, Plumwood and Routley, while agreeing that military and corporate interests have pushed for Amazonian development, also draw attention to international capital, especially in highway construction.[69] Despite Brazil's transition to democracy, the military remain a powerful voice in Brazilian politics with respect to the Amazon.

Poverty

Only brief consideration can be given here to what is a highly complex subject. Poverty is one of the most hotly-contested issues, not just for forests, but in development and environment policy-making in general. It is not suggested here that the poor *cause* deforestation. However, it is quite a different proposition to argue that the poor are agents of deforestation, and that the causes of poverty lie in deeper structural factors. Once again the cause–agency distinction is important.

One view is that the issue of poverty cannot be considered without reference to the high consumption levels of the North, for example of timber and beef, which, it is alleged, drives deforestation more than poverty in the South.[70] This argument was presented several times by governments from the South during the UNCED forest negotiations (Chapter 4). Other views focus on South to North financial transfers. Reports from the United

Nations Fund for Population Activities (UNFPA) and the United Nations Development Programme (UNDP) both suggest that net South to North financial transfers and other global inequities contribute to poverty. The UNFPA's 1992 annual report called for a direct attack on the roots of poverty, including unfair trade systems and international debt.[71] The UNDP revealed in 1992 that the richest fifth of the world's population receives 150 times more income than the poorest fifth,[72] a state of affairs that South to North financial flows inevitably exacerbates. SAPs, which insist on cuts in government spending in order to reduce budget deficits, divert funds that could be spent on the welfare needs of the poor.

Poverty in tropical forest countries contributes to deforestation due to, for example, increased demand for woodfuel and land for agricultural smallholdings. Clearly, certain causes of poverty lie at the country level, such as national development policies that displace populations or give low priority to rural areas. High levels of expenditure on the military or on prestige projects in the South exacerbate poverty by diverting funds from rural areas and welfare programmes. Poverty in the South therefore has its origins in both national policy and global dynamics. However, poverty is not just a function of global economic forces and of government development policies. It is also related to two other factors to which attention will now turn, namely population growth and land inequities.

Population growth

There are two poles to the population debate. At the one extreme are the neo-Malthusians who argue that population growth will lead to environmental and human catastrophe. Following Thomas Malthus, who saw population growth as the central variable in economics, the neo-Malthusians see population growth as the critical factor in environmental degradation. For example, Paul and Anne Ehrlich see forest clearing in the tropics as principally driven by agriculture to feed increasing populations and to meet the needs of expanding urban areas for firewood and timber.[73] At the other extreme of the debate are the ecologists. Here it is necessary to distinguish between ecologism as a science and ecologism as a belief system or ideology.[74] For example, Haas considers ecologism to be a framework that 'assimilates other scientific disciplines'.[75] However, Sachs differentiates between two views of global ecology, the first of which approximates to ecologism as a science, whereas the second inclines to ecologism as ideology. To Sachs ecologism is either 'a technocratic effort to keep development afloat [or] a cultural effort to shake off the hegemony of ageing Western values'.[76] Ecologism as ideology provides a political critique of the status quo, whereas ecologism as science does not.[77] Here the term 'ecologism' is employed to refer to an ideology. Ecologists tend to deny that population growth is a factor in deforestation, and they focus instead on land inequities (see next section).

Other analysts, like Myers and Harrison,[78] adopt a position between the neo-Malthusians and the ecologists, and argue that population growth, acting in conjunction with other factors, contributes to deforestation. Myers notes that the populations of tropical forest countries expanded by 15 to 36 per cent during the 1980s while deforestation expanded by 90 per cent, largely because the populations of shifted cultivators, displaced from traditional farmlands, 'are often increasing at rates far above the rates of nationwide increase'.[79] Harrison links the fate of marginal people living on the edge of the economy with marginal environments, which are the least likely to be conserved in a market economy: 'Marginal people and marginal environments are chained together and become the agents of each other's destruction.'[80]

The relationships between population growth and the extensification or intensification of agriculture are very complex. Extensification may be the initial response to population pressure, but intensification occurs where access to land is restricted or where only peripheral areas remain.[81] Poverty is partially a function of population growth, particularly in peripheral areas where resources are overstretched. Similarly, population growth is partially a function of poverty; it is well documented that family size declines as family income increases. Although demographic shifts alone do not cause deforestation, they do serve to magnify and reinforce other deforesting forces, such as migration into forests and the demand for timber. Demand for woodfuel is also a factor. Leach and Mearns estimate that woodfuel collection accounts for between 60 and 95 per cent of total national energy use in sub-Saharan Africa.[82] Munslow estimates that in countries of the South African Development Coordination Conference, woodfuel provides four-fifths of total energy consumption.[83]

The International Planned Parenthood Federation (IPPF), like Myers and Harrison, sees population growth as a factor, though not the most important one, in environmental degradation. In an edition of the IPPF publication *Earthwatch*, José Serra-Vega argued that there is a linkage between population growth and deforestation in the Amazon and that 'the expansion of family planning services into the Amazon will help bring down population growth rates in a region that is being inexorably eaten away by people and their needs'.[84] The *Harare Declaration on Family Planning for Life*, issued by a conference cosponsored by the IPPF and the Deutsche Gesellschaft für Technische Zusammenarbeit, noted the role of population growth in tropical forest despoliation, and also drew attention to 'inappropriate lifestyles in developed countries'.[85] The latter part of this statement was a clear reference to high consumption patterns in the North.

All agents of deforestation are human, and it follows that the more human beings there are the more agents there are. However, such agents often partake in deforestation on behalf of, or as the result of policies implemented by, wider economic interests. Such policies frequently result in landlessness for those with no economic or political power. It is to the subject of landlessness that we now turn.

Inequities in land tenure

Harrison, who it was noted above occupies a position between the neo-Malthusians and the ecologists, does not see population growth and land tenure as mutually exclusive issues: 'the two accounts are not alternatives. Both are needed for a complete view. Population growth and inequality work together in a destructive synergy.'[86] A view held by many ecologists is that inequities in land tenure result in rural poverty, which drives the landless poor into tropical forests in search of land for agriculture. Dorner and Thiesenhusen see land reform as essential if deforestation is to be curbed in the Amazon,[87] a view shared by Lutzenberger who argues that colonists enter the forest as land elsewhere is concentrated in the hands of economic interests.[88] The necessity for land reform is a common theme that emerges from a series of studies, organized by the World Rainforest Movement, where Colchester concludes that 'inequitable patterns of land use and ownership in the tropics have been exaggerated by the incorporation of the third world into the global market'.[89] Lohmann supports this point, arguing that expanding international markets deprive poor farmers of power and rights over their land.[90] Elsewhere, Friends of the Earth adopts a similar analysis; inequitable land distribution restricts the poor to the least fertile agricultural lands, 'whilst traditional élites reap most of the benefits of their countries' natural resources'.[91] It is clear that land concentration strengthens local economic élites and marginalizes the poor, and the view that land reform is a prerequisite to curbing deforestation is gaining currency. This is due in large measure to NGO pressure. Indeed, the World Rainforest Movement has published on this subject with the specific intention of bringing the issue of agrarian reform into international debate.[92]

Natural phenomena

A final cause of deforestation is natural disasters such as storms, hurricanes and drought. Such factors lie beyond direct human control. However, it should be noted that the question arises to what extent such phenomena are, to some extent at least, human made as the result of, for example, altered landscapes and climate change.

Concluding remarks

The objective of this section has been to argue that not only do the forces of deforestation lie outside the forest, but that they are frequently transnational in nature. In the final section of this chapter it will be argued that finding a solution to the causes of deforestation is a problem that cannot be considered in isolation from other factors; it is linked to two other problems, namely the need to devise new institutions and the need to arrive at a new notion of forest proprietorship.

THE EFFECTS OF DEFORESTATION

Global warming

The effects of deforestation range from the global to the local level. One of the most important ramifications of deforestation is its effect on the degradation of a global common, namely the atmosphere. Deforestation contributes to global warming, which occurs from increased atmospheric concentrations of greenhouse gases leading to net increases in the global mean temperature. The atmospheric process by which greenhouse gases contribute to temperature increases is known as radiative forcing. There are four principal greenhouse gases, namely carbon dioxide (CO_2), methane (CH_4), nitrous oxide (N_2O) and chlorofluorocarbons (CFCs). The effects of global warming are potentially calamitous. It has been predicted that severe global warming will result in changes in global patterns of agricultural productivity, a melting of the Arctic and Antarctic ice caps, thermal expansion of the oceans and a net rise in sea levels which would threaten coastlines worldwide. Not only does deforestation contribute to global warming, but global warming will, in turn, pose a renewed threat to nature conservation; the warming of the oceans will threaten marine life, while climatic zones will migrate towards the poles, thus placing stress on fragile ecosystems.

Scientific consensus has emerged on the role of greenhouse gases in global warming, including the relationship between CO_2 levels in the atmosphere, deforestation and global warming. A major research project, which contributed to scientific consensus on the linkage between deforestation and global warming, was the twenty-ninth report of the Scientific Committee on Problems of the Environment (SCOPE) of the International Council of Scientific Unions (ICSU). SCOPE, established at the twelfth general assembly of the ICSU in Paris in 1968, has a close working relationship with other organs within the ICSU family, most notably the International Union of Biological Sciences and the International Union of Geodesy and Geophysics.[93] The report — SCOPE 29 — was published in 1986 and dealt mainly with the impact of global warming, including the role played by deforestation in increased concentrations of atmospheric CO_2.[94] However, the forum that made the greatest contribution to scientific consensus on the

linkage between deforestation and climate change was the Intergovernmental Panel on Climate Change (IPCC). This published its first report in
1990, and noted that for the period 1980–90, 'Deforestation, biomass burning including fuelwood, and other changes in land-use practices release
CO_2, CH_4 and N_2O into the atmosphere and together comprise about 18
per cent (with an uncertainty range of 9–26 per cent) of the enhanced radiative forcing.'[95] An IPCC Working Group report noted that, in addition to
global climatic effects, local climatic variations may occur, with potential
effects including changes to the hydrological cycle and increases in the
reflectivity of deforested land.[96] The IPCC presented its findings to the
Second World Climate Conference (SWCC) which was held in Geneva in
1990. Unlike the First World Climate Conference (1979), which was
primarily a scientific forum, the SWCC was attended by both scientists and
governmental delegations. Some NGOs attended as observers. The conference statement of the SWCC, endorsed by both scientists and politicians,
acknowledged that scientific consensus existed *inter alia* on the relationship
between deforestation and global warming: 'Emissions resulting from
human activities are substantially increasing atmospheric concentrations of
the greenhouse gases. These increases will enhance the natural greenhouse
effect, resulting on average in an additional warming of the Earth's surface.
The Conference agreed that this and other scientific conclusions set out by
the IPCC reflect the international consensus of scientific understanding of
climate change.'[97] However, although few scientists doubt the existence of
global warming, a degree of scientific uncertainty remains on the precise
nature of the phenomenon. This uncertainty arises from the complexity of
the world's climatic system, and insufficient knowledge on the precise
contributions of the individual greenhouse gases to global warming.

Loss of biodiversity

Tropical forests also serve as storehouses of biodiversity, and consequently deforestation destroys plants, fauna and insect species, some of
which have still to be catalogued. Three international environmental NGOs,
namely the World Resources Institute (WRI), IUCN, WWF–US and Conservation International have collaborated with the World Bank on biodiversity
conservation.[98] WRI and IUCN have also collaborated with UNEP, in
consultation with FAO and UNESCO, in the preparation of the Global
Biodiversity Strategy.[99] Many of the world's medicines have been developed
from species discovered in tropical forests, and in destroying the forest and
its biodiversity, possible future medicines, still to be discovered, are irretriev-
ably lost. For example, the rosy periwinkle plant has helped in the production of two anti-cancer drugs. The importance of biodiversity for medicine
has led to at least one example of international cooperation between the

government of a tropical forest country and a pharmaceutical TNC; in September 1991, the world's largest drugs company, Merck and Company (USA), signed an agreement with the Costa Rican government. Under this agreement any profits Merck make from products developed from the biodiversity of Costa Rica's tropical forests will be shared with the Costa Rican government.[100]

Local and regional effects of deforestation

Deforestation destroys other forest values and functions. In Brazil, deforestation in the northeast of the Amazon has increased vulnerability to droughts in surrounding areas.[101] Forests fill ecological functions such as natural watershed management. They are rich sources of timber and may serve as a supply of woodfuel for local communities. In addition, tropical forests provide a wide range of non-wood products. These include rattan, leaf products, bamboo, honey, resins, tannins, fruits, mushrooms and nuts. Many of the local functions that the forests provide are not appreciated until deforestation has occurred. Declining soil fertility and soil erosion are two local and regional effects of deforestation in the Amazon, while deforested land is more prone to flooding after heavy rains. Indigenous forest peoples regard the forests as their ancestral habitat and they may attach a cultural value to specific areas of forest.

All actors benefit to some extent from forest conservation; as we saw earlier, forests play a vital role in the regulation and servicing of a global common, namely the atmosphere. (This is not to say that forests are a global common; there are cases for and against such a proposition as we shall see below.) Some actors gain in other ways from forest conservation; indigenous peoples will retain their habitats, and companies harvesting non-wood products will be assured of the supply of their products. But other actors may lose, such as those TNCs that profit from the clear felling of forests for timber, the national treasuries that accrue tax revenues from timber exports and the consumers who buy the products. The multiple functions that forests perform result in contention over forest use, contention which is in itself an impediment to prompt and effective political action.

In short, the wide range of goods that forests provide makes deforestation an especially acute political issue. The fact that forests provide so many functions results in conflicts between actors over the use to which the forests should be put. Furthermore, forest destruction may provide a good. Deforestation releases land for agriculture, either for the local landless poor or for large-scale industrial agriculture, and for mining. There are therefore numerous potentialities for conflict over the goods that both the forest and deforested land provide. Hence a vast diversity of actors — at the international, national and local levels — have a stake in forest conservation, while other actors have a stake in forest destruction. It is at the interface between political economy and political ecology that the many substantive and sensitive political conflicts over forests are to be found.

THE EMERGENCE OF FOREST CONSERVATION AS AN INTERNATIONAL POLITICAL ISSUE

In international politics, agenda formation may be seen as the process by which an issue becomes salient to important international actors. This section adopts a pluralist perspective to explain the emergence of deforestation as an international issue. A pluralist approach emphasizes interaction and dialogue among a diversity of actors and explains agenda formation by factors such as the spread of ideas, the role of global communications and NGO lobbying.

The media and the power of images

Deforestation emerged as an international issue during the 1980s; by 1990 it was firmly established as an issue of global concern, and has remained so since. One explanation for this concerns the role of the media. The powerful images of rainforest destruction, particularly in Amazonia, contributed to growing concern about the issue. The media have disseminated photographic and video images of forest destruction, including the satellite imagery of the Brazilian Amazon, which particularly shocked many politicians and citizens on 9 September 1987 when the NOAA–9 satellite photographed, according to one estimate, 7603 fires.[102] The same satellite monitored 11,904 forest fires in a seven day period in August 1992.[103] Politically, the fires contributed to the visibility of the issue among the general public, particularly in the developed North.

Local community organizations

Deforestation in the Amazon led to a mushrooming of grass-roots groups in Brazil in the 1980s.[104] This included the Rubber Tappers' Association of Brazil, led by 'Chico' Mendes until his murder on 22 December 1988.[105] Mendes's death fuelled a sense of outrage among the general public in several countries, which led to renewed lobbying by international NGOs, some of which formed alliances with the newly-emerging Brazilian NGOs. Grass-roots activity, such as the Chipko (tree hugging) movement in India and the blocking of logging roads by the Penan in Sarawak, have also received attention from the international media and international NGOs.

Mass-action campaigns

A third explanation centres on mass-action campaigns, such as the petitioning and lobbying of policy-makers and international institutions. A significant example of such activity is the campaign launched in July 1987 by the European NGOs' network ECOROPA. Two years later, in September 1989, a petition of 3.3 million signatures calling for an emergency session of the UN General Assembly on tropical rainforest destruction was presented to the UN Secretary-General, Pérez de Cuéllar, in New York.[106] Although the campaign was unsuccessful — no emergency General Assembly session was convened — it did help promote the continuing visibility of the issue.

Global deforestation surveys from 1980 to 1993

A further factor that explains increasing concern among international policy-makers on the issue was evidence that deforestation was accelerating. Table 1.1 compares four surveys for the period 1980–93. The figures in columns 1 to 3 inclusive are extracted from Myers's 1989 survey. The figures in column 4 are extracted from FAO's *Forest Resources Assessment 1990 Project*, which was finalized in 1993.[107] The project's provisional findings were released in June 1992 to coincide with the UNCED in Rio de Janeiro.[108] The final figures provide country percentage rates of deforestation.

It is not possible directly to compare the figures of Myers with those of the FAO. First, the two sets of survey use different base years. Second, the methodologies for compiling the two sets of data differ. Myers's 1989 survey was compiled principally from desk surveys and from questionnaires circulated to tropical country forestry departments, while FAO's 1993 survey used high resolution satellite data and computer modelling. Hence a comparison of columns 5 and 6 reveals some major discrepancies, particularly with the figures for Bolivia, Cambodia and Venezuela.

Nonetheless, the following can be discerned from Table 1.1. First, FAO's 1993 figures indicate that rates of deforestation are more severe than their 1980 survey indicated. Second, and notwithstanding some of the discrepancies between Myers and the FAO, if one compares only the FAO's two sets of figures, or only Myers's, in both cases the overall trend is of accelerated deforestation across all continents with tropical forests.

A final factor, which helps explain the emergence of deforestation as an international issue in the 1980s, was the emergence at approximately the same time of other global environmental issues such as ozone destruction, marine pollution, and desertification. In addition, the occurrence of major environmental disasters, such as the chemical factory explosion in Bhopal,

Table 1.1 Tabulation of deforestation figures for selected tropical
forest countries, 1980–93

	Column 1 Myers (1980) Annual area deforested km²	Column 2 FAO (1981) Annual area deforested km²	Column 3 Myers (1989) Annual area deforested km²	Column 4 FAO (1993) Annual deforested km²	Column 5 % increase or decrease of Myers (1989) from Myers (1980)	Column 6 % increase or decrease of FAO (1993) from FAO (1981)
Bolivia	750	870	1500	10097	100	1061
Brazil	14500	14800	50000	34434	245	133
Cambodia						
(Kampuchea)	600	250	500	2631	−17	952
Cameroon	1200	800	2000	1272	67	59
Colombia	4600	5100	6500	3709	41	−27
Congo	200	220	700	656	250	198
Ecuador	2200	1650	3000	2696	36	63
Gabon	200	150	600	1397	200	831
India	2600	1430	4000	2689	54	88
Indonesia	6600	6000	12000	11861	82	98
Ivory Coast	3800	2900	2500	1180	−34	−59
Laos	800	1000	1000	1567	25	57
Madagascar	2000	1500	2000	2167	0	44
Malaysia	2900	2550	4800	3838	65	51
Mexico	6100	5950	7000	9125	15	53
Myanmar						
(Burma)	1800	1000	8000	4923	344	392
Nigeria	3100	2850	4000	2618	29	−8
Papua New						
Guinea	700	220	3500	No data	400	No data
Peru	2900	2700	3500	5679	21	110
Philippines	4600	2900	2700	2370	−41	−18
Thailand	3400	2450	6000	2742	76	12
Venezuela	1100	1250	1500	8282	36	563
Vietnam	1800	650	3500	1590	94	145
Zaire	2600	1800	4000	7023	54	290

Sources and Notes: Columns 1, 2 3 and 5, Norman Myers, *Deforestation Rates in
Tropical Forests and their Climatic Implications* (London: Friends of the Earth,
December 1989), p. 34. [Figures in column 1 were originally produced in Norman
Myers, *Conversion of Tropical Forests* (Washington DC: National Research Council,
1980), with some updating in 1984 and 1985. Figures in column 2 were originally
produced in Food and Agriculture Organization, *Tropical Forest Resources*, (Rome:
FAO, 1981).] Column 4 compiled by FAO's *Forest Resources Assessment 1990
Project* and presented to FAO's Committee of Forestry in March 1993. The original
figures show estimates of forest cover change for the period 1981–90 in millions of
hectares. These have been converted into km² by the present author. Column 6
calculated by the author.

India (1984) and the nuclear power station explosion in Chernobyl, Ukraine (1986), fundamentally changed perceptions of the global environment and engendered a general feeling among many actors of a pervasive global environmental crisis requiring an urgent and concerted international response. Deforestation was seen by many actors as a symptom of this crisis.

Concerted pressure group and lobbying activity, along with the global dissemination of visual images, was instrumental in drawing attention to the problem of deforestation. However, it was necessary for politicians from the powerful economic countries to recognize the size and significance of the problem before large-scale global political discussion on the issue occurred. Bramble and Porter credit the emergence of the idea of a global forest convention (GFC) on the international agenda to NGO lobbying at the G7 Houston summit of 1990.[109] As we shall see in Chapter 4, the United Nations Conference on Environment and Development (UNCED) failed to produce a GFC, despite having the support of many Northern governments, most of which have continued to call for a GFC since the UNCED (Chapter 6).

Deforestation as an ethical issue

Deforestation is not just an international issue, it is also an ethical one. The ethical dimension can be traced to a series of international publications that appeared throughout the 1980s. The World Conservation Strategy of 1980[110] popularized the notion of sustainable development.[111] In 1987 the World Commission on Environment and Development (the Brundtland Commission) argued that development should only occur at a rate that can be sustained by the earth's regenerative capacity and defined sustainable development as 'development that meets the needs of the present without compromising the ability of future generations to meet their own needs'.[112] This concept does not apply solely to forests, but has had a strong impact on the forest debate, leading to the emergence of concepts such as 'sustainable forest management' (see Chapters 3 and 6). Embedded in the idea of sustainable development is the notion of intergenerational equity, which holds that present generations have a duty to pass the planet on to future generations in the same state as they inherited it.

THE FOREST CONSERVATION PROBLEMATIC

Deforestation emerged as a global issue because it was perceived to be of global concern, with global effects necessitating global action. However, it is a central argument of this work that there has been no coherent and generally accepted formulation of the problem. While the issue was considered worthy of global action, there was no agreement as to precisely what that action should be. Clearly problem formulation is an essential prerequisite to problem solution. This section will outline the author's views of the problems peculiar to deforestation, collectively referred to here as the forest conservation problematic, of which three dimensions are identified. The first of these, the causal dimension, hinges on the identification and arrest of the transnational causes of deforestation, which crosscut the intergovernmental system and have their loci in the global political economy. The second dimension of the forest conservation problematic is institutional in nature. It is argued that the development of new polities, and of new political processes and structures, is necessary both to address fully the causes of deforestation and to incorporate the views and knowledges of local and indigenous peoples whose participation is essential to forest conservation initiatives. Finally, there is the proprietorial dimension which results from the absence of a commonly-accepted notion of forest proprietorship.

The causal dimension

It is argued in this chapter that many of the causes of deforestation are transnational in nature. Deforestation, as a global phenomenon, cannot be solved if global economic processes are ignored. Some causes may be tackled within countries, by governments and local agencies. Intergovernmental cooperation may arrest other causes. However, the constraints of the intergovernmental political system mean that tackling economic causes is far more difficult. There remains an incongruence between the international economic system, composed of a diversity of actors, and the dominant international political system, composed of governments. Hence NGOs have called for greater public accountability for TNC activities.[113] A coordinated international response is necessary, with norm-governed behaviour being observed not only by state actors, but by all actors with a stake in forest use in the global system. Simply recognizing the importance of forest conservation as a global norm is an insufficient response. For forest conservation to be attained, it is necessary for all actors with a stake in forest use to make forest conservation their single most important concern.

The institutional dimension

Since forest conservation has become an international issue, there is what may be seen as a triangular tension of interests: global interests; national interests; and local interests. Furthermore, other actors are drawn into forest conservation policy-making on a vast range of issues from a multitude of disciplines and from every level of international society, from the United Nations General Assembly down to the village level.

Present international institutions have been unsuccessful in tackling deforestation. As Litfin observes, the 'strongest indictment of existing institutions comes from the recognition that, despite the flurry of institution building over the past two decades, the quality of the global environment has degenerated over the same period'.[114] To Blowers and Glasbergen, a major constraint on effective environmental policy making 'is the process of policy making itself'.[115] Similar concerns were previously presented in 1987 by the World Commission on Environment and Development which, in a quote that is worth citing at length, noted that:

> *Separate policies and institutions can no longer cope effectively with ... interlocked issues. Nor can nations, acting unilaterally. The integrated and interdependent nature of the new challenges and issues contrasts sharply with the nature of the institutions that exist today. These institutions tend to be independent, fragmented, and working to relatively narrow mandates with closed decision-making processes. Those responsible for managing natural resources and protecting the environment are institutionally separate from those managing the economy. The real world of interlocked economic and ecological systems will not change; the policies and institutions must.*[116]

Forest conservation cannot be achieved without the willing and effective participation of people at the local level, such as indigenous people, villagers and local community groups. Such peoples may possess knowledges essential to effective conservation policies, and these knowledges should receive full and fair consideration by other actors. Indeed, many campaigners assert that conservation can only be achieved through local control, or at the very least local endorsement of control, by local people. Environmentally benign policies by those national companies and TNCs whose policies impact upon, often with calamitous consequences, the world's forests is also necessary. Politically there is an imperative for all actors to communicate and agree upon policies guided by the norm of forest conservation.

Hence, the second dimension of the forest conservation problematic is the need to devise institutions and political structures that will integrate the views of local peoples, at the lowest level of international society, with government departments, TNCs and with those actors at the highest level of international society, such as UN organs. A range of such institutions — at the international, national and local levels — is necessary. Such institutions would serve two functions. First, they would scrutinize the activities of those actors with a stake in the forest, ensuring they adhered to conservationist norms. Second, they would serve to ensure the active participation of actors from the local level, ensuring that their views are heard with equal status alongside those of other actors from international society.

It is emphasized that there is a qualitative difference between participation and consultation; a consultative arrangement carries with it only the understanding that views *may* be expressed (in either written or verbal form), whereas a participative arrangement carries with it a genuine opportunity to influence decisions and outcomes. In intergovernmental fora, NGOs and other actors have been granted only consultative status, with the decision-making actors being government delegations. In order to overcome the institutional dimension of the forest conservation problematic a qualitatively new type of institution is necessary in which all actors may effectively participate.

The proprietorial dimension

The fact that forest depletion is now an issue on the international agenda leads to questions such as who 'owns', or has a legitimate stake in, the world's forests. Three competing claims can be identified. First, many actors concerned about the global environmental ramifications of forest destruction have inclined towards, although often stopped short of, asserting that forests are a global common. This has been resisted by governments of the South, who have asserted a counter claim, namely that forests are a national resource to be used in line with national policy. Third, many local peoples, especially indigenous peoples, acting with the help of international NGO networks such as the World Rainforest Movement, have asserted that the forests are local commons belonging to local peoples.

The existence of these three competing claims is the proprietorial dimension of the problematic. This section begins by discussing the tension that exists from the competing claims to forests of global common and national resource. We will then consider claims that forests are a local common.

If actors in the wider international community were to undergo a behavioural transformation, and to adopt the norm of forest conservation, would they then have a stake in those forests? Do forests cease to be a national resource and instead become a global common? At present, the case for viewing forests as a national resource is a strong one. In the 1970s the concept of sovereignty came to be widely accepted as applying to the sovereignty of states over their natural resources. The 1972 United Nations Conference on the Human Environment, held at Stockholm, married this right with the responsibility of states to avoid transboundary environmental damage. Stockholm Principle 21 states that:

> *States have, in accordance with the Charter of the United Nations and the principles of international law, the sovereign right to exploit their own resources pursuant to their own environmental policies and have the responsibility to ensure that activities within their jurisdiction do not cause damage to the environment of other States or of areas beyond the limits of national jurisdiction.*[117]

The latter part of the Principle reaffirmed the notion of *sic utere tuo ut alienum non laedas* (that one must use one's property in such a way as not to injure the property of another) which originates from the Trail Smelter Arbitration on transboundary air pollution between the USA and Canada.[118] In 1972 the UN General Assembly adopted a self-denying resolution which stated that no subsequent General Assembly resolutions may affect Stockholm Principle 21.[119]

The ambiguity of Stockholm Principle 21 was noted in the Preamble to the 'Hague Recommendations on International Environmental Law', which were adopted by the participants at the International Environmental Law Conference convened by IUCN–Netherlands in August 1991. The Hague Recommendations noted that the principle of national sovereignty is often interpreted so as to neglect the interdependence of the global ecosystem, and argued that it should be broadened: 'It should be acknowledged as a rule that the principle of sovereignty implies the duty of a state to protect the environment within its jurisdiction, the duty to prevent transboundary harm and the duty to preserve the global commons for present and future generations.'[120] The term 'global commons' is not defined in the Hague recommendations and should be explored. What do we mean by the term 'global common'? The World Commission on Environment and Development considered there to be only three global commons: the oceans, outer space and Antarctica.[121] Caldwell identifies these three plus the atmosphere.[122] Vogler notes that some countries have claimed outer space, especially the geostationary satellite orbit, as a global common.[123] To Porter and Brown, the global commons include 'natural systems and resources, such as the atmosphere and oceans, that belong to all living beings rather than to individual nations'.[124]

So far it would appear that forests cannot be considered a global common. Indeed governments of most tropical forest countries have proved unwilling to entertain such a notion. As UNCED Secretary-General Maurice Strong notes, the concept of sovereignty poses a problem in international environmental diplomacy: 'national sovereignty has been an immutable, indeed sacred, principle of international relations ... that will yield only slowly and reluctantly to the new imperatives of global environmental cooperation.[125] Sovereignty is a reality that those who wish to assert the international community's stake in the world's forests must overcome. With Stockholm Principle 21 widely accepted, most actors who wish to assert a stake by the international community in the world's forests choose not to dispute national sovereignty over forest resources, but instead they seek to redefine the nature of that sovereignty. This has been done in two ways.

The first challenge has been to redefine forests so that previously accepted notions of proprietorship are questioned. Romm notes that the globalization of forest uses has challenged the traditional territorial definition of forests 'as territories that display certain forms of vegetation, use and jurisdiction'. Forests, in the eyes of some members of the international community, have been redefined in terms of the functions, including global functions, they provide, such as biodiversity, carbon sinks, wildlife and beauty, to yield a functional definition. To Romm, a functional definition of forests is defined by what an area of forest does.[126] Those in the international community who wish to establish a stake in the world's forest point out that many of these benefits accrue to all countries, and not just to those with forests. Those who reconceptualize forests to yield a functional definition thus assert, implicitly or explicitly, that the international community has a stake in the conservation of the world's forests.

A second challenge made to the nature of state sovereignty over forests is the moral argument that forests should be preserved for the common good of humanity. This view is questioned by McCleary who, using Brazilian Amazonia as a case study, and drawing upon Kant's Formula of Humanity,[127] argues that the international community has an obligation to help Brazil and Brazilians if the latter are to be expected to help the international community: 'the claim asserted against Brazil to preserve the rainforest ... is not a genuine moral claim unless the international community acknowledges its duty derived from the Formula of Humanity'. McCleary proceeds to argue that the international community would only have a justified stake in Brazilian Amazonia if it were to 'institute just practices in international regimes governing activities of trade, development and debt-servicing'.[128]

McCleary's insights suggest that an interesting future research avenue for political scientists could be an investigation into possible linkages between global environmental conservation and normative theories of distributive justice. However, the main purpose of this discussion has been to illustrate that views on whether it is the state or the international community that has a legitimate stake in forests are sharply polarized, but not necessarily mutually exclusive. As Hurrell argues, tropical forests provide benefits for all humanity; they are 'both a global "commons" providing a collective good from which all benefit and the "property" of an individual state'.[129] The gap between the two claims may also be seen as the space between Romm's functional and territorial definitions: international law favours a territorial definition of forests, whereas global environmental concerns adhere to a functional definition.

The discussion so far has dealt with two competing claims. But a third aspect to the proprietorial dimension exists, namely the assertion that forests are a local common. This view is adhered to by deep ecologists and local community organizations. Forest destruction has sometimes been referred to as a tragedy of the commons. Developed by Garrett Hardin,[130] the tragedy of the commons model argues that environmental degradation will ensue in local commons as individual users will have a short-term incentive to over-exploit the common resource. The collective result of individual exploitation is the degradation of that resource.

However, the model has been the subject of criticism, for it ignored the question of open access. Monbiot and Harrison have each elaborated their own version of the 'real' tragedy of the commons. Both emphasize the importance of effective local ownership of land in avoiding degradation. Monbiot argues that Hardin's thesis works only when there is no ownership of land, whereas traditional commons are closely regulated by local people.[131] Harrison arrives at a very similar conclusion, noting that most pre-modern societies tend to develop common rules to govern sharing of local resources; such forms of ownership 'are perfectly appropriate in situations of low population density'.[132] To Harrison, an example of the real tragedy of the commons is when the state assumes ownership of the land and local communities lose the power to control and benefit from forest conservation. The state has proved unable to take care of forest lands sustainably. Ghai notes a similar phenomenon in Africa; when the state has assumed responsibility for the commons, traditional systems of care have been undermined and the result is 'uncontrolled and shortsighted exploitation of common property resources that further accelerated environmental degradation'.[133]

Indigenous forest peoples have become increasingly better organized as international pressure groups and have presented their claims to land rights to fora such as the ITTO and the UNCED. The increased cohesion of forest peoples was evident in February 1992 when the 'Charter of the Indigenous–Tribal Peoples of the Tropical Forests' was issued from a forest peoples' conference hosted by the World Rainforest Movement in Penang. This brought together forest peoples from Africa, Asia and the Americas. In Article 7 of the Charter, forest peoples demand 'respect for our autonomous forms of self-government, as differentiated political systems at the community, regional and other levels. This includes our right to control all economic activities in our territories.' The Resolutions from this conference stated that those responsible for forest destruction 'are united and coordinated at the international policy-making level regarding both natural resources and the denial of the right to self-determination of our peoples'.[134] Furthermore, two conventions of the International Labour Organization (ILO) recognize the rights of indigenous and tribal peoples. Article 11 of ILO Convention No. 107 of 1957 stipulates that 'the rights of ownership, collective or individual, of the members of the populations concerned over the lands which these populations traditionally occupy shall be recognized'.[135] Article 7 of ILO Convention No. 169 of 1989 states that indigenous and tribal peoples 'shall have the right to decide their own priorities for the process of development as it affects their lives, beliefs, institutions and spiritual well-being and the lands they occupy or otherwise use, and to exercise control, to the extent possible, over their own economic, social and cultural development'.[136] These two conventions give indigenous peoples a right in international law, a right which is in clear and direct tension with Stockholm Principle 21. The claims by all indigenous peoples, not just those from the forest, are receiving increasing attention at the international level. For example, 1993 was designated the International Year of the World's Indigenous People by the UN General Assembly.

Effective forest conservation cannot be achieved while three competing claims to the world's forests exist simultaneously. With much of the contention between actors focusing on forest ownership, the chances of effective forest conservation will be increased if actors can agree upon a common formula or metaphor, acceptable to both state and non-state actors, on the proprietorial status of forests. Such a formula or a metaphor would successfully bridge and reconcile the three claims of global common, national resource and local common.

Concluding remarks

This section has presented the author's formulation of the forest conservation problematic. It is asserted that a solution to the three dimensions of the problematic will increase the chances of effective forest conservation initiatives. The three dimensions should be seen as interrelated.

First of all the causal and proprietorial dimensions are interlinked. As noted above, many of the causes of deforestation are transnational in nature, and include international trade and external debt. McCleary, it has been noted, argues that if the international community wishes the Brazilian Amazon to be conserved, it must first implement just practices in trade and external debt. McCleary seems to suggest that, on moral grounds, the international community would have a legitimate stake in tropical forests if, and only if, some of these transnational causes of deforestation[137] were to be addressed by the wider international community. To this extent two of the dimensions may be seen as interlinked: the proprietorial dimension can be bridged if transnational causes are dealt with.

This linkage also applies in reverse. The causal dimension is unlikely to be addressed if forests remain, in accordance with Stockholm Principle 21, a resource to be exploited by a country in line with national development policy. International society will be unwilling to address the issues McCleary identifies, such as debt relief, unless forest countries make firm commitments and guarantees to conserve their forests. The causal dimension and the proprietorial dimension can therefore be seen as twin problems that must be solved in tandem.

In turn the institutional and causal dimensions are also linked. A new type of institution, with oversight powers over state and non-state actors with a stake in the forest, would play an instrumental role in dealing with the transnational causes of deforestation, as well as ensuring that local views are heard. Here the institutional and proprietorial dimensions become linked: improved liaison between all actors, in a new type of institution, is unlikely to be successful unless actors from the international, national and local levels manage to reconcile their presently conflicting views on forest proprietorship. In short, the three dimensions of the forest conservation problematic should be seen as interlinked questions requiring a common solution.

Chapter 2

The Tropical Forestry Action
Programme

On a conceptual level there are four different applications of the name Tropical Forestry Action Programme (TFAP, until September 1990 known as the Tropical Forestry Action Plan). First, there is the TFAP as a global strategy, namely to save the tropical rainforests. (In fact TFAP has always accepted applications from non-tropical governments in the developing world.) Second, there is the TFAP as a document, the *Tropical Forestry Action Plan*. (There have been two such publications, in 1985 and 1987.) Third, the TFAP may be seen as an evolving process involving a wide range of actors in the international system. Finally, there is the TFAP as a national process, which is referred to in the policy literature as either National TFAPs or National Forestry Action Plans/Programmes (NFAPs).

In this work, the operative mode of the term TFAP, used in isolation and without qualification, will denote the TFAP as a developing international process. Where the two editions of the *Tropical Forestry Action Plan* are referred to these will be italicized. National Forestry Action Plans/Programmes will be referred to as NFAPs. Where the acronym TFAP is employed to denote a global strategy, this will be emphasized in the text. The origins of the TFAP as an international process, the history of this process up to 1990, the international organization of the TFAP and the national level processes that lead to the formulation of NFAPs are analysed in the pages that follow.

THE ORIGINS OF THE TFAP

The TFAP was created in the 1980s following the publication of evidence of widespread tropical deforestation (Table 1.1, Chapter 1). There are two roots to the TFAP as a developing international process. The first originates within the FAO and the second within the Washington-based NGO, the World Resources Institute (WRI). This section describes the creation and early history of the TFAP, which involved a combination of formal and informal processes.

The FAO was created in 1945 as a United Nations specialized agency and its Forestry Department was created shortly thereafter. In 1982 an Experts' Meeting on Tropical Forestry, convened by UNEP, UNESCO and FAO, suggested that FAO's Committee on Forest Development in the Tropics (CFDT) take a more active coordinating role in tropical forestry affairs. In 1983 the statutes of the CFDT were amended to give it responsibility for reviewing international cooperation on the conservation and development of tropical forests.[1] Also in 1983, the sixth session of the CFDT recognized the need to identify high priority areas for tropical forest conservation. Accordingly, the CFDT recommended that FAO establish *ad hoc* groups to elaborate action programmes at the regional or global level. The recommendation was made after a UNEP delegate stated that UNEP's Governing Council, noting that the CFDT is a centre of international tropical forestry collaboration, 'expected the elaboration of an integrated programme of activities and not the expression of mere intentions',[2] a statement which suggests that the UNEP felt it necessary to exercise its catalytic mandate in order to propel the FAO into action.

FAO subsequently established five *ad hoc* groups. In March 1985 a FAO expert meeting reviewed the findings of these groups and recommended their proposals become five action programmes.[3] The proposals were subsequently endorsed by the seventh session of the CFDT in June 1985, which recommended that they be presented to the ninth World Forestry Congress (Mexico, July 1985)[4] where endorsement was subsequently given. The action programmes, which are shown in Box 2.1 below, have since formed the conceptual backbone for NFAPs formulated under the TFAP umbrella. The FAO subsequently published the first version of *Tropical Forestry Action Plan* in October 1985.[5] In February 1986, the International Conference on Trees and Forests, hosted by the French government, recommended the adoption of TFAP as a platform for strengthening and harmonizing international tropical forestry cooperation.[6] In April 1986 the eighth session of FAO's Committee on Forestry (COFO) endorsed the TFAP.[7]

The second root of the TFAP originated from the WRI. In December 1984 the WRI convened an International Task Force to devise a programme for reversing tropical rainforest destruction. The membership of the Task Force included representation from invited bilateral development agencies, the World Bank and the United Nations Development Programme (UNDP). The Task Force's report, *Tropical Forests: A Call for Action*, was published in the same month as the FAO's report, namely October 1985. The WRI did not see the Task Force as competing with FAO's initiative, as the WRI president made clear when stating that the report 'contributes to the continuing efforts of the [FAO]'.[8] Indeed the WRI initiative adopted FAO's five action programmes as its framework. Thus there was a conceptual linkage between what were then two separate, but complementary, initiatives.

Box 2.1 THE FIVE ACTION PROGRAMMES OF THE
TROPICAL FORESTRY ACTION PLAN

1. Forestry in Land Use:
This programme is at the interface between forestry and agriculture and aims at conserving the resource base for agriculture, at integrating forestry into agricultural systems and, in general, at a more rational use of land.

2. Forest-based Industrial Development:
This programme aims at promoting appropriate forest-based industries by intensifying resource management, promoting appropriate raw material harvesting and developing the marketing of forest industry products.

3. Fuelwood and Energy:
This programme aims at restoring fuelwood supplies in the countries affected by shortages through global assistance and support for national fuelwood and wood energy programmes.

4. Conservation of Tropical Forest Ecosystems:
This programme aims at conserving, managing and utilizing tropical plants and wild animal genetic resources through the development of national networks of protected areas.

5. Action Programme on Institutions:
This programme aims at removing the institutional constraints impeding the conservation and wise use of tropical forests by strengthening public forest administrations and related government agencies.

FAO, World Bank, WRI and UNDP,
The Tropical Forestry Action Plan (Rome: FAO, June 1987), p8.

In 1987 the two initiatives formally merged. A new document was published, also entitled *Tropical Forestry Action Plan*, based on the earlier FAO and WRI publications. The five action programmes were endorsed as the framework for national level action in the 1987 document which was presented to the Bellagio Strategy Meeting on Tropical Forests (July 1987). This was the first international meeting on tropical forestry that brought together the four cofounders of the TFAP (FAO, UNDP, World Bank and WRI).[9]

One important change from the original FAO document, and one that can be attributed to the influence of the WRI, was the emphasis on NGOs: 'Local communities must be involved in managing and utilizing the forests, and be convinced that this is in their interests. In this respect, non-governmental organizations (NGOs), working at the grass-roots level, have an important role to play.'[10] The WRI played a significant role in obtaining NGO participation at Bellagio, and has since persistently promoted local NGO participation in NFAPs.[11] An NGO statement to the Bellagio meeting concluded that 'NGOs are prepared, and express their strong desire, to participate fully in the Tropical Forestry Action Plan... We demand equal responsibility and participation in all stages of implementation'.[12] In September 1988 a second meeting in Bellagio, referred to as 'Bellagio II', considered forestry research needs in developing countries.[13]

By now the TFAP process was consolidated. The FAO and WRI initiatives were formally merged and TFAP, as a process, was developing into a broad-based coalition between UN agencies, governments, international NGOs and local NGOs. FAO had assumed the coordinating role as the TFAP's lead international agency.

THE ORGANIZATIONAL STRUCTURE OF THE TFAP AT THE INTERNATIONAL LEVEL

The role of the Forestry Advisers Group

Prior to the merging of the two processes some important developments occurred. In November 1985, just one month after the publication of the WRI and FAO documents, the government of the Netherlands launched an initiative on behalf of the donor community. Forestry advisers from 32 donor agencies, including governmental aid agencies and international NGOs, attended a meeting at The Hague. The meeting endorsed the TFAP document as 'the framework to guide future multilateral and bilateral development cooperation activities in tropical forestry'.[14] The meeting also led to two initiatives that helped to shape the TFAP process at the national and international levels. First, the meeting proposed that the *Tropical Forestry Action Plan* be translated into national TFAPs 'in harmony with national priorities and development plans'.[15] Second, the meeting led to the birth of the TFAP Forestry Advisers Group (FAG). The Hague meeting became, in effect, the first FAG meeting. The FAG has since met at six monthly intervals to discuss donor responses and strategies. Representatives from developed country government departments, principally those from Canada, Finland, France, Germany, the Netherlands, Sweden, Switzerland and the United Kingdom, and from NGOs such as the WRI, WWF and IIED, are among those to have attended FAG meetings. The FAG did not finalize agreement on a role and mandate for itself until its ninth meeting in Washington DC in 1989. Describing itself as 'an informal assemblage of forestry advisers', the FAG defined its role as the promotion of 'increased international support to the implementation of the TFAP process within the framework of the decisions of the Committee on Forest Development in the Tropics (CFDT) and the Committee on Forestry (COFO)'.[16]

The FAG has no formal institutional relationship with TFAP organs at the FAO, although it has a widely recognized functional role, with its input to the TFAP process widely respected. The Deutsche Gesellschaft für Technische Zusammenarbeit views the FAG as 'a valuable clearing house between the donors'.[17] Although it has no executive authority, and despite its informal status, the FAG has become an integral part of the TFAP process at the international and national levels. Initially intended to serve purely as a donors' forum, the FAG frequently invites representatives from other interested actors, such as representatives from the FAO Forestry Department, NFAP coordinators and representatives from the World Bank, UNESCO and UNDP. While coordination of donor support remains its principle role, the FAG has become more than a donors' forum, although it remains dominated by members from the developed North. It has a small steering group which considers broader policy issues on the FAG's role and tropical forestry issues in general, whereas full FAG meetings deal with detailed donor-related issues.

The role of the Food and Agriculture Organization

Since its inception, FAO has been the TFAP's lead international agency. Statutory authority[18] for the TFAP is located within the FAO's two intergovernmental forestry committees, namely the CFDT and the COFO. As its name suggests, only tropical forest countries are represented in the CFDT, which has a membership of up to 60 Member States and associate Member States. The CFDT reports to the Director-General, and through him to the FAO Council. COFO, which reports to the FAO Council, has oversight functions with respect to the TFAP.

The NFAP Support Unit, which until 1995 was called the TFAP Coordinating Unit, is located within the FAO's Forestry Department. Formed in 1986, the unit is headed by the TFAP coordinator who is responsible to the assistant director-general of FAO's Forestry Department. The unit's principal function is to respond to country requests for TFAP assistance and it is the 'main institutional link between national and international TFAP efforts'.[19] A seat is reserved for the NFAP Support Unit at all FAG meetings. In 1986 the FAO created a TFAP Steering Committee, which has since been superseded by a Multidisciplinary Support Group comprised of FAO officials. The TFAP coordinator attends Multidisciplinary Support Group meetings. As the CFDT meets only every two years, key policy decisions in the interim are taken by the Multidisciplinary Support Group, with day-to-day operations the responsibility of the Coordination Unit. The FAG fills an informal but crucial role in this process. Figure 2.1 below is an organization chart of the TFAP at the international level.

THE FORMULATION AND EVOLUTION OF A NATIONAL FORESTRY ACTION PROGRAMME

Neither of the two editions of *Tropical Forestry Action Plan* constitutes a global plan. As a strategy, the TFAP is best viewed as an aggregation of national plans that are conceptually linked (through the five action programmes) and institutionally linked (through the NFAP Support Unit and the FAG). In line with the recommendation of the 1985 Hague donors' meeting, the five action programmes are translated into National Forestry Action Plans. With the exception of two changes (which will be outlined below), the basic procedure for a NFAP exercise has remained essentially unchanged since the TFAP's inception. In 1989 FAO issued a set of guidelines for the formulation of NFAPs.[20] Given that forest types and socioeconomic conditions vary widely between, and sometimes within, countries, the guidelines are not binding rules or procedures, although they are intended to provide the conceptual framework within which such rules and procedures can be enunciated.

An NFAP begins with a request for assistance from the host government to the FAO.[21] In practice, no such request has been refused. If donors are prepared to fund NFAP projects in the host country the FAG will, in liaison with the FAO, agree on a lead donor agency (the core support agency, or CSA). The CSA is the lead executive, but not necessarily the lead *funding*, agency (the core support funding agency, or CSF).

The CSA will appoint an international team leader to work with the national coordinator who is appointed by the lead agency at national level, the national lead institution (NLI). An NFAP office will be established by the national coordinator. Table 2.1 details the governments, departments or ministries assigned the role of NFAP national lead institution by tropical forest governments as at March 1991.

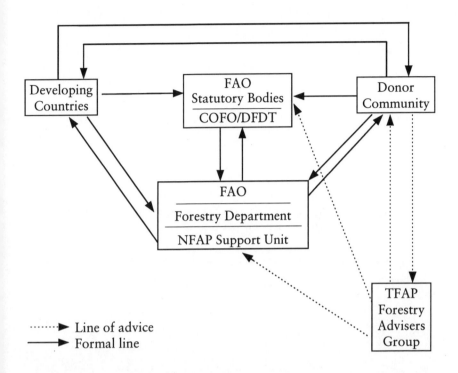

Source: Herman Savenije, *BOS–Document 11, Tropical Forestry Action Plan: Recent Developments and Netherlands Involvement*, (Wageningen, Netherlands: Ministry of Agriculture, Nature Conservation and Fisheries, 1990), p. 14.

Figure 2.1 *The organization of the TFAP at the international level*

The host government may also establish a NFAP steering committee (sometimes known as coordination committees), consisting of government, private sector and local peoples' NGOs. The intended function of a national steering committee is to contribute during the NFAP planning process when decisions are needed requiring a broad-based consensus. Table 2.2 details information concerning NFAP steering committees. Note that 76 countries are listed, whereas the total number of countries with tropical forests is in the region of 120.

The next stage is the convening of a seminar or workshop known as Round Table 1. Ideally, Round Table 1 should bring together all interested national actors — government ministries, private companies and NGOs — to define the procedures for compilation of the NFAP. Next, donor-sponsored consultants will carry out field missions with a view to identifying suitable projects for sponsoring.

Forest Politics

Table 2.1 *National lead institution for NFAPs as at March 1991*

National lead institution	Latin America	Asia	Africa	Total
National Planning Agency	1	–	–	1
Department/Ministry of Forestry	14	7	6	27
President's Office	2	–	–	2
Ministry of Agriculture	3	2	4	9
Ministry of Environment	1	2	3	6
Ministry of Natural Resources	3	–	–	3
Ministry of Primary Industries	–	1	–	1
Ministry of Rural Development	–	1	5	6
No Information	1	4	16	21
TOTAL	25	17	34	76

Source: FAO document FO:TFAP/92/2, Annex 3, p15.

The findings of all parties to date will then be incorporated into the Mission Report, which forms the basis for the NFAP. Round Table 2 is then convened to debate and amend, if necessary, the NFAP Mission Report. Usually this stage develops into a series of meetings. The finished NFAP document is then circulated to donors. The NFAP will provide details of intended projects and outline a long-term forest management strategy.

Round Table 3 is the international round table at which donor agencies are expected to identify proposed NFAP projects that they are prepared to fund. By now the NFAP process will have been in progress for about 18 months. (An indication of the value some tropical forest governments have attached to the TFAP was seen at Round Table 3 of the Nicaraguan NFAP in November 1992, where the opening ceremony was chaired by President Violeta Chamorro.[22]) From Round Table 3 there should emerge a commitment on funding from the CSA and from other donors (supporting agencies or SAs). Table 2.3 details the roles (as CSAs, CSFs or SAs) which bilateral and multilateral donors have filled in NFAP exercises as at March 1992. A study of Table 2.3 reveals some points that are worthy of comment.

First, of the cofounders of the TFAP, FAO is the major CSA for NFAP exercises, while UNDP is the major CSF. These actors have played additional important roles, with FAO the largest, and UNDP the second-largest, SA. The World Bank is the second-largest CSA, with involvement in nine NFAPs, while it has also been involved (as CSF or SA) with six others. Note that in addition to WRI, other international NGOs have acted as SAs, namely IIED, IUCN and WWF. These actors have participated in the FAG,

Table 2.2 *Steering committees in NFAP host countries as at March 1991*

	Latin America	Asia	Africa	Total
Steering Committee				
Steering committee formed	17	11	13	41
Steering committee not formed	3	–	6	9
No information	5	6	15	26
TOTAL	25	17	34	76
NGO involvement				
NGOs involved	15	11	12	38
NGOs not involved	3	–	8	11
No information	7	6	14	27
TOTAL	25	17	34	76
Industry involvement				
Industry involved	16	10	11	37
Industry not involved	1	–	6	7
No information	8	7	17	32
TOTAL	25	17	34	76
People's involvement				
People involved	14	9	10	33
People not involved	3	–	5	8
No information	8	8	19	35
TOTAL	25	17	34	76

Source: FAO document FO:TFAP/92/2, Annex 2, p14.

as have the regional development banks. Of the latter category, the Asian Development Bank has become an important CSA in Asia, while the African Development Bank and the Inter-American Development Bank have played a lesser role as SAs. Note also the role of other actors within the United Nations system.

The table indicates that the WRI has been an SA for one project. However, WRI disputes this figure and claims to they have been involved with NFAP exercises in Burkino Faso, Central America, Ecuador, Guatemala and, in cooperation with IIED, in Cameroon and Zaire.[23] IIED claims to have been involved in NFAP exercises in Costa Rica, Papua New Guinea, Tanzania and Vietnam and, in cooperation with WRI, in Cameroon and

Table 2.3 Bilateral/multilateral donor participation in NFAPs (March 1992)

Agency/country	CSA	CSF	SA	Total
TFAP Co-sponsors				
FAO	23	–	31	54
UNDP	3	18	11	32
World Bank	9	1	5	15
International Development Banks				
African Development Bank	–	–	1	1
Asian Development Bank	5	–	–	5
Inter-American Development Bank	–	–	2	2
International NGOs				
IIED	–	–	1	1
IUCN	–	–	2	2
WRI	–	–	1	1
WWF	–	–	1	1
Other International Actors				
EC	–	1	2	3
ITTO	–	–	1	1
UNEP	–	–	1	1
UNFPA	–	–	2	2
UNSO	–	–	1	1
World Food Programme	–	–	3	3
Donor Countries				
Australia	–	–	2	2
Austria	–	–	1	1
Belgium	–	1	1	2
Canada	4	–	3	7
Denmark	–	–	2	2
Finland	3	3	4	10
France	4	–	–	4
Germany	1	–	9	10
Italy	–	–	4	4
Japan	–	1	1	2
Netherlands	2	2	8	12
New Zealand	–	–	2	2
Sweden	1	–	2	3
Switzerland	–	–	3	3
United Kingdom	1	1	2	4
USA	2	–	3	5
USSR	–	–	1	1
TOTAL	58	28	113	199

Source: FAO document FO:TFAP/92/2, Annex 1, p13.
Notes: These figures are extracted from a FAO source. The WRI and the IIED each dispute that it has acted as an SA only; see main text above. UNSO stands for United Nations Sudano–Sahelian Office. A UN programme based in New York, the UNSO deals with the problems of the Sudano–Sahelian region. It was established in the mid-1970s.

Table 2.4   Status of NFAPs as at November 1994

	Africa	Asia/Pacific	Latin America/ Caribbean
		NFAP formulated	
Original NFAP implemented; new NFAP being formulated	Ghana Guinea	Nepal Sri Lanka Bhutan	Honduras Belize
NFAP being implemented	Cameroon Cote d'Ivoire Equatorial Guinea Guinea Bissau Mauritania Senegal Sierra Leone Tanzania	Fiji Indonesia Laos Pakistan Papua New Guinea Philippines Vietnam	Bolivia Colombia* Costa Rica Ecuador Guatemala Jamaica Nicaragua Panama Peru
NFAP formulated but not yet implemented	Benin Cape Verde Congo Ethiopia Kenya	Bangladesh China (Simao District) Thailand	Caribbean Community Chile Mexico
NFAP implementation interrupted	Burkina Faso Rwanda Somalia Sudan** Togo Zaire		Cuba Dominican Republic El Salvador Guyana
		NFAP not formulated	
Countries currently preparing NFAPs — Final phase	Niger Nigeria	Malaysia	
Inter. phase	Madagascar Malawi	India Solomon Islands Western Samoa	Argentina
Initial phase	Gabon Uganda Zambia	Cambodia Myanmar (Burma) Tonga Vanuatu	Paraguay
NFAP planning phase not yet started	Angola Burundi Central African Rep. Chad Gambia Lesotho Liberia Mali Mauritius Mozambique Zimbabwe		Haiti Surinam Uruguay Venezuela

Source: FAO, *TFAP Stocktaking*, (Rome: FAO, November 1994), pp19–20.
Notes: *Colombia totally reformulated its NFAP after two years. **Sudan's NFAP was formulated in 1986 following a World Bank review of the Sudanese forestry sector. However, implementation has yet to begin.

Zaire. It is, however, misleading to include IIED in a table of donors, as IIED is not a donor agency, although it may be contracted to handle money on behalf of governmental and non-governmental donors.[24]

The next stage of the NFAP is where the linkages between the NFAP (the national processes) and the TFAP (the international process) are most evident. There will be contact with the NFAP Support Unit to review progress in NFAP implementation. Monitoring of NFAP implementation may also occur in the FAG. For example, at the thirteenth FAG meeting (Rome, December 1991) seven NFAP round table exercises were considered, namely those from Cameroon, Congo, Myanmar (Burma), Vietnam, Laos, Ecuador and the Caribbean.[25] Of these, representatives from all but Congo and Laos attended.

Table 2.4 above details the status of NFAPs up to November 1994; the information was compiled as part of an FAO TFAP stocktaking exercise. Note that countries with non-tropical forests, such as Argentina and Nepal, are listed; as noted at the start of this chapter, since its inception, TFAP has been open to non-tropical forest countries from the developing world.

THE 1990 LEGITIMACY CRISIS

By 1990 it could be argued that TFAP was a successful international mechanism. First, a high rate of participation among tropical forest countries had been achieved. Seventy-nine tropical forest countries had started, or had expressed an interest in, an NFAP by June 1990. Second, through the FAG, developed-world governments and NGO donors had given support. The early support by the donor community at the 1985 meeting in The Hague was an important step in the international acceptance of the TFAP. Third, other donors, such as regional development banks and various UN organs, had established relations with the TFAP. The TFAP, as a strategy and an international mechanism had, it would seem, attained a position of widely-accepted legitimacy and authoritative status among key actors.

But by 1990 the TFAP was attracting criticism. Deforestation rates in tropical forest countries had risen (Table 1.1, Chapter 1). Environmental and peoples' NGOs had warned that the TFAP could accelerate forest loss.[26] WWF had called for a fundamental revision of the TFAP in August 1989.[27] Subsequently the FAO initiated an independent review. In March 1990 an NGO 'retreat' hosted by WWF–US in Washington concluded that the 'original conception of TFAP was flawed by an internal contradiction: it recognized many causes of deforestation, but offered only forestry solutions'.[28] Then in 1990 the TFAP's legitimacy was severely undermined with the publication of three reviews.

Criticism from the independent review, the WRI and the WRM

The World Rainforest Movement published the first review in March 1990. The FAO-initiated independent review was published in May 1990, and the WRI published a third review in June. The independent review team was composed of three people: Ola Ullsten, a former prime minister of Sweden, Salleh Mohammed Nor, and Montague Yudelman. The composition of the team attracted NGO criticism that the team was far from independent. George Marshall, then of *The Ecologist*, UK, noted that two of the team had previous close involvement with the TFAP.[29] Nor was a member of the WRI task force, while Yudelman was formerly employed by the World Bank.[30] Marshall's criticism therefore has a certain validity.

The authors of the three reviews used different data and documents to conduct their research. The WRI review provides a lengthy bibliography of sources covering the international and national levels of the TFAP. The independent review team relied heavily on desk studies of a questionnaire circulated to actors in host and donor countries. The WRM review team obtained the documentation from nine NFAPs and drew conclusions from these.

Despite their different methodologies, many of the conclusions of the three reviews are similar. Five commonalities linked the three reviews. First was the acknowledgement that the TFAP, and the NFAPs it spawned, had not contributed to slowing rates of deforestation. The WRI report noted that opportunities for slowing deforestation by launching policy and institutional reforms were being neglected.[31] To the independent review 'the rate of deforestation appears to have accelerated in spite of the TFAP'.[32] To the WRM deforestation had accelerated in part *because* of the TFAP; it accused TFAP donors of 'facilitating substantially increased financing of unsustainable projects'.[33] Such allegations were made elsewhere by other NGOs.[34]

A second area of agreement among the reviews was that the TFAP had been unable to reconcile country level interests with international concerns. The independent review observed 'a need to find a consensus approach that bridges the differences between national and international views'.[35] But whereas the independent review noted only the importance of national interests, the WRI and WRM also emphasized local interests. To the WRI, the TFAP had 'apparently assumed that there would be few conflicts between local and national interests'.[36] A separate WRI report argued that the TFAP may have contributed to cultural destruction due to an attitude which 'in varying degrees penetrates all [N]FAPs analysed — that indigenous peoples are "backward" less productive members of society, and as such are obstacles to "progress" as defined by national economic goals and international market forces'.[37] The WRM shared this concern and criticized the 1985 TFAP document for paying insufficient attention to the needs and rights of forest dwellers. The WRM also called for a funding moratorium until the TFAP was restructured to yield 'a democratic development process in which local people have a decisive voice in the formulation of policy'.[38]

The third feature that united the reviews was that, by concentrating solely on the forest sector, NFAPs had ignored the causes of deforestation that lie outside the forest. The independent review implied this by noting that there are preconditions necessary for successful forest development,[39] while the WRI and WRM reports stated it directly. The latter considered that the nine NFAPs examined were 'narrowly focused on the forestry sector'[40] and concluded that deforestation is likely to increase under the TFAP. The WRI argued that NFAPs have 'a focus too narrow to adequately assess the root causes of deforestation, much less to affect them significantly'.[41] Although the three reviews agreed that the causes of deforestation lay outside the forest, beyond this there was little consensus between them on this subject.

The fourth area in which the reviews found common ground was that at the international level the TFAP required institutional reform. The WRI argued that 'a broadly representative international steering committee' was urgently needed to provide overall guidance on the TFAP's implementation'.[42] This proposal was one of the most important to emerge from the three reviews and led to major efforts both within and outside the FAO to establish a consultative group.[43] WRI also urged the development of a new management structure for TFAP, outside the FAO Forestry Department. This was similar to an independent review recommendation which urged that the FAO Forestry Department 'should be freed from direct responsibility ... so that the TFAP becomes a distinct administrative unit under the FAO umbrella'.[44] The WRM urged that the TFAP be restructured and in a letter circulated to TFAP funding agencies later in 1990 urged that an essential step in the reform process 'must be to remove the TFAP from FAO's overall control'.[45]

The fifth area in which the three reviews agreed was on NFAP decision-making procedures. In particular, there was concern that international agencies and tropical forest governments had a disproportionate say in the planning and implementation of NFAPs. The independent review concluded that NFAPs were donor-driven and project-oriented and criticized the NFAP process for being based conceptually on the practices of development banks. The WRI endorsed this view when noting that development banks have preferred to fund industrial projects, with the priority for bilateral aid agencies being land-use and institution-building projects.[46] Hence, within a given country, a preponderance of one type of donor resulted in an unbalanced NFAP. These observations relate to the supply side of aid flows. The independent review noted a demand side when stating that some tropical forest country governments 'have been drawn in by the lure of aid'.[47] The WRM notes both the supply and demand sides to aid flows: first 'governments will push forward their favoured projects'; and second, 'donors will pick out projects that suit them rather than which benefit local peoples'.[48] The end result was NFAPs driven by the interests of host governments and donors.

Criticism from other actors

Pressure on the TFAP and its funding agencies grew throughout 1990. On 2 May the chairman of the United States Senate Committee on Foreign Relations sent an open letter to the president of the World Bank repeating two of the points in the WRM review: first, there was 'mounting evidence that the implementation of the TFAP on the national level ... may in fact cause rates of deforestation to increase'; second, the letter urged the World Bank to suspend funding for 'forestry projects through TFAP pending completion of a thorough review of the TFAP process'.[49]

Two months later the TFAP crisis was recognized by the Group of Seven Industrialized Countries. The G7, meeting at Houston in July 1990, urged that the TFAP 'be reformed and strengthened, placing more emphasis on forest conservation and protection of biological diversity'.[50] Problems accumulated for the TFAP when two prominent international NGOs announced that they would make no further financial contributions to the TFAP. WWF announced its withdrawal on 3 October 1990 because, in the words of one spokesman, 'We had to respond to our affiliates in the Third World who fear that, in its present form, the plan may do more harm than good.'[51] The withdrawal of the second international NGO was even more damaging for the TFAP. In April 1991 one of the TFAP's cofounders, the WRI, announced it would cease making financial contributions.

By this time NGOs and green parties in several countries had publicly called for either a cessation to, or reform of, the TFAP. Following the fall of the Berlin Wall, the first meeting of west and East European green parties (December 1989, Brussels) had passed a resolution opposing the TFAP, and supporting the campaign for an emergency session of the UN General Assembly on tropical deforestation (see Chapter 1).[52] NGO pressure was maintained in September 1990 with a joint letter to the president of the World Bank signed by 19 NGO representatives.[53]

Environmentalists — both green parties and NGOs — thus played a significant role in advocating TFAP reform, although the role of major actors such as the G7 and the US Senate Committee on Foreign Relations was also important. In short, a major broad-based coalition advocating either abandonment or, at a minimum, reform of the TFAP had emerged, a coalition which FAO found impossible to ignore.

THE TFAP RESTRUCTURING PROCESS

Since 1990 FAO has striven to retrieve its previously leading role in international forestry. This section will trace the efforts by FAO and other actors to restructure the TFAP; it will be seen that the period since 1990 has been fraught with further complications and controversy.

The first meeting of a FAO statutory body after the publication of the three reviews was the tenth session of the COFO in September 1990. Delegates recommended that new guidelines be drafted for NFAP implementation, and supported an independent review recommendation that country capacity projects be introduced.[54] The notion of country capacity projects is premised on the assumption that certain institutional and human resource conditions are necessary for effective forest conservation. The COFO also approved a minor semantic change, namely that the name of the TFAP should henceforth be known as the Tropical Forestry Action Programme. (In fact many actors, including some individuals within the FAO, now refer to the Tropical Forests Action Programme.) The TFAP restructuring process has taken far longer than originally anticipated, and at the time of writing (July 1996) is only now nearing completion. Table 2.5 below details the key events and meetings that contributed to the restructuring process. The main debate during the restructuring process was whether an independent TFAP consultative group should be established. In the pages that follow we will first consider this debate before attention turns to the role of the Forestry Advisers Group and to other aspects of the restructuring process.

The consultative group debate

The restructuring process began with an *ad hoc* meeting of experts in Geneva (March 1991) hosted by the four cofounders. Individuals attended the meeting in a personal capacity, and not as representatives of any group or government. Seven members from six NGOs attended, namely the World Rainforest Movement (Marcus Colchester), Friends of the Earth–US, the Environment Liaison Centre International (Kenya, ELCI), the Indonesian Environmental Forum (WALHI), IUCN and the Amazonian indigenous-peoples' group, COICA. In line with a WRI recommendation, the goals and objectives of the TFAP were reconsidered.[55] The meeting agreed upon the revised goals of the TFAP, namely 'to curb tropical forest loss by promoting the sustainable use of tropical forest resources to meet local and national needs through fostering international and national partnerships to manage, protect and restore forest resources and lands in tropical countries for the benefits of present and future generations throughout the world'.[56] Six functions for a consultative group, to be established in line with the WRI recommendation for an international steering committee, were agreed upon,

Table 2.5 Chronology of key events and meetings during the TFAP restructuring process*

Date	Place	Details
1989		
August	Gland, Switz	WWF calls for 'fundamental revision' of TFAP
16–17 Dec	Brussels	Meeting of West and East European green parties
1990		
2 March	Washington DC	NGO retreat on TFAP (hosted by WWF–US)
6–8 March	Oxford	3rd meeting of FAG steering committee
March	Penang/Dorest, UK	Publication of World Rainforest Movement review
May	Kuala Lumpur	Publication of independent review
2 May	Washington DC	US Senate Foreign Relations C'ttee calls for funding moratorium
June	Washington DC	Publication of World Resources Institute review
11–16 June	Rome	10th session of Forestry Advisers' Group
11 July	Houston	G7 call for TFAP to be reformed and strengthened
24–28 Sept	Rome	10th session of FAO's Committee on Forestry
3 October	Switzerland	WWF announces it will cease funding TFAP activities
9–14 Dec	Helsinki	11th session of Forestry Advisers' Group
1991		
6–8 March	Geneva	Expert meeting to discuss revamping of TFAP
15–16 Apr	New York	Meeting of four cofounders – FAO, World Bank, UNDP and WRI. WRI announces it will cease funding TFAP activities
27–31 May	Ottawa	12th session of Forestry Advisers' Group
10–21 June	Rome	99th session of FAO Council**
13–14 Sept	Paris	Paris consultation on TFAP restructuring
17–26 Sept	Paris	10th World Forestry Congress
5–8 Nov	Rome	100th session of FAO Council**
25 Nov	Washington	WRI renews call for independent consultative group
2–6 Dec	Rome	13th session of Forestry Advisers' Group
9 Dec	Rome	1st session of Ad Hoc Group on TFAP
10–13 Dec	Rome	10th session of FAO committee on forest development in tropics
1992		
4 May	Rome	1st session of FFDC
5 May	Rome	2nd session of Ad Hoc Group on TFAP
18–22 May	Dublin	14th session of Forestry Advisers' Group
15–17 July	Rome	Meeting of sub-group of Ad Hoc Group on TFAP
3 Sept	Rome	2nd session of FFDC
4 Sept	Rome	3rd session of Ad Hoc Group on TFAP
9–20 Nov	Rome	102nd session of FAO Council**
29 Nov–4 Dec	Costa Rica	15th session of Forestry Advisers' Group
1993		
8–12 Mar	Rome	11th session of FAO's COFO
25–28 May	New York	16th session of Forestry Advisers' Group
14–25 June	Rome	103rd session of FAO Council**
1994		
June	Rome	105th session of FAO Council**

Notes: *The chronology is not exhaustive. Numerous meetings have taken place on the TFAP; the intention of this table is to itemize the main ones. **Only those FAO council sessions at which TFAP was an agenda item have been listed.

namely 'to provide broad strategic advice, establish priorities for action, review adherence to [the] TFAP goal and objectives, undertake periodic progress and impact reviews, promote dissemination of information on TFAP, and help identify funding needs and sources'.[57] The meeting agreed that the group would be an independent (not an intergovernmental) body, would accomplish its objectives through consensus and 'not be subject to ratification by other bodies'.[58]

This raised hopes among many NGO campaigners that the consultative group would be an innovative polity, ensuring genuine participation by local communities and indigenous peoples. Marcus Colchester of the World Rainforest Movement collaborated with Friends of the Earth–US to conceive, and lobby for, a vision of such a group. They urged the establishment of a consultative group completely independent from the TFAP cofounders with the capacity to review NFAPs to ensure that they adhered to genuinely democratic modes of action. A two-tier reform process was envisaged: first, there would be an international consultative group, serviced by an independent secretariat; second, at NFAP level, there would be a separate consultative group for each country. Comprised of affected local peoples' and indigenous peoples' groups, the national consultative groups would be integral to the NFAP process. The international group would have some oversight functions over NFAPs, and would be tasked with ensuring transparency and accountability at the international level.[59] However, the type of consultative group eventually decided upon, but which has yet to meet, falls far short of this vision. Indeed it does not meet even the recommendations of the Geneva meeting.

One month after the Geneva meeting the four cofounders met in New York at the UNDP offices (April 1991) where the WRI announced that it would cease sponsoring TFAP activities while continuing to contribute to the restructuring dialogue. It was agreed that NGOs would be included in the membership of the consultative group, which should hold its first meeting in 1991.[60] This schedule was not met.

The FAO Council subsequently recommended that a further meeting consider the question of a consultative group. This met in Paris prior to the tenth World Forestry Congress (September 1991).[61] Known as the Paris consultation, the meeting had five agenda items,[62] only one of which was considered, namely a revision of the Geneva text on the functions of the consultative group.[63] Deliberations were slowed by protests from the Malaysian delegation that a consultative group should respect the sovereignty of states over their natural resources.[64] It became clear after the Paris consultation that, by invoking sovereignty, tropical forest government delegates to FAO had seized control of the restructuring debate from the NGOs.

In November 1991 WRI issued a statement repeating its call for a consultative group to be established 'composed of representatives from all of the major TFAP actors [and] responsible to no single entity'. The statement urged that should such a group not be established the TFAP should be terminated 'so that something new and more hopeful can be created at the international level in its stead'.[65]

With the Paris consultation having considered only one agenda item, the 100th session of the FAO Council (Rome, December 1991) established an intergovernmental TFAP Ad Hoc Group. This met in Rome on three occasions between December 1991 and September 1992. Representatives from interested UN agencies and NGOs 'who had demonstrated their commitment to the TFAP' were invited as observers.[66] The FAG chairman also attended the Ad Hoc Group. The FAG took an active interest in the TFAP restructuring process, and initially supported the formation of a consultative group.[67] The Ad Hoc Group's first meeting of December 1991 reached no conclusions and requested that the FAO Forestry Department outline the options for a consultative mechanism. One day later the tenth session of the CFDT convened where the FAG chairman voiced concerns over the length of time the reform process was taking.[68]

What were the reasons for this delay? A study of FAO and other available documents reveals three contentious questions: the role of NGOs; the relationship of a consultative group with FAO; and the number of members the group should have. Debate over these questions became protracted. This was partially because deliberations slowed as delegates awaited the outcome of the UNCED (June 1992), and partially because of the contentiousness of the questions themselves. An important factor influencing the debate was the formation by the Group of 77 (G77) of the Forestry Forum for Developing Countries (FFDC), a group which may be seen as a TFAP caucus for developing countries. The FFDC's inaugural meeting, which took place in May 1992 the day before Ad Hoc Group's second meeting, advocated that existing FAO bodies be used for the consultative group with FAO maintaining a leading role in TFAP.[69]

At the second Ad Hoc Group meeting FAO presented five possible options for a consultative mechanism:

- Option I Committee on Forestry
- Option II Committee on Forest Development in the Tropics
- Option III Forestry Advisers' Group
- Option IV Forestry Forum for Developing Countries
- Option V A new consultative mechanism.

Central to this debate was the relationship of the proposed group with FAO organs. Senior FAO officials were aware that the 1990 legitimacy crisis posed a threat to the organization's leading role in tropical forestry.[70]

Options I and II would have resulted in the establishment of a consultative group firmly within FAO, and would have broken the Geneva consensus for an independent group. Options III, IV and V would have resulted in the creation of a group outside FAO's formal structure. Option V, a new mechanism, was preferred by the developed countries while, as noted above, the FFDC supported the assumption of a consultative group's functions by an established FAO body (Options I and II).[71]

The second meeting of the Ad Hoc Group decided that subsequent discussion should focus on Options I and V. A sub-group was established in an attempt to reach a compromise formula. This proposed a consultative group 'within the institutional structure of FAO, linked in a looser way with COFO but still reporting directly to FAO's governing bodies'. Broad rights of participation should be accorded to IGOs and NGOs, although the 'final decisions (and voting rights) . . . would remain with Member Governments'; the group should be open to all FAO governments 'which express their desire to become members'.[72]

The status of NGOs remained a contentious point. Delegates from tropical forest countries, concerned about national sovereignty, expressed reservations about the proposed involvement of NGOs alongside government representatives. The first meeting of the Ad Hoc Group had underlined that the group 'should not be a body for supervizing the activities of sovereign states'.[73] Subsequently, the sub-group emphasized that any consultative mechanism 'should be based on the concept of national sovereignty and should thus allow for the final decisions to be taken by governments alone'; the views of NGOs 'should not be placed on a parity with those of Member Nations'.[74] Malaysia, Indonesia and Brazil led the opposition of the tropical countries to the establishment of a consultative group outside the FAO.[75]

At the Ad Hoc's Group third and final meeting in September 1992 the FAG chairman stated that the compromise proposal 'did not respect the fundamental principles of the Consultative [Group] agreed upon at the Geneva meeting. Especially the two most important features of independence and informality would be lost if it is situated within FAO or any other international organization'.[76]

Despite this intervention, the Ad Hoc Group endorsed the recommendations.[77] FAO's programme committee subsequently forwarded the proposal to the 102nd session of the FAO Council (November 1992),[78] over which the director-general expressed reservations. He estimated that the annual cost of servicing meetings would be in excess of US$ 2 million and stated that 'the allocation of scarce resources of the magnitude estimated can hardly be justified'. Furthermore, the number of participants would be in excess of the FAO Council, 'which calls into question the basic concept that an advisory body should be smaller than the entity receiving its advice'.[79] He proposed a membership size of 30 and the council requested its chairman to prepare a fresh proposal on membership.[80]

Following discussions with the FAO secretariat, the OECD group[81] and the G77, the chairman reported to the FAO Council's 103rd session. A membership size of 36 with five categories was recommended; category 1 would be the executive heads of the FAO, World Bank, UNDP, UNEP and ITTO; category 2 would be 14 developing countries; category 3 would be seven developed countries; category 4 would be one representative each from the WRI, FFDC and FAG; while category 5 would be seven NGO representatives, of which two should represent the forest industry and five should represent environmental and developmental interests.[82] In this way, the type of NGO that could participate was heavily proscribed, with no proviso made for local community groups or indigenous peoples. Although it was suggested that NGOs select their own representatives, an added proviso that angered many NGO campaigners was that 'this procedure would require the consent of the member country in the case of national NGOs'.[83]

After five years of debate it is by no means certain that the group will convene. The original meeting of the group was scheduled for December 1994 but FAO postponed this to take account of the April 1995 session of the UN Commission on Sustainable Development (CSD), which *inter alia* considered implementation of the Agenda 21 provisions on forests. At its 1995 session, the CSD agreed to establish an Intergovernmental Panel on Forests; this is due to report in 1997 (Chapter 6). FAO will await the panel's recommendations before considering whether it is appropriate for the group to be established. This is unlikely as the type of consultative group decided upon by the Ad Hoc Group has not been fully accepted by all stakeholders, with FAO rules not facilitating the full participation of NGOs, indigenous peoples and the private sector.[84]

The role of the Forestry Advisers' Group during the TFAP restructuring process

Throughout the period 1991–3 the FAG became increasingly concerned at the direction and slow pace of the restructuring process. Initially the FAG supported the TFAP restructuring process and stated that it would 'act in an advisory capacity to the proposed Consultative Group and other interested partners'.[85] In December 1991, as noted above, the FAG chairman voiced concern to the CFDT on the slow pace of reform. One year later, in December 1992, the FAG still supported the Geneva conception of a consultative group but, as discussions had moved away from this idea, the FAG 'decided to draw back from further involvement'.[86] Then in May 1993 the chairman of the FAG sent a letter to the FAO Council stating that 'The vast majority of the members of the Forestry Advisers' Group, taking into consideration the fruitless efforts over three years to establish a consultative [group] agreeable to all parties, strongly recommends that Council *not pursue this matter any further*' (emphasis in original).[87] Among the reasons cited for this view were that NGOs would not participate unless accepted as equal members, and the proposed group would be the third FAO forestry body and would not have added value over the COFO or the CFDT.

The FAG also indicated support for a post-UNCED initiative, namely the World Commission on Forests and Sustainable Development (Chapter 6). This initiative is referred to in the letter sent to the FAO Council. The final comments of this letter are scathing, coming from a group that was specifically established to support the TFAP; the FAO's proposal for a consultative group 'has been overtaken' by the World Commission on Forests and Sustainable Development proposal, it 'is "too little and too late" and should not be allowed to distract from efforts being made towards more fruitful undertakings'.[88] Following their disillusionment with the TFAP, the FAG redefined its mandate at its sixteenth meeting in New York (May 1993). The FAG no longer focuses exclusively on TFAP affairs. Now called simply the Forestry Advisers' Group (and no longer the TFAP Forestry Advisers' Group), the redefined role of the FAG is 'to support the formulation and implementation processes of national forest programmes and international support frameworks, and their interactions'.[89] The FAG is now an advisory group for all national forestry programmes (and not solely NFAPs) and other international initiatives, and at its sixteenth meeting it agreed to offer its services to the UN Commission on Sustainable Development and the Global Environment Facility.[90] The FAG continues to debate and coordinate donor responses for NFAPs, although this is no longer its sole task.

Other developments during the TFAP restructuring process

The debate on the consultative group was just one part, although an extremely important one, of the restructuring process. In line with a recommendation of the independent review, new operational principles have been published.[91] The principles emphasize that an integral approach to forest management is necessary, with linkages between the forestry sector and other social and economic sectors receiving full consideration. Three further changes have occurred since 1990, namely one at the regional level, and two at the national level.

There is now a greater emphasis on regional and sub-regional cooperation. By June 1993, six TFAP sub-regional exercises to enhance cross-border forestry conservation had been established.[92] This process has paralleled a movement towards regional cooperation among NFAP coordinators. In March 1992 the first meeting of the Latin American and Caribbean

Regional Group of NFAP coordinators took place in Santa Cruz, Bolivia. NFAP regional advisers are no longer based at the FAO in Rome, and since January 1996 they have all been posted to the FAO regional offices in Accra, Bangkok and Santiago. However, the NFAP Support Unit remains in Rome and continues to support countries during the preparation of their NFAPs.[93]

Two significant changes have taken place at the national level. The first is a change to the NFAP process. Some actors considered that review and implementation was weak in the NFAP process. In September 1990 the WWF, noting a need for ongoing review of NFAPs, called for the introduction of Round Tables 4 to ensure continuing donor coordination.[94] In June 1991 the FAG noted problems of implementation, further observing that donor participation was limited to Round Tables 3.[95] The FAO has since introduced Round Tables 4, the objective of which is to review progress in NFAP implementation.[96]

The second major change at national level is the introduction, in line with an independent review recommendation, of country capacity projects. UNDP will take the lead for the TFAP on national capacity building. After the UNCED, UNDP launched a Capacity 21 programme to complement the UNCED's Agenda 21. As Oksanen et al note, 'Under [the Capacity 21] programme, and initially to assist with delivery of TFAP, UNDP has proposed a more systematic and consistent approach to the support and management of national forestry programmes. The Country Capacity for National Forest Programmes (CCNFP) ... is widely supported by developing countries, and consistent with the Agenda 21 objective of strengthening national capacities.'[97] The intention is that CCNFPs will improve institutional and human resource support for NFAPs and help to implement participatory planning by the establishment of national steering committees involving key stakeholder groups. An objective of Capacity 21 projects is to integrate environmental considerations into other activities.[98]

The TFAP restructuring process has seen some of the original tenets of the 1985 and 1987 publications of *Tropical Forestry Action Plan* questioned. The implicit assumption that concentration on the forestry sector alone will solve the problem of deforestation has been abandoned. There is less of an emphasis on international planning, a greater stress on decentralization and a notable shift to the regional level.

FAO has been criticized for the lengthy time of the TFAP restructuring process, and in particular for the decision that the consultative group, if it meets, will be established within the FAO bureaucracy. However, such criticism is not justified. It is important to distinguish here between the FAO Forestry Department and FAO statutory bodies. Once it became clear that the TFAP was experiencing difficulties, the staff of the FAO Forestry Department were prepared to listen to, and to act upon, the advice of other actors. Indeed they launched the restructuring process. Their efforts were thwarted by the consensual decision-making procedures of the FAO's inter-governmental statutory bodies, where the real levers of power within FAO are to be found. It was in these organs, and also in the intergovernmental Ad Hoc Group, that the opposition was led by the FFDC delegates due to their reservations on NGO participation and where the attempts to create an independent consultative group were stifled.

In many respects the concept of TFAP has now been superseded, with Chapter 11 of UNCED's Agenda 21 calling for all governments of the world, and not just tropical governments, to develop and implement forestry 'plans and programmes'.[99] FAO believes that NFAPs, renamed National Forestry Programmes in 1995, are suitable for forest management outside the tropics and that the process can, with amendments, be adapted by developed countries and countries with economies in transition. In early 1996 the NFAP Support Unit was assisting the governments of Armenia, Bulgaria, Lithuania and Slovakia with the preparation of NFAPs.[100] With the recent decentralization of the TFAP's organization from Rome, the future is likely to see fewer references to the Tropical Forestry Action Pro-gramme, and more reference to national programmes and regional coordination in both tropical and non-tropical countries.

Chapter 3

The International Tropical Timber Organization

The work of the International Tropical Timber Organization (ITTO) is the formal, institutionalized part of a complex system of interactions occurring in the international tropical timber economy. This chapter will not deal with the full range of these activities, although due consideration will be given to those that have affected the work of the ITTO. Consideration will first be given to the background to and contents of the International Tropical Timber Agreement, 1983 (ITTA 1983), the first commodity agreement to include a conservation component, which gave birth to the ITTO. This agreement would have expired on 31 March 1994 if the negotiation of a successor agreement had not been concluded. Shortly before that deadline, in January 1994, the negotiation of the International Tropical Timber Agreement, 1994 was concluded. The negotiations leading to this agreement will be considered in Chapter 5. This chapter will analyse the history of the ITTO. At the time of writing (July 1996) the ITTA 1994 has yet to enter into legal effect, and until such time as it does the ITTA 1983 may remain in effect by special vote of the International Tropical Timber Council.

INTERNATIONAL TROPICAL TIMBER AGREEMENT, 1983

Origins

The International Tropical Timber Agreement, 1983 emerged from a protracted series of preparatory meetings and negotiations held under the auspices of the United Nations Conference on Trade and Development (UNCTAD). Resolution 93(IV) of the fourth session of the UNCTAD (Nairobi, 1976) agreed to establish an Integrated Programme for Commodities (IPC) and requested UNCTAD's secretary-general to organize negotiating conferences for individual commodity agreements.[1] Pursuant to this resolution, a proposal for an international tropical timber agreement was tabled by Japan in 1977.[2] The first preparatory meeting opened in Geneva in May 1977 where it was agreed that a tropical timber agreement should aim to improve the incomes of tropical timber-producing countries through increased export earnings.[3]

The fifth preparatory meeting was not convened until July 1980. Disagreement on finance helps explain the lengthy time the preparatory process took. The fifth meeting considered establishing a global fund based on a 1 per cent levy on internationally-traded tropical logs. However, the proposal did not attract general support.[4] The meeting established two intergovernmental groups of experts, one on research and development and the other on improving tropical timber market intelligence. The reports of these groups were subsequently considered at the sixth preparatory meeting (June 1982, Geneva), along with reports by the FAO and UNCTAD secretariats on reforestation and forest management and on increased processing of tropical timber in developing countries.[5] The meeting agreed that the texts of these four elements represented the final preparatory phase of the negotiations for an international tropical timber agreement.[6]

The United Nations Conference on Tropical Timber, 1983

Prior to the main negotiating conference (the United Nations Conference on Tropical Timber, 1983) an intergovernmental Meeting on Tropical Timber met to consider remaining questions where a significant intervention was made by the IUCN. The IUCN observer emphasized 'the symbiotic relationship between conservation and development', stressing that 'in the long run neither could be achieved in isolation'.[7] As a result of this intervention, which was supported by other NGOs, the meeting agreed that due regard should be given to ecological considerations. The meeting concluded that the conference should establish an International Tropical Timber Organization (ITTO), the highest authority of which would be the International Tropical Timber Council (ITTC), and further establish committees to assist in the implementation of the four elements agreed upon at the sixth preparatory meeting. Hence, by the time the UN Conference on Tropical Timber opened in Geneva in March 1983 agreement had been reached on some of the important features of an agreement. However, the conference was unable to conclude its work, with the producers emphasizing the production of tropical timber while the consumers concentrated on trade-related questions.[8] A second part of the conference was held in November 1983 where agreement was finally reached on the text for the International Tropical Timber Agreement, 1983 (ITTA 1983).

Some 70 countries (36 producer countries and 34 consumer countries)[9] had taken part in the conference. The ITTA 1983 was opened for signature on 2 January 1984. However, ratification proceeded slowly, and the governing council of UNEP exercised its catalytic mandate in calling upon governments to become parties to the agreement.[10] As outlined in Chapter 2, the UNEP also exercised its catalytic mandate in the same year with respect to the FAO initiative that led to the Tropical Forestry Action Plan.[11]

The Objectives of the International Tropical Timber Agreement, 1983

Among the objectives of the ITTO, as stipulated in Article 1 of the ITTA 1983, are to provide a framework for cooperation and consultation; to promote research and development; to improve market intelligence; to encourage further processing of timber; industrial tropical timber reforestation; and marketing and distribution of tropical timber exports. However, the two objectives that have attracted the most attention within the ITTO are Article 1(b), which aims to 'promote the expansion and diversification of international trade in tropical timber', and Article 1(h) which states that the ITTA 1983 aims to encourage 'the development of national policies aimed at sustainable utilization and conservation of tropical forests and their genetic resources, and at maintaining the ecological balance in the regions concerned'.[12] The IUCN intervention noted above therefore represents a significant example of an NGO altering the course of an intergovernmental negotiating process.

The entry into force of the International Tropical Timber Agreement, 1983

Article 37 of the ITTA 1983 stipulated that the agreement could enter into legal effect on 1 April 1985 if ten or more governments holding at least 50 per cent of the votes for producer countries shown, and fourteen or more governments holding at least 65 per cent of the votes for consumer countries, had acceded to the agreement by 31 March 1985.[13] The ITTO has a complex voting system, with 1000 votes allocated each to producer and consumer countries. Consumer countries' votes are determined by their share of tropical timber imports. Producer countries' votes are determined principally by their share of tropical timber exports with some consideration also given to forest area. Hence, as Colchester argues, the *'net result is that the more a country destroys tropical forests, the more votes it gets* [emphasis in original]'.[14]

In March 1985 a meeting of NGOs hosted by the International Institute for Environment and Development (IIED) in London called for governments that had not done so to ratify the agreement.[15] The agreement entered into effect on 1 April 1985 when it was announced that 12 producer countries holding 58.8 per cent of the producer votes, and 15 consumer countries

Box 3.1 *PRODUCING AND CONSUMING COUNTRIES TO
HAVE ACCEDED TO THE INTERNATIONAL TROPICAL
TIMBER AGREEMENT, 1983 BY 31 MARCH 1985 AND THEIR
VOTES ACCORDING TO THE ANNEXES OF THE AGREEMENT*

Producing countries	Votes	Consuming countries	Votes
Brazil	130	Belgium/Luxembourg	21
Congo	20	Denmark	13
Ecuador	14	Egypt	11
Gabon	21	Finland	10
Ghana	20	France	56
Honduras	9	West Germany	44
Indonesia	139	Greece	14
Ivory Coast (Côte d'Ivorie)	21	Ireland	12
Liberia	20	Italy	41
Malaysia	126	Japan	330
Peru	25	Netherlands	35
Philippines	43	Norway	11
		Sweden	11
		United Kingdom	41
Total: 12 countries	588	Total 15 countries	650

Source: UNCTAD *Bulletin*, No 211, April 1985, p.11.
Note: The ITTO voting rights for Belgium and Luxembourg are grouped together
because they form a customs union.

holding exactly 65 per cent of the consumer votes, had deposited articles of
ratification by the 31 March 1985 deadline; see Box 3.1 above. By 31
March 1994 the membership of the ITTO numbered 50, namely 23 producers and 27 consumers; see Box 3.2 below.

The first session of the International Tropical Timber Council

By the time the agreement entered into legal effect preparations and
negotiations had taken eight years. The first session of the International
Tropical Timber Council opened in Geneva in June 1985 where it became
clear that there would be a further delay before the ITTO became fully operational, with members unable to agree on the choice of the executive
director or the site for the headquarters. Eight countries made bids to host
the headquarters, with the cities offered being Amsterdam, Athens, Brussels,
Jakarta, London, Paris, Rio de Janeiro and Yokohama.

Box 3.2   COUNTRIES TO HAVE ACCEDED TO THE INTERNATIONAL TROPICAL TIMBER AGREEMENT, 1983 BY 31 MARCH 1994

Producing countries	Consuming countries
Bolivia +	Australia #
Brazil *	Austria +
Cameroon +	Canada +
Colombia #	China +
Congo *	Egypt *
Ecuador *	European Community:
Gabon *	Belgium *
Ghana *	France *
Guyana #	Denmark *
Honduras *	West Germany/Germany *[1]
India +	Greece *
Indonesia *	Ireland *
Ivory Coast (Côte d'Ivoire) *	Italy *
Liberia *	Luxembourg *
Malaysia *	Netherlands *
Panama #	Portugal #
Papua New Guinea +	Spain +
Peru *	United Kingdom *
Philippines *	Finland *
Thailand +	Japan *
Trinidad and Tobago +	Nepal #
Togo #	New Zealand #
Zaire #	Norway *
	South Korea +
	Sweden *
	Switzerland +
	USSR/Russian Federation +[2]
	USA +
TOTAL PRODUCERS 23	TOTAL CONSUMERS 27

Sources: UNCTAD Bulletin, No 211, April 1985, ITTO document ITTC(I)/L.10 and UN document TD/TIMBER.2/3.

Key: *Acceded by 31 March 1985 (12 producers, 15 consumers); + acceded between 1 April 1985 and the end of the first session of the International Tropical Timber Council (1 August 1986) (6 producers, 8 consumers); # acceded between the end of the first session of the International Tropical Timber Council and 31 January 1993 (5 producers, 4 consumers).

Notes: (1) West Germany and East Germany were reunified on 3 October 1990. (2) After the dissolution of the USSR on 31 December 1991 the Russian Federation assumed the obligations of the Soviet Union in the ITTO.

The first session eventually met in three separate parts (June 1985, October 1985 and July 1986). The discussion on the choice of headquarters led to prolonged debate. At the third part of the first session a secret ballot decided that the ITTO would be based in Yokohama. A major contributory factor in the choice of Yokohama was the generous subsidies offered by the Japanese government towards the running of the ITTO. A Malaysian, Dr Freezailah bin Che Yeom, then deputy director-general of the Malaysian Forestry Department, was chosen as the executive director. NGO campaigner Marcus Colchester reports there was a deal whereby the Japanese supported Freezailah's nomination in exchange for producer support from Southeast Asia for Yokohama's nomination to host the headquarters.[16]

The ITTO's complicated voting system played a major role in the choices of both a headquarters and an executive director. Japan has the most votes of any tropical timber importer, with the EC second. (EC countries are required to vote as a bloc.) Malaysia, Brazil and Indonesia are the biggest tropical timber exporters. The number of votes each country receives is recalculated annually based on the latest trade data. However, it is noteworthy that no vote has been taken at an ITTC session since the choices were made for a site for the headquarters and an executive director, with Article 12 of the ITTA 1983 emphasizing that, 'The Council shall endeavour to take all decisions and to make all recommendations by consensus.' However, although votes are not taken, awareness of the differences between the member countries can influence how a consensus develops.

THE INSTITUTIONAL STRUCTURE OF THE ITTO

Formal institutional elements

The ITTO is composed of six elements: an executive director, a secretariat, a council and three permanent committees. The executive director's position is essentially administrative, with full decision-making powers resting with the ITTC. However, as shown below, since 1991 the ITTC has granted the executive director powers with respect to project expenditure.

The ITTC meets in full session once every six months. The November meeting takes place in Yokohama, with the May meeting hosted by a producer country rotating between Africa, Asia and Latin America. The three permanent committees fulfil the functions of three of the four elements agreed upon at the sixth preparatory meeting. These are the Permanent Committee for Reforestation and Forest Management, the Permanent Committee for Forest Industry and the Permanent Committee for Economic Information and Market Intelligence. The fourth element, namely research and development, is a common function of all committees. The committees held their first sessions alongside the ITTC's third session and have since met, along with the ITTC, at six-monthly intervals. Chairmanship of the ITTC and the three committees alternates between producer and consumer countries.

The role of environmental NGOs and timber traders

Article 15 of the ITTA 1983 allows observer status to be granted to NGOs at ITTC and permanent committee sessions. In practice, any NGO expressing an interest in the ITTO's work that arrives at the start of a council session will, upon presentation of its credentials, be granted observer status.[17] It is one of the most open arrangements offered by an IGO for NGO access. Some heads of individual delegations have appointed NGO representatives to serve as conservation advisers on their national delegations. Examples include the UK, Denmark and Malaysia (WWF), the Netherlands (IUCN), and the USA (National Wildlife Federation). National delegations may also appoint representation from the timber trade. The case of the UK is illustrative.

Up to and including 1989 the lead agency for the British delegation was the Department of Trade and Industry (DTI), and the British delegation consisted of DTI delegates and timber-trade advisers, with occasional representation from the Overseas Development Administration (ODA). Since then two important changes to the composition of the British delegation have occurred. First, Francis Sullivan of the WWF, who had previously attended ITTC sessions as an NGO observer, was appointed to the British delegation as a conservation adviser at the ITTC's eighth session. And second, at the ITTC's ninth session, the lead British agency for the ITTO switched from the DTI to the ODA.[18] These two changes represented a recognition by the British government that tropical timber is essentially a conservation and development, rather than purely a trade, issue. From the ninth to the twelfth sessions a typical British delegation to ITTO was composed of three ODA delegates, one timber-trade adviser and one forest conservation officer from WWF. At the thirteenth session, WWF withdrew its forest conservation advisers from all national delegations (see Chapter 5).

In addition to those NGOs that have secured representation on national delegations, many others, whether unwilling to serve or unable to gain a place on national delegations, have attended the ITTC with observer status. Examples include Friends of the Earth (FoE), Survival International and the Japan Tropical Forest Action Network. Individual NGO representatives serving on national delegations act as a useful two-way conduit of information, conveying opinions from the wider NGO community to individual national delegations and vice versa.

Only occasionally do NGOs make individual statements to the ITTC, although they are always permitted to make joint statements. Hence, as

Bramble and Porter note, the 'ITTO is an institution for which a North–South NGO alliance is crucial'.[19] NGO unity has generally been preserved at the ITTO, with joint positions agreed upon behind the scenes, and statements made to the ITTC by a prearranged spokesperson acceptable to all NGOs.

THE FINANCING OF THE ITTO

The ITTA 1983 provides for two accounts, the administrative account and the special account. The former funds the secretariat and costs incurred in holding ITTC sessions. Contributions by members to the administrative account are in proportion to the number of votes held for each financial year. There are two sub-accounts to the special account, namely the pre-project sub-account, and the project sub-account. The role of the ITTO on project work will be considered below. There are three possible sources of funding for the special account, namely regional and international financial institutions, voluntary contributions and the Second Account of the Common Fund for Commodities.[20] The ITTC's eleventh session noted that the Second Account of the Common Fund for Commodities had been opened for use in 1991. By March 1993 the Common Fund had received five ITTO project proposals which were considered by the fund's technical advisory body, the consultative committee. None of these were recommended to the executive board of the Fund for approval.[21] Most projects have been paid for by voluntary contributions from consumer governments, with the Japanese government being the largest single donor.[22] When governments donate money to the special account they usually stipulate projects on which it should be spent (earmarked funds), although governments may make donations without specifying a particular project or project type (unearmarked funds).

THE HISTORY OF THE ITTO

After the ITTC's first session the ITTO broke its orbit around the UNCTAD and became an autonomous IGO. It was only at the ITTC's second session, with the decisions for the executive director and headquarters settled, that members could concentrate on the operational aspects of the ITTO. Hence the real work of the ITTO began in 1987, ten years after the preparatory process began, and two years after the ITTA 1983 came into effect. A chronology of the events in the negotiation of the ITTA 1983, and of the regular ITTC and permanent committee sessions held until 31 March 1994, is shown in Table 3.1 below.

The project work of the ITTO

An important feature of the ITTO is its emphasis on project work to achieve the objectives of the ITTA 1983. The main work of the three permanent committees is vetting and approval of pre-projects and projects. A pre-project is the preparatory phase of a project and is essentially concerned with background research and information collation. Not all projects go through a pre-project phase; this occurs only if exploratory research or preparatory activity is required. Proposals may only be submitted by ITTO members or the secretariat. The ITTC may reject proposals, return them for amendments to be made by the permanent committees, or accept them. In principle, inappropriate or poorly-designed projects should not pass through the permanent committees. Examples of projects, to name just a few, include the building of databases on the tropical timber trade, research on incentives, reforestation projects and projects to improve the further processing capacity of producer countries.

Pre-project and project proposals must fall within the remit of one of the three permanent committees. As Table 3.2 below details, two-thirds of ITTO project expenditure has centred on the work of the Permanent Committee on Reforestation and Forest Management. By 31 December 1992 the ITTO had also undertaken 55 approved pre-project studies, the total budget of which was $US 5.5 million. Of these, 28 were completed, 15 operational, 11 awaiting contracts or finance and 1 suspended.[23]

At the ITTC's fifth session in 1988 NGOs called for the designation of a secretariat staff member as an NGO liaison officer to facilitate secretariat–NGO communications on projects and policies.[24] The secretariat did not act upon this proposal. Nonetheless, NGOs have continued to offer their assessments on pre-projects and projects at ITTC sessions. In 1987 the ITTC initiated the drafting of a project cycle, namely the rules of procedure to be adhered to by members in the drafting and submission of pre-projects and projects, and the procedures to be followed by the permanent committees and the council in approval of the same. One year later, and after several amendments, a project cycle was approved by a joint session of the three permanent committees.[25]

Since then, and following pressure from the governments of the UK and the Netherlands for a more efficient project approval process,[26] the ITTC has passed several decisions to expedite the project cycle. In 1990 it established an Expert Panel for Technical Appraisal of Project Proposals as an interim measure; the ITTC subsequently decided to continue the panel on a permanent basis. Composed of 12 representatives, with equal representation from producer and consumer countries, the panel first met in February 1991 in Kuala Lumpur, and now meets regularly between ITTC sessions. It assists the secretariat with project screening to ensure that only well-designed

Table 3.1 Chronology, May 1977–March 1994: the International Tropical Timber Agreement, 1983 and ITTC and permanent committee sessions held during the lifespan of this agreement

Date	Place	Meeting
May 1977–July 1980	Geneva	First five ITTA Preparatory Committee Meetings
16–20 Nov 1981	Geneva	Meeting of Intergovernmental Group of Experts on Research and Development for Tropical Timber
23–27 Nov 1981	Geneva	Meeting of Intergovernmental Group of Experts on Improvement of Market Intelligence on Tropical Timber
1–11 June 1982	Geneva	6th Preparatory Meeting on Tropical Timber
29 Nov–3 Dec 1982	Geneva	UNCTAD Meeting on Tropical Timber
14–31 March 1983	Geneva	UN Conference on Tropical Timber (Part I)
7–18 Nov 1983	Geneva	UN Conference on Tropical Timber (Part 2)[1]
2–6 July 1984	Geneva	Preparatory Committee, 1st ITTC session
1 April 1985		Entry into force of International Tropical Timber Agreement, 1983
17–28 June 1985	Geneva	1st session of ITTC (Part I)
25–29 Nov 1985	Geneva	1st session of ITTC (Part II)
28 July–1 Aug 1986	Geneva	1st session of ITTC (Part III)[2]
January 1987	Yokohama	Establishment of ITTO secretariat
23–27 March 1987	Yokohama	2nd session of ITTC
16–20 Nov 1987	Yokohama	3rd session of ITTC, 1st of committees
22 June–1 July 1988	Rio de Janeiro	4th session of ITTC, 2nd of committees
9–16 Nov 1988	Yokohama	5th session of ITTC, 3rd of committees
16–24 May 1989	Abidjan	6th session of ITTC,[3] 4th of committees
30 Oct–7 Nov 1989	Yokohama	7th session of ITTC, 5th of committees
16–23 May 1990	Bali	8th session of ITTC, 6th of committees
16–23 Nov 1990	Yokohama	9th session of ITTC, 7th of committees
29 May–6 June '91	Quito	10th session of ITTC,[4] 8th of committees
28 Nov–4 Dec 1991	Yokohama	11th session of ITTC, 9th of committees
6–14 May 1992	Yaounde	12th session of ITTC, 10th of committees
16–24 Nov 1992	Yokohama	13th session of ITTC, 11th of committees
11–19 May 1993	Kuala Lumpur	14th session of ITTC, 12th of committees
10–17 Nov 1993	Yokohama	15th session of ITTC, 13th of committees
31 March 1994		Date International Tropical Timber Agreement, 1983 would have expired had a successor agreement — International Tropical Timber Agreement, 1994 — not been concluded

Notes:
1. Conclusion of negotiations for International Tropical Timber Agreement, 1983.
2. Decision on headquarters site and election of executive director.
3. First extension of the ITTA 1983 for the period 1 April 1990–31 March 1992.
4. Second and final extension of ITTA 1983 for period 1 April 1992–31 March 1994.

Table 3.2 *Project work of the International Tropical Timber Organization as at 31 December 1992*

Permanent committee	Approved projects	Total budget US$ m	ITTO budget US$ m
Reforestation and forest management	96	99.9	71.2
Forest industry	60	39.5	25.7
Economic information market intelligence	23	10.4	9.6
Total	179*	149.8	106.5

* Of these 179 projects, 28 were completed, 64 were operational, 30 were pending contracts, 33 were pending finance and 24 had been set aside under the ITTO's sunset provision.

Source: UN document TD/TIMBER.2/3, 'Background, Status and Operation of the International Tropical Timber Agreement, 1983, and Recent Developments of Relevance to the Negotiation of a Successor Agreement', 26 February 1993, paras 46–7, pp9–10.

projects meeting the stated objectives of the ITTA 1983 are forwarded to the permanent committees. In 1991 the ITTC passed a 'sunset provision': approved pre-projects or projects that have not received funding within 20 months will lose their approved status; those that have received funding, but for which implementation has not begun, will lose their approved status after 26 months. Also in 1991, the ITTC granted the executive director the power to approve pre-projects and projects up to the value of US$ 50,000;[27] this was increased to US$ 75,000 in 1992.[28]

FoE alleges that 'political considerations apparently override environmental, social, technical [and] financial ... considerations' in the project approval process.[29] Two projects that attracted particular NGO criticism at the ITTC's eleventh session in 1991 were those for multipurpose tree planting and forest nursery development in Egypt.[30] The expert panel recommended that technical changes be made before concluding, with identical comments for both proposals, 'The Panel was divided in its view on the relevance of the Proposal for ITTO support, and therefore felt that the decision on relevance should be taken by the Committee itself.'[31] FoE reported that pressure was exerted on the expert panel by the secretariat not to reject the proposals, despite the fact that they did not meet the objectives of the ITTO as specified in Article 1 of the ITTA 1983. The chairman of the Permanent Committee on Reforestation and Forest Management subsequently concluded for one of the projects 'that while he considered that the project did not contribute directly to major ITTO objectives ... *there was no justification in either the ITTA or past project approval practices of ITTO to preclude approval and funding of the project*' (emphasis added).[32] The two projects were subsequently approved by the ITTC leading FoE to suggest that approval was given to avoid possible withdrawal by the Egyptian government from the ITTO. FoE has documented other examples of interference in the project approval process. For example, when the ITTC meets in a producer country it has become commonplace for the host country to table several project proposals, 'approval of some of which at least is deemed to be recompense for the expenses incurred in hosting meetings'.[33] Despite these criticisms, there is general agreement that the expert panel has rationalized and improved the efficiency of the project screening process.

The IIED report on sustainable forest management

In 1988 an ITTO-commissioned study by the IIED presented its findings to the ITTC's fifth session on the sustainability of forestry management practices worldwide. The study team, led by Duncan Poore, concluded that, 'The extent of tropical moist forest which is being deliberately managed at an operational scale for the sustainable production of timber is, on a world scale, negligible.'[34] The report found that less than 1 per cent of the global tropical timber trade, namely from Queensland, Australia, came from what the investigation team considered to be sustainable sources.

There were two main ramifications of the IIED report's findings. First, as detailed below, the report led to initiatives from both the ITTO membership and the NGO community to promote sustainable forest management. Second, the report contributed to a long-running debate on precisely what constitutes sustainable forest management. The question of whether sustainable forest management is possible is one that has frequently occupied NGOs. Colchester has distinguished between two types of NGOs: those that believe that sustainable logging of tropical forests is possible and those that do not. In the former category he includes IIED, IUCN and FoE.[35] However, even within NGOs the debate may continue. For example, a representative of Sahabat Alam Malaysia, the Malaysian branch of FoE, has argued that there is 'no management solution' to deforestation caused by logging.[36]

What is 'sustainable forest management'?

At the heart of the debate is the question of what criteria are to be included in any definition or measurement of sustainable forest management. Article 1(h) of the ITTA 1983 emphasizes the maintenance of the ecological balance in forest regions, and the IIED report also emphasized genetic resource conservation and respect for the rights of indigenous peoples as elements of an 'ideal policy' for forest conservation. Sustainable forest management is just one 'sustainable' concept to have been floated at the ITTO. 'Sustainable timber production' is another. In the report Poore notes that representatives from some countries have interpreted this concept to mean 'continuity of [timber] supply from the natural forest, implying that when one source is exhausted, another will be found'.[37] Not surprisingly, this is an interpretation with which Poore and environmental NGOs disagree. Poore declines to provide a firm definition of sustainable forest management or sustainable timber production, noting that criteria will vary according to, for example, market and silvicultural conditions. He does, however, note that 'If production of timber is to be genuinely sustainable, the single most important condition to be met is that nothing should be done that will *irreversibly reduce the potential of the forest to produce marketable timber* — that is there should be no irreversible loss of soil, soil fertility or genetic potential in the marketable species [emphasis in original].'[38] Utting notes that others interpret 'sustainable forest management' to mean that 'the volume of timber extracted in a period of years should not exceed the volume of new growth'.[39] In 1991, following the report of an ITTO expert panel, which considered possible methods for defining general criteria for and measurement of sustainable tropical forest management,[40] an ITTC decision adopted an approved ITTO definition of sustainable tropical forest management: 'Sustainable forest management is the process of managing permanent forest land to achieve one or more clearly specified objectives of management with regard to the production of a continuous flow of desired forest products and services without undue reduction of its inherent values and future productivity and without undue undesirable effects on the physical and social environment.'[41] In 1992 FAO promulgated a further definition of how sustainable development applies to forests (and other resources coming under the FAO's domain):

> *Sustainable development is the management and conservation of the natural resource base and the orientation of technological and institutional change in such a manner as to ensure the attainment and continued satisfaction of human needs for the present and future generations. Such sustainable development (in the agriculture, forestry and fisheries sectors) conserves land, water, plant and animal genetic resources, is environmentally non-degrading, technically appropriate, economically viable and socially acceptable.*[42]

In many ways the concept of sustainable forest management is similar to that of sustainable development. First, it is a concept that attempts to synthesize developmental and conservationist objectives, in this case Articles 1(b) and 1(h) of the ITTA 1983. Second, although it has attracted popular (though not universal) approval, those actors who subscribe to the concept of sustainable forest management have so far proved unable to agree upon a clear unambiguous definition for it, still less agree upon precise criteria by which it can be gauged. It can be seen that in the two definitions provided above, there is a wide degree of latitude within which actors' interpretations may vary. In the ITTO's definition, terms such as 'clearly specified object-ives' and 'undue desirable effects' are not defined. Similarly, in the FAO's definition, value judgements such as 'economically viable' and 'socially acceptable' are left unclear. (One may ask 'viable' in what way and 'accept-able' to whom?) In short, there is a wide margin for interpretation in both definitions. It is within these margins that contention between actors occurs, with disagreement arising as to what criteria are to be included, and how they are to be measured. It is clear that once the social and economic dynamics of sustainability enter the forest management discourse, the range and number of actors within the frame of forest management is vastly increased, as is the possibility of competing interpretations. In such a con-text it has been argued that 'sustainable forest management' can only be defined in relation to a defined group of actors and interests. As we shall see in Chapter 6, several fora have sought to agree upon criteria and indicators for 'sustainable forest management' since the UNCED.

The ITTO's Target 2000 and the WWF's 1995 target

The findings of the IIED report catalysed action in the NGO community. In August 1989 WWF adopted the target date of 1995 by which time it was intended the entire tropical timber trade should come from sustainable sources.[43] The WWF target has since been extended to non-tropical timbers. At the ITTC's seventh session in November 1989 the WWF called for the ITTO to adopt a target date[44] with the intention that the ITTO would meet its own 1995 target. The ITTO members resisted pressure to match WWF's date, but at the ITTC's eighth session in Bali in 1990 they took steps to establish the target date of the year 2000. The origins of this date, known as Target 2000, can be traced to a 'Draft Action Plan and Work Program' prepared by the Permanent Committee on Forest Industry in May 1990, which noted that the target date of 2000 had received the support of both producers and consumers. This document contributed to the ITTO action plan, which endorsed the year 2000 as the target date by which 'all tropical timber exports should come from sustainably managed forests'.[45] The action plan was presented to the ninth ITTC session at Yokohama, November 1990, where WWF unsuccessfully lobbied for the ITTO to match its 1995 date.[46] The ITTC formally endorsed Target 2000 at its tenth session in Quito when an ITTC decision invited members to provide a paper on their proposed progress towards the target to the eleventh session.[47] Target 2000 has attracted controversy with contention centring on whether the target should refer to the trade in tropical timber or the sustainable management of all tropical forests; the ITTO action plan stipulates that Target 2000 refers only to the tropical timber trade, while many NGOs consider that the target should refer to all tropical forests, irrespective of whether or not they contribute to the trade. During the negotiation of the International Tropical Timber Agreement, 1994 Target 2000 was renamed Objective 2000 (Chapter 5).

Meanwhile, the WWF continued to work towards sustainable forest management by 1995. In the UK, WWF has formed a working relationship with the Timber Trade Federation on this issue. A second prong to WWF's UK policy is the '1995 Group', consisting of companies that have publicly endorsed the WWF's target date.[48] In February 1996 WWF's '1995 Group' was relaunched as the '1995 Plus Group'. Only one government formally endorsed the WWF 1995 target date, namely that of the Netherlands.[49] The Dutch timber industry supported the 1995 campaign and after two years of negotiation a framework agreement was signed between Dutch government ministries, trade unions, the Netherlands Timber Trade Association, WWF and IUCN.[50] In February 1994 this consensus was broken when the WWF and IUCN withdrew from the agreement in protest at Dutch government attempts to include as 'sustainable' timber from conversion forests (that is forests that are in the process of being converted to an alternative form of land use such as agriculture).[51] Once again, the confusion that arises between different actors with different conceptions of sustainable forest management proved divisive.

The ITTO as a normative organization

Three sets of guidelines have been adopted at ITTC sessions. The first set, the *ITTO Guidelines for the Sustainable Management of Natural Tropical Forests*, was adopted in Bali in 1990 where the ITTC commended them as 'an international reference standard to Members'.[52] The guidelines contain 41 principles for the sustainable management of natural tropical forests, with 36 recommended possible actions as to how these principles may be realized. The ITTO's executive director recommended that the guidelines be 'shaped into more specific guidelines which are compatible with regional and national forestry practices'.[53] The guidelines reflected inconsistencies between conservationist and developmental objectives. For example, 'possible action 33' notes that environmental impact studies should 'assess compatibility of logging practices with *declared secondary objectives such as conservation and protection, and with the overall principle of sustainability*' (emphasis added).[54] Here the expansion of the tropical timber trade [Article 1(b) of the ITTA 1983] clearly has a higher priority than conservation (Article 1(h)].

In June 1991 the ITTC approved a second set of guidelines, the *ITTO Guidelines for the Establishment and Sustainable Management of Planted Tropical Forests*,[55] containing 66 principles and 75 recommended possible actions. The *ITTO Guidelines for the Conservation of Biodiversity in Tropical Production Forests*,[56] consisting of 14 principles and 20 recommended possible actions, were approved by the ITTC in November 1992.

The mechanics of the drafting processes for these guidelines are worth brief consideration. In each case a small expert panel was established, composed of representatives from producer and consumer countries, UN specialized agencies, NGOs, the timber trade and the ITTO secretariat. A draft document was prepared in advance of the expert panels' first meetings. In the case of guidelines for the natural tropical forests, the UK ODA assisted the ITTO secretariat to prepare a paper. The German government offered to finance the initial draft on planted tropical forests,[57] a gesture which gave them a say in the content of the resulting guidelines; the initial draft was produced by the Research Institute for Forestry and Forest Products in Hamburg.[58] Finally, the biodiversity guidelines were based on the outputs of a workshop held alongside the eighteenth General Assembly of the IUCN (Perth, Australia, November–December 1990).[59]

Although the ITTC decisions adopting the guidelines 'invited' members to take them into account when deciding national policy, not one country to date has openly admitted to using the ITTO guidelines as the basis for producing national guidelines. As noted above, the ITTC's tenth session invited members to report at the eleventh session on proposed country measures to be taken to realize Target 2000. At the eleventh session, only seven of ITTO's members (47 at the time) submitted a report of any form. Information submitted by ITTO members is not used to assess compliance with the agreement, although it is useful for compiling data on the global tropical timber economy. The poor level of national reporting has given rise to a sustained campaign by NGOs, which have criticized member governments for not implementing ITTC decisions, for failing to review progress on projects, and for not demonstrating greater commitment to Target 2000 and to the ITTO's objectives and guidelines.

The ITTO and indigenous forest peoples

Indigenous peoples' groups are becoming increasingly well-organized, both at the ITTO and other fora, where they have put forward their claims that forest dwellers should be granted title to their customary and ancestral lands. Many environmental NGOs have forged a close relationship with, and lobbied the ITTC on behalf of, indigenous peoples' groups. For example, at the ITTC's tenth session environmental NGOs urged that ITTO projects be developed with the full participation of affected forest peoples'.[60] However, the principle that indigenous peoples and other local communities should be accorded full rights of participation in the design and implementation of projects has yet to secure the agreement of all ITTO members, with producer delegations concerned that such a principle could lead to an erosion of sovereignty.

At the ITTC's eleventh session, environmental NGOs drafted a resolution which, if accepted, would have affirmed a commitment from the ITTO to respect the rights and secure the livelihoods of forest-dwelling peoples and would have instructed the ITTO secretariat to commission an independent study of the impact of timber extraction on forest-dwelling peoples. The draft was discussed informally outside council, where the EC voiced support, but was not adopted due to producer opposition.[61]

Despite the lobbying of environmental NGOs and indigenous peoples no ITTC decision has recognized the land rights of forest peoples. FoE and the World Rainforest Movement have reported that, despite the emergence of evidence that ITTO projects have caused social problems and although the ITTO's natural tropical forests guidelines recommend that indigenous peoples be consulted and customary rights respected,[62] ITTO projects infringe upon the customary lands of indigenous forest peoples who are included in neither decision-making nor project management.[63] On the rare occasions when ITTO projects have provided for consultation with indigenous peoples, such as the Chimanes project in Bolivia,[64] this has been possible only after NGO pressure and prior approval of the host government. Producer governments have been prepared to recognize only the importance, but not the claimed land rights, of these groups. The rights to timber (as asserted by the timber traders and producer countries) and the rights to land (as asserted by the alliance between indigenous peoples' groups and environmental NGOs) has therefore been one of the most acute points of conflict within the ITTO, with the former adopting a narrow economistic definition of sustainable forest management, and the latter arguing that sustainability cannot ignore broader social concerns.

Indigenous forest peoples' groups have used their statements to the ITTC to challenge narrow definitions of sustainability. For example, the ITTC's tenth session at Quito was addressed by a spokesman for the Coordinating Body for the Indigenous Peoples' Organizations of the Amazon Basin (COICA) who stated that 'we insist that one cannot speak of sustainability without sustaining the livelihoods of those who live in the forests'.[65] Similar points have been made by other NGOs. An interesting case concerns the ITTO Mission to Sarawak.

The ITTO Mission to Sarawak

In 1989 the ITTO initiated, at the request of the Malaysian government, an expert mission under the leadership of the Earl of Cranbrook to investigate the sustainability of forest management in Sarawak. This is the only occasion that the ITTO has investigated the forest management practices of one of its members. The Mission's report was presented to the ITTC in 1990. It concluded that Sarawak's forests would disappear by 2001 if current logging rates continued and called for a 30 per cent reduction in the timber harvest,[66] a figure criticized by WWF as too low.[67]

The Mission's report brought into renewed focus the debate on sustainable forest management. FoE and the World Rainforest Movement criticized the mission for adopting a narrow technical interpretation of its terms of reference, investigating only the extraction of timber and 'thereby marginalizing not only human considerations but also alternative forms of forest use'.[68] The Malaysian government's response to NGO concerns on indigenous peoples' claims to ancestral land was that such claims must be dealt with under Malaysian law, and not by the ITTO. The Malaysian government further asserted that its invitation to the ITTO to establish the Mission 'was in no way diminishing the exercise of its sovereignty and independence of action'.[69]

The incentives and labelling debates

Perhaps the debates that best highlight the gulf between the environmental NGOs and the performance of the ITTO are those on labelling and incentives. These two debates became entangled with each other and by 31 March 1994 the ITTC had made no firm decision or commitment with respect to either issue.

The labelling issue was first debated at the fifth session of the Permanent Committee on Economic Information and Market Intelligence at Yokohama in 1989 when a pre-project proposal, *Labelling Systems for the Promotion of Sustainably-Produced Tropical Timber*, was tabled by the British delegation.[70] Prepared by FoE with some input from the Oxford Forestry Institute (OFI), the proposal was opposed by Malaysia.[71] The committee's judgement was that the pre-project 'was a veiled attempt to install ... an incentive to encourage the current campaign of boycott against the import of tropical timber products'.[72] This passage strongly implies that the proposal threatened the interests of the producer countries and the timber trade. The DTI, then the lead agency for the UK delegation, subsequently redrafted the proposal later in the session without consulting FoE. In the new version, *Incentives in Producer and Consumer Countries to Promote Sustainable Development of Tropical Forests*,[73] all references to labelling had been excised. At this stage FoE withdrew its support.[74]

The revised proposal was not forwarded to the ITTC and instead became the subject of further research. The ODA, by now the lead British agency, engaged the UK Timber Research and Development Association (TRADA) which together with the OFI drafted a pre-project proposal on possible financial and non-financial incentives for sustainable forest management. Financial incentives include funding of forest management services by, *inter alia*, tax transfers, debt-for-nature swaps and grants, while non-financial incentives include security of land tenure, certification schemes for good forest management practice and the development of non-timber forest products. The OFI/TRADA report[75] was debated at the ITTC's tenth session in Quito in 1991 at a specially convened round table. The chairman of the round table considered there was a 'need to define an acceptable compromise between the environmental value of the forest and the economic value of trade in tropical timber'.[76] Again the tension between the developmental and conservation components of the ITTA 1983 is evident, a tension reflected in the subsequent ITTC decision. This invited members 'to enhance their ability to attain the Year 2000 Target *by investigating liberalized trade in tropical timber*' (emphasis added),[77] a view that clearly accepts that conservation can be achieved without trade restrictions.

Prior to the Quito round table, the British ODA initiated a study on the economic linkages between the tropical timber trade and sustainable forest management. The London Environmental Economics Centre (LEEC) acted as consultants for this study, which became an ITTO project. Its report, *The Economic Linkages Between the International Trade in Tropical Trade in Tropical Timber and the Sustainable Management of Tropical Forests*,[78] considered how environmental and social costs could be internalized into pricing mechanisms, and was debated at the ITTC's thirteenth session in November 1992. Its findings with respect to labelling are considered below.

A related development at this time was the Austrian parliament's vote for legislation that all tropical timber imports should be labelled. At the ITTC's thirteenth session an Austrian delegate asserted that the legislation was not discriminatory or protectionist, nor a restriction to trade.[79] The measure was applauded by environmental NGOs at the ITTO[80] but met with opposition from producer countries, especially Malaysia and Indonesia. At the General Agreement on Tariffs and Trade (GATT) Council meeting of November 1992 the ASEAN contracting parties, namely Indonesia, Malaysia, the Philippines, Singapore and Thailand, expressed 'serious concern' about the action and claimed it was discriminatory as it did not apply to temperate timbers.[81] Austrian industrialists, concerned about the possibility of a trade war, lobbied the Austrian parliament following which the law was amended, but only after Green MPs staged a 30-hour filibuster.[82]

The creation of the Forest Stewardship Council

WWF signalled disillusionment over the ITTO's performance on labelling in 1988 when stating that if 'the ITTO fails to actively promote tropical forest conservation ... then conservation organizations will have to seek other mechanisms to achieve this'.[83] Three years later WWF carried this threat into effect. The failure of the ITTO to deal effectively with labelling led to a shift in WWF labelling policy. From advocating government-backed labelling schemes, which can be considered GATT-illegal (see below), WWF promoted a global private-sector scheme which, as all the relevant actors concerned are non-governmental, cannot be GATT-illegal. After a series of meetings with timber traders and other NGOs, agreement was reached to found a Forest Stewardship Council (FSC). In October 1993 the founding assembly of the FSC was held in Toronto. The FSC focuses on activity at the forest-concession level as opposed to the intergovernmental level. The rationale behind the FSC is that cooperation with those firms granted forest concessions is more likely to ensure sustainability than intergovernmental activity at the ITTO. FSC membership is voluntary. The FSC aims to accredit national certifying authorities to issue certificates where the management practices of forest managers applying for certification meet the FSC's criteria for well-managed forests. The FSC brings together an unusual coalition of conservation NGOs and timber traders. These two groups have previously disagreed at the ITTO, and it remains to be seen how well they will cooperate in the FSC. However, there have already been splits in the NGO community, with Greenpeace and FoE withdrawing from the process (while agreeing to remain as observers) in protest at the founding assembly's decision to give those with an economic interest representation on FSC organs.[84]

The FSC initiative and the LEEC report rekindled the labelling debate within the ITTO. The LEEC report recommended that 'ITTO should encourage the establishment of a *country certification* scheme' (emphasis in original).[85] It was precisely such a scheme that FoE had promoted in 1989. However, since then many NGOs have moved their support from country schemes to arguing that only a voluntary non-governmental approach, such as the FSC, can be credible.[86] The FSC receives further consideration in Chapter 6.

In May 1993 Chris Elliott of WWF–International, a leading figure in the launch of the FSC, attended the ITTC's fourteenth session in Kuala Lumpur. Labelling was discussed for two days at a session of the Permanent Committee on Market Information and Market Intelligence. The consumer delegations favoured labelling for tropical timber only, while the producers advocated a scheme for all timbers. The NGOs shared the producers' view that a labelling system should cover all timbers, and supported a finding of the LEEC report that labelling was the most effective trade-related incentive for sustainable forest management, while disagreeing with the LEEC's conclusions that labelling be authorized at country level. Consistent with their general support for the FSC, the NGOs argued for a global scheme with certification at the forest-concession level. The permanent committee made no recommendations to the ITTC on this issue.

What started as a proposal by FoE on labelling resulted in protracted, and as yet unresolved, discussion. In 1992 an NGO representative commented that 'ITTO has failed to deal adequately with the whole question of incentives, monitoring, tracing, certification and labelling'.[87] The labelling and incentive debates illustrate how the ITTO's consensual decision-making procedures have stifled original policy initiatives. In effect, every member, or at least every member with a large number of votes, holds the power of veto. This has resulted in the blocking of decisions on any form of market intervention. Bramble and Porter have commented that 'the objective of ITTO is to promote the timber trade and no decision-maker in it has a primary interest in conservation'.[88] This may be rather cynical, and perhaps exaggerated, but certainly it is the case that conservation has not been allowed to threaten trade interests at the ITTO.

THE INTERNATIONAL RELATIONS OF THE ITTO

The role of the FSC clearly overlaps to a degree with that of the ITTO. Indeed, there are several international institutions whose mandate may overlap with that of the ITTO. This section analyses the relations of the ITTO with respect to three such institutions, namely the GATT, the TFAP and the Convention on the International Trade in Endangered Species of Wild Fauna and Flora (CITES).

The ITTO and the GATT

Before analysing the tensions that existed between the GATT and the ITTA 1983, it is necessary to clarify what is meant by the GATT. First, there is the GATT as an international legal agreement, of which there have been two, namely the General Agreement on Tariffs and Trade, 1947 and subsequent amendments, and the General Agreement on Tariffs and Trade, 1994. The latter consists of the GATT 1947; all legal instruments that have entered into force under the GATT 1947 before the date of entry into force of the World Trade Organization Agreement, 1994; the instruments concluded during the Uruguay Round; and the Marrakesh protocol to the GATT 1994.[89] Second, there is the GATT as an institution. This existed from 1947 until December 1995 and embraced the GATT secretariat, the GATT council, GATT standing committees and GATT panels convened to settle trade disputes. Finally, when negotiations, or 'rounds', have been conducted on trade and trade-related issues, the GATT may be seen as an intergovernmental negotiating process. The objective of this section is to evaluate the relationship between the GATT as an international legal agreement and the ITTA 1983. In the paragraphs that follow, where the term GATT is used in isolation, this refers to the GATT 1947.

The conservation objectives of the ITTA 1983 were in direct tension with the GATT in three ways. The first concerned extraterritoriality, that is where one GATT contracting party attempted to conserve natural resources outside its territorial domain. Possibly the most important environmental case to go before a GATT panel, and one with ramifications for the tropical timber trade, was the dolphin–tuna case of 1991 between the USA and Mexico. Prior to this the USA had banned imports of tuna fish caught by Mexican fishermen using nets that ensnared dolphins. The Mexicans made a complaint, and a GATT panel subsequently ruled that GATT Article XX(g) on the conservation of exhaustible natural resources could not be invoked by one GATT contracting party to ensure the protection of the environment or of natural resources beyond its territorial boundaries.[90] As WWF notes, 'This may be logical in the context of a free trade agreement, but does not further the objective of ensuring that any trade liberalization resulting from the agreement is sustainable.'[91]

The second tension concerned GATT clauses prohibiting discrimination between like products on the basis of their manufacture.[92] These clauses effectively gave unsustainably-produced timber the same status in the international market as sustainably-produced timber. Various analysts concluded that unilateral bans on imports of tropical timber from unsustainable sources would have contravened the GATT as GATT signatories would have been prohibited from using tariffs or quotas to favour timber from sustainable sources.[93]

The third instance where GATT and ITTA clauses were in tension centres on Article 1(e) of the ITTA 1983 on the further processing of timber. This is designed to encourage producers to add value to timber prior to export, thus increasing their export earnings by selling processed planks and plywood as opposed to unprocessed logs. In 1990 the WWF reported that the EC disputed a ban on the export of raw tropical logs by Indonesia on the basis that it infringed GATT Article XI, which aimed at the general elimination of quantitative restrictions. The Indonesians claimed that the measure was a conservationist one. The EC denied that the ban was a genuine conservationist measure as it was not applied, as stipulated by GATT Article XX(g), 'in conjunction with restrictions on domestic production or consumption'. It can be argued that, irrespective of whether or not domestic levels of timber production and consumption are reduced, a ban on log exports is a conservation measure inasmuch as it allows a country's export earnings to be maintained for a lesser volume of timber. (This is the high-value/low-volume concept to which many conservationists refer.) Despite its disapproval, the EC did not request the establishment of a GATT panel with respect to Indonesia's measures on logs, and neither did it raise a formal complaint on the matter at a GATT council meeting.[94] However, if such a complaint had been made and upheld, a GATT panel could have ordered Indonesia to recommence the export of raw logs, a factor which would have increased the rate of deforestation of Indonesian forests if Indonesia's export earnings were to be maintained.

The Indonesia–EC case is the second example where ITTO members have cited the GATT. In the Austrian labelling case (see above) reference to the GATT by the ASEAN governments was a factor that contributed to the reversal of Austrian policy. This was not the case with the Indonesian–EC case, which one legal expert attributes to developed states being more willing to provide developing states 'with some leeway in fulfilling their GATT obligations than they would be willing to provide each other'.[95]

The WWF recommended that the ITTO secretariat seek a waiver from the GATT for any trade measures deemed necessary to contribute to the conservation and the sustainability of tropical forests, a position reiterated by TRAFFIC International and the updated version of the WWF/IUCN/ UNEP's World Conservation Strategy published in 1991.[96] To date, the secretariat has not done this. With none of the possible contradictions between the ITTA 1983 and the GATT having been tested in a GATT panel, and with the ITTO having no dispute settlement procedures of its own, the tensions between the GATT and the ITTA 1983 were never resolved.

On 1 January 1995 the World Trade Organization was created. (The closing session of the GATT Contracting Parties was held in Geneva in December 1995; there was therefore a period of one year when the GATT and the WTO existed in parallel.) The ITTA 1994 has yet to enter into force and, as noted earlier in this chapter, until such time as it does the ITTA 1983 may remain extant by special vote of the ITTC. However, all of the clauses in the ITTA 1983 that clashed with clauses in the GATT 1947 remain in the ITTA 1994, some of them in a modified guise. Furthermore, the GATT 1994 provides no clarification of those areas of dispute between the GATT 1947 and the ITTA 1983. In short, despite the major transition that has occurred in the multilateral trading system, the contradictions identified in this section remain. Future researchers will wish to consider how these issues are dealt with by the WTO, the ITTO and other actors. In this respect it is significant that although the GATT 1994 and other instruments agreed upon in the Uruguay Round make no explicit provisions for environmental protection, the environment is an issue to which the WTO will give consideration. The first meeting of the WTO General Council established a Committee on Trade and the Environment. Presumably this committee will seek to deal with the contradictions between the GATT 1994 and those legal instruments with a conservation mandate, such as the ITTA 1994.

The ITTO and the TFAP

There are four ways of considering the nature of the ITTO–TFAP relationship: the degree of cooperation on projects; the formal institutional linkages at the international level; the linkages between the Forestry Advisors Group (FAG) and the ITTO; and the role NGOs and other actors have played in providing linkages between the two processes.

The second action programme of the *Tropical Forestry Action Plan* is forest-based industrial development, while the fourth action programme is conservation of tropical forest ecosystems (Chapter 2). These action programmes correspond very closely to Articles 1(b) and 1(h) respectively of the ITTA 1983. However, the horizontal linkages between the ITTO and the TFAP have been particularly weak, despite their overlapping mandates. By March 1992 the ITTO has acted as a supporting agency for only one NFAP (Table 2.3, Chapter 2). This was the Papua New Guinean NFAP where the ITTO funded two projects, with the main donors being the Australian government's aid agency, the World Bank and the UNDP.[97] At the ITTC's tenth session in 1991 the delegate for Papua New Guinea emphasized 'the difficulties experienced by host countries in adhering to different guidelines for different donors'.[98] The eighth session of the Permanent Committee on Forest Industry, which approved one of the projects, amended the project design to avoid duplication with other NFAP projects in the country.[99] In short, the horizontal linkages between ITTO projects and NFAPs have been either non-existent or weak.

There are certain formal linkages between the ITTO and the TFAP at the international level. Members of the ITTO secretariat, sometimes the executive director, have attended sessions of the FAO's Committee on Forestry and its Committee on Forest Development in the Tropics. A member of the FAO's Forestry Department has attended all ITTC sessions as an observer, and representatives from the other TFAP cofounders, namely the World Bank, United Nations Development Programme and the World Resources Institute have frequently, although not always, attended ITTO sessions.

The relationship between the ITTO and the FAG has not been strong, although this is an issue to which the FAG is now paying increasing attention. The FAG chairman has occasionally attended the ITTC as a nongovernmental observer as have NFAP national coordinators attending ITTC sessions. However, representation from the FAG and NFAP offices at ITTC sessions has very much been the exception rather than the rule. In 1988 an NGO statement to the council noted that NGOs were 'somewhat disturbed at the lack of joint activities' between the FAG and the ITTO and called for 'an effective functional partnership between the ITTO and other institutions'.[100]

A fourth way of gauging the strength of the ITTO–TFAP relationship is by identifying NGOs and other actors that have relations with both processes. In an interdependent world with multiple channels of communication NGOs can play a role in promoting closer cooperation between institutions, even when formal horizontal linkages are poor or non-existent. NGOs that have had a working relationship with both the ITTO and the TFAP include IUCN, WWF, WRI, FoE and IIED. The IIED, through its participation in the FAG, has taken a particularly significant interest in the ITTO–TFAP relationship. At the fifteenth meeting of the FAG in Costa Rica in 1992 Caroline Sargent of the IIED noted that although the ITTO's Target 2000, strictly interpreted, referred only to the international trade in tropical timber, a broader interpretation saw the target as very close to the TFAP's new goal and objectives.[101] Yet the majority of ITTO projects are formulated without reference to NFAPs. Since then the IIED has conducted a study on the ITTO–TFAP relationship which argues that work within the ITTO 'must be nested within national forest planning and management mechanisms; whilst TFAP would benefit from the establishment of mutually determined guidelines and agreed standards as introduced by the ITTO'. The study concludes that 'the objectives of each initiative would benefit from closer cooperation'.[102] However, some NGO assessments are less optimistic. A WWF commissioned study argued that the TFAP, with a general forestry mandate, has become focused on industrial wood production, while the ITTO, a commodity agreement, has a conservationist objective; hence there is no complementarity between the two initiatives and the 'level of effort and funding that goes into development of industrial forestry via TFAP renders nugatory ITTO's ... experiments in sustainable forest management'.[103]

The weak nature of the ITTO–TFAP relationship has attracted comment. In November 1990 the WWF called for the replacement of the TFAP with an appropriate mechanism with better linkages to the ITTO.[104] In September 1992 a German–Japanese Expert Meeting on Tropical Forests stated that the TFAP and ITTO should be 'complementary instruments'.[105] In December 1992 the government of the Netherlands stated that it favoured the establishment of national long-term forest management plans by producer countries within the ITTO, which 'may help in the establishment' of NFAPs.[106] Future researchers who wish to assess changes in the nature of the ITTO–TFAP relationship may wish to consider the number of NFAPs for which the ITTO acts as a supporting agency (in other words the number of ITTO projects coming under the auspices of separate NFAPs), the number of NFAPs that refer to ITTO guidelines and/or the number of ITTO projects that refer to TFAP guidelines, and other indications of increased cooperation at the international, regional, national and local levels.

The ITTO and CITES

At the ITTC's third session in 1987 the WWF called for ITTO members to identify endangered tropical tree species.[107] Following this the ITTO commissioned the World Conservation Monitoring Centre to produce a pre-project study on the conservation status of traded tropical timber. (Based in Cambridge, the World Conservation Monitoring Centre monitors bio-diversity conservation worldwide and is a joint venture of the IUCN, UNEP and the WWF.) The study found that over 300 tropical timber species are threatened with extinction in Africa and Southeast Asia, many of them in ITTO countries.[108] The study was presented to the ITTC's tenth session in Ecuador, together with a proposal for a tropical timber conservation database. Subsequently, this proposal was reviewed at an ITTO workshop in Cambridge UK which discussed the possibility of establishing a global endangered timber species database under the auspices of the ITTO and in cooperation with the World Conservation Monitoring Centre. The govern-ments of the UK and the Netherlands offered to finance the project. Scientific experts from Brazil, Indonesia, Malaysia and Ghana were among the delegates. However, following opposition from producer delegates the meeting decided against a global database, but did call for research on the development of national databases 'under the ownership of the country concerned', separate from the ITTO secretariat, with 'no access to databases ... permitted to any other party'.[109]

The recalcitrance of the ITTO and its member delegations to become involved with the conservation status of traded tropical timber led to pressure for the CITES to list endangered species. At the eighth meeting of the parties to CITES in Kyoto, Japan in March 1992 commercially-traded tropical tree species were listed as requiring protection for the first time since the inception of the ITTO. One species was listed on CITES Appendix I, which bans international trade, while three were placed on Appendix II listing, which requires the monitoring of international trade.[110] Delegations were divided as to whether the ITTO or CITES should have the authority to restrict the trade in endangered tropical tree species. A proposal by the Netherlands to list two further species was resisted by Malaysia, which argued that the ITTO was the responsible intergovernmental organization for such measures.[111]

Following these disagreements, the ITTC session held after the Kyoto meeting passed a decision designed to improve ITTO–CITES cooperation. While recognizing that ITTO and CITES 'are separate institutional entities with distinct mandates and separate membership', the decision recognized that 'the roles of the two entities are potentially complementary in some areas related to internationally-traded tropical timber'.[112] The first draft of the decision was presented by the Netherlands. TRAFFIC Oceania lobbied during the redrafting stages for the ITTO to undertake ongoing activities on tropical tree species conservation, as the result of which a paragraph was inserted in the decision inviting ITTO members to 'improve the information base regarding the conservation status of internationally traded tropical timber species'.[113] However, the decision was loosely-worded and made no concrete provisions to improve the ITTO–CITES relationship following producer protests. Indeed there has been general reluctance by the ITTO producers to pass any decision which could be viewed as a trade restriction, be it incentives, labelling or limiting the trade in endangered species.

In November 1994 the ITTO–CITES relationship improved when the parties to the CITES created a Timber Working Group. The group will make recommendations on Appendix II listings of timber, review the status of listed timber species and appraise the relationship of the CITES with other institutions, including the ITTO. Membership will principally be government-nominated experts, although IUCN and TRAFFIC have also been invited to attend. The ITTO secretariat is also represented on the working group, which held its first meeting in London in November 1995.[114]

Summary

The relations of the ITTO with other international institutions with similar or overlapping mandates have been poorly-defined and unclear. Despite their similar mandates, ITTO cooperation with the TFAP and the CITES has been poor, although in both cases the relationship is improving. However, the ITTO did not take any steps to deal with the contradictions between the ITTA 1983 and the GATT, and these tensions remain in the new multilateral trading system that has followed the creation of the World Trade Organization.

Chapter 4

The Forest Negotiations of the UNCED Process[1]

In this chapter the negotiation of a global forests instrument (GFI) is analysed during the preparatory process of the United Nations Conference on Environment and Development (UNCED). A GFI is defined as an international agreement between governments on the issue of forest conservation. It may be a legally binding instrument, such as a global forests convention (GFC) or a protocol to another convention, or it may be a non-legally binding instrument. The negotiation of a non-legally binding GFI is examined during the UNCED forest negotiations, namely the 'Non-legally binding authoritative statement of principles for a global consensus on the management, conservation and sustainable development of all types of forests'. The question is also considered here of why, despite the efforts of the governments of the developed North and of the UN's Food and Agriculture Organization (FAO), no GFC was opened for signature at the UNCED.

PROPOSALS FOR A GLOBAL FORESTS INSTRUMENT

The year 1990 was an eventful one for international forestry politics. As well as seeing the launching of the TFAP restructuring process and the publication of the ITTO's *Guidelines for the Sustainable Management of Natural Tropical Forests*, the year also witnessed several proposals for a GFI following United Nations General Assembly Resolution 44/228 (December 1989). This resolution announced that the UNCED would be convened in Rio de Janeiro in June 1992 and noted several environmental issues of major concern including, 'Protection and management of land resources by, *inter alia*, combating deforestation, desertification and drought'.[2] Resolution 44/228 set in motion nine proposals for a GFI.

The first proposal came from the Intergovernmental Panel on Climate Change (IPCC) workshop on Agriculture, Forestry and Other Human Activities (AFOS). The workshop was one of four sub-groups established by IPCC Working Group III, the Responses Strategies Working Group, which was tasked with formulating policy recommendations to limit greenhouse gas emissions. At its January 1990 meeting in São Paulo the workshop recommended a forestry protocol to a climate change convention. Its statement, the São Paulo Declaration, asserted that 'Forests cannot be considered in isolation, and solutions must be based on an integrated approach which links forestry to other policies, such as those concerned with poverty and landlessness. . . . Deforestation will be stopped only when the natural forest is economically more valuable than alternative uses for the same land.'[3] The São Paulo Declaration thus sought to broaden the forests component of the global warming debate, moving it from purely climatic considerations and introducing socioeconomic criteria.

The second recommendation for a GFI, and the first specifically to recommend a GFC, was made in May 1990 by the TFAP independent review (see Chapter 2). The review considered that the FAO could host the negotiations for an 'International Convention on Forests', and suggested that the proposal be put before the tenth session of FAO's Committee on Forestry (COFO) in September 1990.[4]

The third recommendation for a GFI was also made by a review of the TFAP, namely that of the World Resources Institute (WRI). This recommended that an 'international convention and protocols should be negotiated on a range of TFAP-related and parallel actions that are needed to address global deforestation issues, in order to achieve net afforestation within a decade'.[5]

The fourth recommendation was made in June 1990 at the European Council summit held in Dublin which recommended 'the early adoption of a Climate Convention and associated protocols, including one on tropical forest protection'.[6] This is the only proposal to call for a tropical, as opposed to a global, forestry instrument.

Fifth, the G7 Declaration at Houston in July 1990 stated that G7 leaders 'are ready to begin negotiations, in the appropriate fora, as expeditiously as possible on a global forest convention or agreement'.[7] (It will be recalled from Chapter 2 that this declaration also called for reform of the TFAP.) The G7's emphasis on a GFC differs substantially from the European Council's proposal for a tropical forests protocol to a climate change convention, despite the overlapping membership between these two fora. Two factors help explain this: international NGOs lobbied strongly on the forests issue at the summit;[8] and, according to a US government press release, President Bush played an important role in the G7's GFC proposal.[9]

The sixth proposal for a GFI, a European Parliament resolution, urged the Commission of the European Communities to 'give priority to advocating and working internationally for the preparation and implementation of a Worldwide Convention on the Protection of Forests'.[10] The resolution differed markedly from the European Council's proposal and signifies a shift within EC circles from a position supporting a tropical forests protocol towards the G7 proposal for a GFC.

The seventh source produced two proposals for a GFI, namely the Second World Climate Conference (SWCC) convened in Geneva, October 1990. Three statements were issued at the end of the conference, namely the conference statement, issued by the final scientific and technical plenary, the statement of non-governmental organizations and the ministerial declaration. The two former recommended a GFI. The conference statement recommended an international forests instrument linked with climate change and biodiversity conventions.[11] The NGO statement, endorsed by the 57 NGOs present, noted a need for 'a comprehensive global forestry agreement dealing with the conservation and sustainable management of boreal, temperate and tropical forests on a fair and equitable basis'.[12] The third SWCC statement, the ministerial declaration, did not recommend a GFI,[13] although ministers demonstrated consensus on other issues, especially the need for a framework convention on climate change. This provided an early indication that a global political consensus for a GFI was lacking.

The ninth proposal for a GFI came from the eighteenth session of the general assembly of the International Union for the Conservation of Nature and Natural Resources (IUCN), held at Perth, Australia. A draft resolution forwarded to the general assembly from one of its workshops urged state members of IUCN to negotiate a protocol on forests protection to a framework convention on climate change. The resolution was adopted by a majority vote, although the US and Canadian governmental delegations voted against as they supported a separate GFC.[14]

These nine proposals are summarized in Table 4.1 below. Note that only governments from the developed North had demanded a GFC. No government from the South had made or supported a proposal for a GFC or other type of GFI.

THE DRAFT GLOBAL FORESTS CONVENTION OF THE FAO

Following criticism for its handling of the TFAP, and in the wake of the above proposals for a GFI, FAO attempted to regain some legitimacy by seeking to gain for itself the leading role in the negotiation of a GFC. FAO considered that a GFC was necessary to realize the objectives of the biodiversity and climate change conventions then under negotiation. At the first

Table 4.1 Nine proposals for a global forests instrument

Date (1990)	Place	Source	Proposed instrument
11 January	São Paulo, Brazil	Agriculture, Forestry and Other Human Activities (AFOS) workshop of Working Group III of the Intergovernmental Panel on Climate Change	Forestry protocol to a climate change convention
May	Kuala Lumpur	Independent review of the Tropical Forestry Action Plan	International convention on forests
June	Washington DC	World Resources Institute review of the Tropical Forestry Action Plan	International convention on forests and protocols on parallel actions
26 June	Dublin	European Council	Tropical forests protocol to a climate change convention
11 July	Houston	Group of Seven Industrialized Countries	Global forests convention or agreement
25 October	Strasbourg	European Parliament	Worldwide convention on the protection of forests
7 November	Geneva	Conference statement of the Second World Climate Conference	International forests instrument linked with climate change and biodiversity conventions
7 November	Geneva	NGO statement of the Second World Climate Conference	Comprehensive global forestry agreement
5 December	Perth, Australia	18th session of the general assembly of the International Union for the Conservation of Nature and Natural Resources	Forests protection protocol to a framework convention on climate change

session of the UNCED Preparatory Committee FAO offered to provide the forum for UNCED-related forest negotiations.[15] The following month, in September 1990, the tenth session of the COFO resulted in ambiguity. The meeting report does not accord with the accounts of some observers. To Jeff Sayer of IUCN, the COFO gave 'a clear signal that it did not want [FAO] to proceed too far with the convention negotiations'.[16] Chris Elliott of WWF–

International also noted reservations by delegates, with Malaysia and Colombia making strong statements against continued FAO involvement with respect to a GFC.[17] There must be a suspicion that the meeting report did not faithfully reflect the proceedings. The report states that the COFO 'supported the concept of an international instrument on the conservation and development of forests' and that 'FAO would naturally play a leading role in the preparation of proposals for the envisaged instrument'.[18]

Although the FAO had no mandate from the COFO or within the UNCED process to take the lead role in any international forest negotiations, it responded by issuing a draft document entitled 'Possible Main Elements of an Instrument (Convention, Agreement, Protocol, Charter, etc) for the Conservation and Development of the World's Forests', hereafter referred to as the FAO draft (see Annex A). Despite the title there are strong indications that the draft was intended as the basis of a GFC; the first section, 'Preamble', is the standard beginning of a legal instrument and, more significantly, the draft refers to itself as a 'Forest Convention'.[19]

The 'Preamble' outlined three basic principles. First, the sovereignty of states over their forest resources is affirmed. The second basic principle is affirmation of the 'stewardship of those resources in such a manner as to ensure the attainment and continued satisfaction of human needs for present and future generations'. Finally, there is the notion of burden-sharing, namely 'an equitable sharing by the international community of the burden of forest conservation and development'.[20] Burden-sharing is to be achieved *inter alia* by increasing resource flows from the developed to the developing world and by trade policies encouraging forest conservation.

The linkage of the three basic principles of sovereignty, stewardship and burden-sharing can be seen as complementary ideas in a global bargain: countries with tropical forests would undertake to act as global stewards of their forests on behalf of international society, which in turn undertakes to share the burden of conservation.[21] The relevance of these three articles will become clear as this chapter unfolds.

The obligations of the parties are outlined in Section V of the FAO draft where three clauses reproduce language from the ITTO's *Guidelines for the Sustainable Management of Natural Tropical Forests*. As Chapter 3 notes, these guidelines consist of 41 principles by which sustainable tropical forest management may be achieved, and 36 'possible actions' by which the principles may be realized.

Part of Article V.2 of the FAO draft, 'Formulation of National Forest Policy', is identical to Possible Action 2 of the ITTO guidelines: 'A national forest policy, forming an integral part of the national land use policy, assuring a balanced use of forests, should be formulated by means of a process seeking the consensus of all the actors involved: government, local population and the private sector.'[22] Two other articles in the FAO draft reproduce language from the ITTO guidelines.[23] The wisdom of this must be questioned. It suggests that the principal focus for a GFC negotiated under the auspices of the FAO would be tropical forests. The use of ITTO text called into question the underlying motives of the FAO, an error compounded by the fact that many tropical timber producing countries, whose support would be needed for a GFC, have not signed or ratified the ITTA.

The FAO undertook the draft at a time when it was under severe criticism for its handling of the TFAP. In fairness to the FAO, it should be noted that one of the implied criticisms of the independent review was a lack of leadership within FAO, and the draft convention can be seen as an attempt to reclaim that leadership. However, hopes within the FAO that the organization could take a lead role in a GFI steadily evaporated. At the SWCC, the Director-General of the FAO announced that he had launched the preparations for the drafting of an international legal instrument on forests,[24] but no support was forthcoming to the FAO at this conference. The final nail in the coffin of the FAO draft came at the second session of the preparatory committee of the UNCED where the UNCED preparatory process assumed the lead role in the negotiation of any GFI.

The rejection of the FAO draft was due to dwindling support for the FAO as the result of criticism of the TFAP, the failure of the FAO to recapture that support, the circulation of the FAO draft prior to proposals to reform the TFAP, the use of clauses from the ITTO guidelines and the eclipse of the FAO by the UNCED process. Nonetheless, the FAO draft is a significant document; it is the first draft of a global forests convention to have been produced, and its contents later impacted upon the UNCED forests debate. FAO Forestry Department officials formed a close working relationship with the UNCED secretariat throughout the preparatory stages,[25] and as a result of this cooperation the three basic principles forming the cornerstone of the FAO's draft GFC were introduced to the UNCED process.

UNCED: AN INTRODUCTION

Four preparatory committee meetings (PrepComs) were held prior to the UNCED in Rio, each of which was divided into three working groups. Working Group I dealt with the atmosphere and land resources, including forests. Working Group II covered oceans, seas and coastal areas, while Working Group III handled legal and institutional matters.

Joint caucus group positions dominated the UNCED negotiations. For the North, the EC, the Nordic countries and the CAN group (Canada, Australia and New Zealand) were the important caucus groups. The USA's position was, to a large degree, synchronized with the EC's through the G7, as was Japan's. For the South the most important caucus was the Group of 77

(G77) developing countries. From PrepCom 3 until Rio the G77 established joint UNCED negotiating positions with China.

Six distinct outputs emerged from Rio. The first output, Agenda 21, is intended to be a blueprint for action by governments, aid agencies and other actors on environmental and developmental issues until the year 2000. Second, there was the Rio Declaration on Environment and Development, originally intended to be the 'Earth Charter'. Outlining the rights and obligations of governments in relation to the environment, the 'Earth Charter' was downgraded to a declaration during the preparatory process following disagreements between North and South. The third output was the establishment of the Commission on Sustainable Development (CSD) as a functional commission of the UN's Economic and Social Council (ECOSOC). The CSD held its first substantive session in June 1993.

The two most significant outputs were the Convention on Climate Change and the Convention on Biological Diversity. These instruments were negotiated on separate tracks from the main negotiations, the former under the auspices of the International Negotiating Committee on Climate Change, and the latter under the auspices of the United Nations Environment Programme (UNEP). Both were thus delinked from the mainstream UNCED preparatory process.

The sixth output was the *'Non-legally binding authoritative statement of principles for a global consensus on the management, conservation and sustainable development of all types of forests'*, hereafter referred to as the Statement of Forest Principles. The principal dynamics in this debate occurred at the intergovernmental level where there were three separate but interacting political processes. The formal negotiations took place in Working Group I (WGI) of the PrepComs. The second process, and the most difficult to document accurately, was informal discussions among delegates at the PrepComs. The third political process was intergovernmental activity outside the PrepComs. In particular, statements made at two sub-groupings from the South were relevant. The Summit Level Group of Developing Countries, also known as the Group of 15 (G15)[26] was attended by heads of state or government and held two meetings prior to Rio, at Kuala Lumpur (June 1990) and Caracas (November 1991). The second sub-grouping consisted of China and various G77 countries and held two ministerial-level meetings. The first (Beijing, June 1991) was attended by 41 countries, while 55 attended the second meeting (Kuala Lumpur, April 1992). It will be seen below that the Beijing declaration contributed to the formulation of joint G77–China negotiating positions for forests and other UNCED issues.

EXPLAINING NORTH–SOUTH DISAGREEMENT DURING THE UNCED FOREST NEGOTIATIONS

The UNCED forest negotiations were marked by extensive and wide-spread disagreement. The remainder of this chapter comprises an analysis of the debate to assess the nature and origins of this disagreement. Attention will centre on three areas of explanation that recurred throughout the debate. All three have a North–South dimension; to a large degree they feed into each other and should not be viewed in isolation.

The first area of explanation centres on the proprietorial dimension of the forests conservation problematic, namely the three competing claims to forest ownership of global common, national resource and local common (Chapter 1). All three claims were visible during the forests debate, but only the two former were made by government delegates; assertions that forests are a local common were made only in the NGO community. According to this explanation, governments from North and South adhered to different views on forest proprietorship as the result of different emphases. The North emphasized environmental issues with a perceived global dimension whereas the South's primary concerns were national environmental and developmental problems. Perceiving deforestation as a global issue, the North inclined towards, but stopped short of, asserting that forests are a global common, and consequently advocated a GFC. However, the South argued that a GFC was unnecessary; viewing deforestation as a national problem, and proclaiming unfettered sovereignty, the South argued that forests should be exploited in line with national development policy.

The second area of explanation is that disagreement between North and South was principally the result of differing views on the causes of deforestation. According to this view, the different policy prescriptions of North and South should be seen primarily as the result of different problem formulations.

According to the third area of explanation, disagreement arose because the North and South each sought to maximize its individual interests. The focus here is on institutional bargaining among actors seeking to achieve relative gains. According to this perspective, disagreement arose during the forest debate because the South sought to extract economic concessions from the North before making any commitment to forest conservation. The North was not prepared to make these concessions, and no bargain was struck.

PREPCOM 1, NAIROBI, 6–31 AUGUST 1990

PrepCom 1 mirrored the lack of consensus prevalent in international society at the time with actors awaiting the outcome of the Second World Climate Conference (29 October–7 November 1990). As noted above, the FAO offered to provide the forum for the negotiation of a GFI. The EC, reflecting the position of the European Council at Dublin, was at this stage committed to a protocol to a climate change convention.[27] The Canadian delegation, noting the G7 proposal for a GFC at Houston, believed 'that work towards a global forests convention should begin as soon as possible'.[28] The only agreement was that the UNCED secretary-general, which in reality meant the UNCED secretariat, should prepare a comprehensive report on the roles and functions of forests (Decision 1/14).

First meeting of the UNCED Working Party on Forests

In between PrepComs 1 and 2, and as a follow-up to Decision 1/14, the UNCED secretariat assembled an expert group, referred to here as the UNCED Working Party on Forests, in Geneva (December 1990). The working party contributed to the drafting of a 31-page report by the UNCED secretariat detailing the roles, functions and values of forests.[29] The working party was not an intergovernmental forum, and was composed solely of UNCED secretariat officials and invited experts.

PREPCOM 2, GENEVA, 18 MARCH–5 APRIL 1991

At PrepCom 2 the EC shifted strategy to advocate a general declaration on forests to include guidance on the text of a future GFC; this should be followed by the negotiation of a GFC after the Rio conference.[30]

In an effort to ensure that the forest debate in PrepCom 2 avoided the stalemate of PrepCom 1 an Ad Hoc Subgroup on Forests[31] was established by WGI under the chairmanship of Mr M S Kismadi (Indonesia). Its terms of reference were 'to consider further the various issues raised in discussions' and 'to submit agreed proposals and recommendations to Working Group I'.[32] Figure 4.1 below charts the organization of forest groups in the UNCED forests debate.

At the sub-group's second meeting an intervention was made by the head of the Malaysian delegation, Ting Wen Lian, who presented a list of points of concern including the relationships between deforestation and poverty, demographic pressures and debt. Ting also asserted the need for transfer of environmentally sound technologies and additional financial resources as 'compensation for opportunity cost foregone'.[33] Other developing country delegations present backed the Malaysian position.[34] Meanwhile, the US delegation stated that it favoured a GFC.[35]

A further indication that deep disagreement existed on the forests issue occurred when the Ad Hoc Subgroup produced a 'Draft Synoptic List'. This document had been drawn up partially on the basis of a statement from a smaller grouping within the Ad Hoc Subgroup.[36] However, the informal nature of this process means that there is no hard knowledge on the exact composition of this group, the manner in which the statement was composed, or its contents. The document offers no concise suggestions and no firm proposals. Ten options for a GFI,[37] many of them mutually exclusive, are listed, a factor that clearly indicates a lack of consensus. Although the Draft Synoptic List fulfilled the mandate of the Ad Hoc Subgroup to 'consider further the various issues' it singularly failed 'to submit agreed proposals and recommendations'. This state of affairs arose despite an intervention from the Brazilian delegate that the sub-group should formulate concrete proposals and not negotiate a GFI by proxy by passing responsibility to the secretariat.[38]

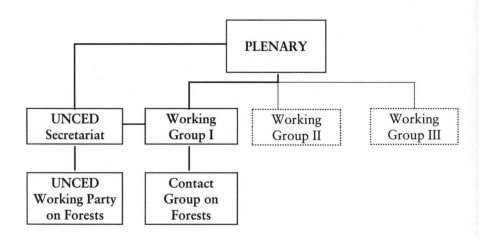

Notes: 1. The Ad Hoc Subgroup on Forests was established during PrepCom 2 as an intergovernmental forum reporting to Working Group I. During PrepComs 3 and 4, and at the UNCED in Rio, it was known as the Contact Group on Forests. At the Rio conference various sub-contact groups were established reporting to this group.
2. The UNCED Working Party on Forests was convened three times between PrepComs. It was composed of UNCED secretariat officials and invited experts. It met three times between PrepComs, namely 17–18 December 1990, 16–17 April 1991 and 13–14 September 1991 (prior to the tenth World Forestry Congress).

Figure 4.1 *Organization chart: forest groups in the UNCED process*

The political balance in the forest debate at the end of PrepCom 2 is summed-up in Decision 2/13 as follows: 'Working Group I will at its third session be in a position to examine all steps towards and options (including at a minimum, taking into account the special situation and needs of developing countries, a non-legally binding authoritative statement of principles) for a global consensus on the management, conservation and development of all types of forests.'[39] The decision also affirmed that 'the UNCED process is the most appropriate forum for conclusive decisions pertaining to global consensus on forests', thus firmly precluding any further possibility of the forests negotiations being passed outside the UNCED process to another organ such as the FAO.

Decision 2/13 reflected a position of compromise between those who wanted a convention, and whose who did not. It clearly indicated that the tide had turned away from a GFC towards a non-legally binding GFI. Regarding the contents of such an instrument, the initiative lay once again in the hands of the secretariat which was requested to analyse and address the items outlined in the Draft Synoptic List. At this stage, the fears of the Brazilian delegation that the PrepCom was effectively empowering the UNCED secretariat with the drafting of a GFI seemed justified.

Second meeting of the UNCED Working Party on Forests

The UNCED secretariat responded to Decision 2/13 by convening the second meeting of the Working Party on Forests (Geneva, April 1991). From this meeting there emerged UN document A/CONF.151/PC/65, 'Guiding Principles for a Consensus on Forests', hereafter referred to as PC/65. It will be recalled that the FAO had undertaken to assist the UNCED secretariat on the forest issue. Evidence of such cooperation is clearly apparent in PC/65, which in places draws from text in the FAO draft GFC. The three basic principles of sovereignty, stewardship and burden-sharing receive prominence in PC/65, which replicates language used in the FAO draft. In the quotes from PC/65 that follow, verbatim wording in the FAO draft is italicized. The PC/65 principle 'Affirm Stewardship' states that a global consensus on forests 'could assert the need for *stewardship of* forest *resources* and forest lands *in such a manner as to ensure the attainment and continued satisfaction of human needs for present and future generations*'.[40] The PC/65 principle, which equates with the FAO draft's principle on burden-sharing, is 'recognize common responsibility'. A global consensus on forests 'could provide for *an equitable sharing by the international community of the burden of forest conservation and development*'.[41]

PC/65 demonstrates that the UNCED Working Party on Forests, and subsequently the UNCED secretariat, had adopted, completely intact, the formula adopted by the FAO in an attempt to create a global bargain. PC/65 was on the table at PrepCom 3 where the introduction of the notion of 'stewardship' was to lead to disagreement.

PREPCOM 3, GENEVA, 12 AUGUST–4 SEPTEMBER 1991

The Ad Hoc Subgroup on Forests was reconvened and renamed the Contact Group on Forests under the chairmanship of Charles Liburd of Guyana. Several key documents were introduced to PrepCom 3. As well as PC/65, a draft Statement of Forest Principles was submitted by the G77 (hereafter referred to as the G77 draft proposal). The title of the G77 draft proposal repeated the text used in paragraph 5 of Decision 2/13, and may be seen as an attempt by the G77 to ensure that any joint declaration on forests at Rio would meet the minimum, and no more than the minimum, required by this decision. Other documents introduced included draft proposals from the USA and Canada and the Beijing Ministerial Declaration on Environment and Development of June 1991.

Developments up to and including PrepCom 3 is now considered to determine if they lend weight to the three areas of explanation outlined above. Turning to the first area, there are strong indications that disagreement on the forests issue was rooted in different views on forest proprietorship. The emergence of the G77–China alliance in Beijing is significant as China has a history of strong assertions of national sovereignty over a wide range of issues, a factor certain to have influenced her G77 allies. Indeed the Beijing declaration asserted that 'environmental considerations should not be used as an excuse for interference in the internal affairs of the developing countries'.[42]

The Canadian and USA proposals bear some similarities to the FAO draft and PC/65. The first principle of the USA proposal was that of 'stewardship' which stated that countries have 'a responsibility to engage in cooperative stewardship to improve global environmental quality for mutual benefit'.[43] The Canadian proposal emphasized the importance of 'responsibility'; states should recognize that forest conservation is 'a common concern of the community of nations which entails corresponding responsibilities'.[44]

By now there was suspicion in the South that the FAO was being manipulated by the North. At the Technical Workshop to Explore Options for Global Forestry Management in Bangkok (convened between PrepComs 2 and 3, hereafter referred to as the Bangkok Workshop) of April 1991,[45] Malaysian delegate Ting Wen Lian stated that her delegation was 'perplexed' as to why FAO was 'being utilized to promote the hasty agenda of some countries to formulate a forest convention'.[46] (Ting occupied a central role during this period and ensured that Malaysia argued a consistent position on the forests issue in all fora; she was the head of the Malaysian delegation to UNCED Working Group I, the Malaysian representative to FAO, and she also represented Malaysia at the Bangkok Workshop.)

To many delegations from the South, the notion of 'stewardship' outlined in PC/65 and in the USA proposal was unacceptable. At PrepCom 3 Malaysia argued that 'stewardship' was embodied in the principle 'reaffirm sovereignty'.[47] Edward Kufuor of Ghana, speaking on behalf of the G77, claimed that the industrialized world was attempting to take control of the resources of the developing countries, while leaving these nations as 'nominal stewards'. Kufuor also accused the North of attempting to assert forests as a global common and stated that, 'We cannot accept the application of such concepts as "global commons" or the "common heritage of mankind" with regard to the territorial domain of developing countries.'[48] Malaysia also objected to the use of 'nebulous terminologies' such as global commons to be an 'assumption of supranational rights' by the North.[49] The G77 draft proposal was consistent with these positions where there was no mention of the contentious principles of stewardship and common responsibility. Developments up to PrepCom 3 therefore provide evidence that differences between North and South on the proprietorial status of forests explain the UNCED forests logjam.

Attention will now turn to the second view, namely that the logjam was the result of differing views on the causes of deforestation. The Malaysian linkage between forests, poverty, demographic pressures and debt made at PrepCom 2 is indicative of this category of explanation. Furthermore, the Beijing declaration asserted that 'the developed countries bear the main responsibility for the degradation of the global environment' due to over-exploitation of natural resources through unsustainable patterns of production and consumption.[50] This assertion was frequently reiterated by delegates from the South during the remainder of the forests debate. Whereas delegations from the North were comparatively reticent on the forces behind deforestation, PrepCom 3 again saw the South offer strong views on the subject. The Indian delegation asserted a relationship between 'the external indebtedness of developing countries and the phenomenon of net transfer of resources from developing countries to developed countries and hence their ability to manage, conserve and protect their forest resources'.[51] This passage raises the question as to whether the Indian delegation regarded debt as a driving force of deforestation or a factor that prevented conservation. Irrespective, the Indians clearly regarded debt as a factor resulting in forest loss.

Meanwhile, Malaysia asserted linkages between: first, indebtedness and the net transfer of resources from South to North; second, the cost of combating deforestation; and third, the need for new and additional resources.[52] Malaysia had previously raised these issues during PrepCom 2 and at the Bangkok Workshop. The G77 draft proposal also stated that debt and South-to-North financial transfers reduced the capacity of developing countries to manage, conserve and develop their resources; unsustainable patterns of consumption and production, 'particularly in industrialized countries', and poverty were other causes mentioned.[53]

The G77 spokesman also attempted to redefine the notion of 'common responsibility' contained in PC/65: 'History tells us that the developed countries must bear greater responsibility for the deforestation that has occurred both in their own countries and in the developing countries. We should therefore be talking of common but differentiated responsibility, not simply of common responsibilities.'[54] In short, the delegations from the South sought to establish a relationship between deforestation and the global economic system.

The third area of explanation is that the South was using forests to bargain with the North for higher stakes. This view cannot easily be separated from the second explanation on causes. Many governments from the South, arguing that the causes of deforestation have their loci in global economic relations, felt justified in making bargaining issue-linkages between forest conservation and issues such as external debt relief.

At the Bangkok Workshop government representative Ting drew attention to the decision taken at PrepCom 2 that UNCED was the most appropriate forum for a global consensus on forests, a statement that strongly implies that this decision was taken following pressure from the South. One possible reason for this could have been to ensure that the debate on forests, rather than being passed outside the UNCED process as happened with the climate change and biodiversity convention negotiations, would be contained in the same forum, and therefore dealt with by the same teams of delegates, as those issues linked by the South to forests. Many of the issues introduced by the South into the UNCED forests debate, such as debt relief and technology transfer from North to South, had previously been advanced in the 1970s in the claims for a New International Economic Order (NIEO). The South now reiterated these claims using forests as a bargaining chip in an attempt to reach a trade-off. Hence the Malaysian intervention in PrepCom 2, as well as lending weight to the view that disagreements on causes was at the heart of the forests logjam, can also be seen as an attempt to introduce forests as a bargaining chip.

Transfer of financial resources and technology were the two issues most frequently linked by the South to forests.[55] The Beijing declaration (without specifically mentioning forests) considered that the key to the success of UNCED depended on whether progress was made on these two issues. This declaration also endorsed a proposal from China for a Green Fund, 'managed on the basis of equitable representation from developed and developing countries',[56] to deal with local environmental and developmental problems in developing countries, in particular forest preservation, tree planting, increasing the supply of fresh water resources and preventing soil degradation.[57] It was envisaged that developed countries and international agencies would provide most of the money for the Green Fund. The proposal ran counter to the preferred financial mechanism of the North, namely the Global Environment Facility (GEF).[58]

In PrepCom 3 the need for new and additional resources was asserted in the Malaysian, Ghanaian (on behalf of the G77) and Indian statements,[59] and the two former also made a claim for access to environmentally sound technology on preferential terms. The G77 draft proposal also included claims for financial and technological transfers.[60] Furthermore a point made by Malaysia in PrepCom 2, and repeated at the Bangkok Workshop, namely that any financial resource transfers from North to South would be 'compensation for opportunity cost foregone', had now been taken up by the G77 as a whole. India used this language in her statement.[61] The G77 draft proposal included an article using very similar wording to the Malaysian and Indian statements suggesting that the governments of these two countries were taking the lead for the G77 in the forests debate.

The increasing cohesion of the positions of the G77 and China was further demonstrated in PrepCom 3 with a joint proposal on financial resources which echoed the claim that transfers from the developed countries would be 'compensatory in nature'. This proposal introduced a new notion, namely 'partnership in additionality', defined as 'a commitment to provide new and additional financial resources to developing countries, for meeting, *inter alia*, the commitments under Agenda 21, and other sustainable development concerns'.[62] The notion of 'partnership in additionality' can be seen as complementary to that of 'common but differentiated responsibility'. The introduction of the concept of 'compensation for opportunity cost foregone' indicated that the forests issue had become entangled with bargaining issue-linkages, and provided a further indication that a united South was using forests to bargain for higher stakes.

Meanwhile, the USA was advocating that the GEF become the mechanism for funding the conservation of the world's forests, a position backed by Australia, the EC and Japan.[63] The North advocated that GEF funding should be dependent on project approval by the donors and on conditionalities such as 'good governance' in the recipient country. The South considered conditionalities to be an interference in the sovereign affairs of independent states. Overall, PrepCom 3 resulted in a solidification of positions on forests and on other issues. The negotiating positions between North and South that had emerged by the end of PrepCom 3 are shown in Table 4.2 below.

Towards the end of PrepCom 3 the UNCED secretariat used PC/65 and the G77 draft proposal, plus comments made in the earlier stages of PrepCom 3, as the basis to produce a draft Statement of Forest Principles for subsequent negotiations. Towards the end of PrepCom 3 the Contact Group chairman submitted the draft to WGI from where it was submitted to the PrepCom Plenary.[64]

Reflecting the numerical superiority of the G77, the G77 draft proposal was used as the overall guiding framework by the UNCED secretariat, and

Table 4.2   The UNCED forests debate and related issues: North and South negotiating positions

North	South
PRIMARY EMPHASIS	
'Global' environmental problems such as ozone depletion, global warming and deforestation.	Environmental and developmental problems of local and regional concern, such as poverty, clean drinking water, land degradation, soil loss and desertification. Deforestation seen as a local, not a global, problem.
SOVEREIGNTY	
Acknowledgement of Principle 21 of the Stockholm Declaration. (Occasional invocation of forests as a global common.) Linkage of sovereignty with the concepts of 'stewardship' and 'common responsibility'.	Strict assertion of unfettered national sovereignty. Forests as a sovereign national resource to be exploited in line with national development policy. Rejection of notion of 'stewardship'. Emphasis not of 'common responsibility', but of 'common but differentiated responsibility'.
CAUSES OF DEFORESTATION	
The North had little to say on this subject although it disputed the South's claims that the normal functioning of global economic relations cause deforestation.	The global economic system is exploitative and is a driving force of environmental degradation, including deforestation. Emphasis on external indebtedness, net financial transfers from South to North, declining terms of trade and high consumption patterns in the North.
PREFERRED FOREST INSTRUMENT	
'Stewardship' and 'common responsibility' require a legally-binding global forests convention.	No need for a global forests convention. The non-legally binding Statement of Forest Principles accepted as a compromise.
FINANCIAL TRANSFERS	
A recognition that new funding should be made available, in line with the notion of 'common responsibility'.	Calls for 'new and additional funding' as 'compensation for opportunity cost foregone'. Funding for local problems, and not just for problems perceived as urgent by the North. Introduction of the concept of 'partnership in additionality'.
FINANCIAL TRANSFER MECHANISM	
Funding to be made available through existing mechanisms, most notably the Global Environment Facility (GEF) administered by the World Bank, UNDP and UNEP. An emphasis on conditionality: good governance and approved projects.	Funding for local environmental problems to be channelled through the Green Fund (to be funded primarily by the North). The GEF to be restructured on more equitable lines. Rejection of conditionality on the grounds that it constitutes an interference in national sovereignty.
TECHNOLOGY TRANSFER	
Technology transfer to be encouraged through existing commercial mechanisms.	Technology transfer at concessional rates or on a grant basis. Subsidies for the transfer of the intellectual property rights of environmentally clean technologies. The cost of such transfers to be borne by the Green Fund.
SUMMARY	
Solve global environmental problems with minimal changes to existing international structures.	Tackle the underlying global causes of underdevelopment and environmental degradation. The burden of adjustment to lie with the North.

consequently the draft overwhelmingly reflected G77 concerns. Some, but not all, proposals from PC/65 were grafted on to the G77 draft proposal, but significantly there was no mention of 'stewardship' or 'common responsibility'. The main thrust of PC/65, that of overcoming the dual claims to forests of global common and national resource, had been discarded. The draft contained square brackets around clauses that remained the subject of wider North–South divisions, including several clauses inserted from the G77 draft on debt, net South-to-North financial flows, compensation for opportunity cost foregone and transfer of technology and financial resources. By analysing those articles without brackets, it is possible to discern those areas where North and South had reached agreement, but these were invariably uncontroversial areas on which North and South could agree without yielding anything of substance. Whereas PrepCom 2 had finished with WGI in a political *cul-de-sac* on forests, PrepCom 3 ended with the G77 in the ascendancy.

This led to a tactical switch by the North, which attempted to insert language on the negotiation of a GFC after UNCED into the Statement of Forest Principles. The EC issued a statement that any declaration on forests opened for signature at Rio should contain 'procedures, including a timetable for the negotiation of a Convention on Forests'.[65] The G77 would resist these efforts in PrepCom 4 and at Rio.

ENVIRONMENTAL DIPLOMACY BETWEEN PREPCOMS 3 AND 4: ABIDJAN AND CARACAS

Two significant intergovernmental conferences were held in the South between PrepComs 3 and 4, namely the Second Regional African Ministerial Conference on Environment and Development (Abidjan, 11–14 November 1991) and the Second Summit Level Meeting of the G15 (Caracas, 27–29 November 1991).

Given the South's position thus far on sovereignty, the Abidjan and Caracas meetings witnessed surprising references to forests as a heritage of mankind. The African Common Position on Environment and Development, adopted at Abidjan, notes that, 'We regret to note that poverty, debt and stringent conditions related to international trade makes it difficult to conserve these forests, this biodiversity and this common heritage of mankind.[66] The second reference was made by the Venezuelan president, Carlos Andres Perez, at the second meeting of the G15. Perez, referring to technology transfer, stated that 'Inordinate responsibility cannot be placed on efforts to preserve the environment, whilst the South has limited access to, or is excluded from, technologies we need to back our national development efforts and fight poverty. . . . If the tropical forests are the heritage of mankind, science and technology should be also.'[67] This is the first known occasion that a head of state from the South had referred, albeit obliquely, to forests as a heritage of mankind. The G15 joint communiqué did not echo Perez's sentiments, but did state that 'it must be recognized that cumulative scientific and technological knowledge and innovation is a heritage of all mankind'.[68]

The real significance of the Abidjan and Caracas statements is that the linking of forest sovereignty to other issues provides the first (and, so far as is known, the only) indications that elements of the South were using forest *sovereignty* (and not merely forest *conservation*) as a bargaining counter for higher stakes. The distinction is important. The position taken by the G77 was that sovereignty could on no account be compromised. However, developments at Abidjan and Caracas suggest that some countries of the South were, despite the stance adopted by the G77, prepared to give up some measure of control over their forests in exchange for a *quid pro quo* from the North.

PREPCOM 4, NEW YORK, 2 MARCH–3 APRIL 1992

The unexpected references made to forests as a common heritage of mankind in Abidjan and Caracas may have suggested that latitude existed for meaningful negotiation at PrepCom 4. However, this did not prove to be the case. The Statement of Forest Principles had arrived at New York with 116 sets of square brackets. The delegations worked through the text, rewording clauses to arrive at compromise language acceptable to all. By the end of PrepCom 4 the reworked text was forwarded to the UNCED in Rio with 73 sets of square brackets remaining. The final PrepCom 4 draft contained no proviso for the negotiation of a future GFC. Attempts by the North to insert such a clause were blocked in the G77.

There was no evidence at New York to support the view that the forests logjam was rooted in different views on forest proprietorship. This is hardly surprising given that PrepCom 3 had settled this issue in favour of the G77. Neither were there any indications that the forest logjam was centred on differing views on causes. However, there were several indications to suggest that forests conservation was being used by the South to lever concessions from the North. As well as the Contact Group on Forests, other contact groups were established to expedite the Agenda 21 negotiations on, *inter alia*, financial resources and technology transfer. So divided was the PrepCom on the financial resources issue that the Contact Group on Financial Resources split into two groups: one, chaired by Australia, debated the GEF; the second, chaired by Sri Lanka, considered possible new financial instruments, including the Green Fund.[69] At the end of PrepCom

4, paragraphs that were fully, or near-fully, bracketed in the Statement of Forest Principles included those on financial resources, technology transfer, external indebtedness and compensation for opportunity cost foregone.[70]

SECOND MINISTERIAL MEETING OF DEVELOPING COUNTRIES, KUALA LUMPUR, APRIL 1992

Further evidence to support the view that the forest debate remained complicated by wider issue-linkages came at the Second Ministerial Meeting of Developing Countries on Environment and Development held in Kuala Lumpur after PrepCom 4. The Malaysian prime minister, Dr Mahathir Bin Mohamad, pleaded for South unity in his opening speech: 'The voice of the individual developing countries will be drowned. It will be different if they speak together with one strong voice in Rio.' Revealingly he enquired, 'What use is there of an Earth Charter if there is no real advance on the critical issues of finance and technology?' and he directly linked forests to finance when stating that, 'If it is in the interests of the rich that we do not cut down our trees then they must compensate us for the loss of income.'[71] The Kuala Lumpur Declaration, signed by 55 ministers, reaffirmed the previously-stated positions of the G77 and China; three references were made to new and additional financial resources, and the need for 'the transfer of technology on preferential and concessional terms' was also repeated. Forests were asserted as a 'part of the national patrimony'.[72] Overall, the Kuala Lumpur meeting was used as a final opportunity to emphasize the bargaining positions of the G77–China alliance in the final period before Rio.

THE UNCED, RIO DE JANEIRO, 3–14 JUNE 1992

The conference was divided into two main bodies. The plenary was the forum for country statements while political negotiations took place in the main committee. Reporting to the main committee were eight contact groups dealing with issues where substantial negotiations were still necessary, namely atmosphere, biodiversity, institutions, legal instruments, finance, technology transfer, freshwater resources and forests. The Forests Contact Group established sub-contact groups to deal with individual paragraphs, while the contact group dealt with less contentious areas.[73]

Although negotiations usually took place in closed session, the forest debate was not separated from other issues at Rio. Forests remained a strong bargaining chip for the South as US chief negotiator Curtis Bohlen observed: 'Some countries are reluctant to take concrete steps to preserve their forests. They are trying to get the money before agreeing to do anything.'[74] Consequently, negotiations on financial resources and technology transfer also proceeded slowly.

Agreement on the Statement of Forest Principles was reached only after Klaus Töpfer, the German Minister for the Environment, assumed responsibility for ministerial level negotiations. A package produced by Töpfer was finally accepted at 3.00 am, 12 June. Negotiations on financial resources lasted until 7.00 pm, 13 June.[75] The slow pace of negotiation on forests, finance and technology transfer suggests that informal behind the scenes manoeuvring took place by negotiators in an effort to reach a 'trade-off' position. This would add weight to our third area of explanation. Although no hard evidence has emerged to support this contention, such developments would be consistent with negotiating positions established earlier in the PrepComs.

The G77 led by India and Malaysia resisted accepting the Statement of Forest Principles until it contained no mention of a future GFC. The following compromise language was adopted in the Statement of Forest Principles: 'In committing themselves to the prompt implementation of these principles, countries also decide to keep them under assessment for their adequacy with regard to further international cooperation on forest issues.'[76]

CONCLUSIONS

The finalized text of the Statement of Forest Principles is appended as Annex B. The main thrust of the text is on policy responses at the national level, with sovereignty asserted at the start of the statement using language reproduced from Principle 21 of the Stockholm Conference on the Human Environment (see Chapter 1).[77] The global ecological role of forests is noted. The G77 had some success in inserting clauses on the causes of deforestation; external indebtedness, net transfers of resources from South to North, trade, industrial and transportation policies leading to deforestation and air pollution are identified as causes, while unsustainable patterns of production and consumption are implied as causes. However, the finalized text does not specify policy prescriptions or frameworks for cooperation.

Overall, the Statement of Forest Principles represents a mean position of the lowest common denominator between North and South, and compares unfavourably with other published principles, such as those of the ITTO. The final text makes no mention of how the document is intended to relate to mechanisms such as the TFAP or ITTO and no indication is given of the intended relationship with forest-related agreements such as the biodiversity and climate change conventions. Despite the complex issue-linkages that surrounded its negotiation, the Statement of Forest Principles gives the impression of having been formulated in a political vacuum. However, the document does represent a first global consensus on forests.

In seeking to explain the logjam that arose in the UNCED forests debate, three lines of enquiry have been pursued in this chapter. The first was that the differences between North and South were centred on forest proprietorship: the South proclaimed unfettered national sovereignty over its forests; the North recognized sovereignty, while at the same time inclining towards a position of forests as a global common. The North attempted to strike a global bargain by linking sovereignty with stewardship and common responsibility (burden-sharing). This was rejected by the South.

The second line of enquiry was that views on the causes of deforestation varied, hence the prescriptions for a cure differed. The South emphasized global economic causes, which the North tended to ignore or downplay.

The third line of enquiry was that both North and South pursued positions of perceived interest. Developments during the UNCED process saw the North trying to extract commitments from the South on forest conservation. The South meanwhile attempted to extract concessions from the North, using forests conservation as a bargaining counter.

Evidence has been provided that supports all three lines of enquiry. However, the third line of enquiry emerges as the strongest single theme. The view that disagreement in the forests debate was due to different conceptions of forest proprietorship has lesser explanatory power. Statements that were made on causes were usually made by the South, frequently while making a bargaining claim against the North. In the final analysis the bargain was not struck. The South did not gain the concessions it wanted from the North, and the North did not extract any binding commitments from the South on tropical forest conservation.

Chapter 5

The Negotiation of the International Tropical Timber Agreement, 1994

An analysis is made in this chapter of the first intergovernmental negotiations on a forest-related issue to occur following the United Nations Conference on Environment and Development (UNCED), namely the negotiation of a successor agreement to the International Tropical Timber Agreement, 1983 (hereafter the ITTA 1983). The main negotiations, like those for the ITTA 1983, were hosted by the United Nations Conference on Trade and Development (UNCTAD). Agreement on a new text — the International Tropical Timber Agreement, 1994 — was reached in January 1994, shortly before the ITTA 1983 had been due to expire on 31 March 1994. A chronology of the most significant meetings, conferences and seminars in the negotiation of the new agreement is shown in Table 5.1 below. Note that gatherings outside the ITTO/UNCTAD negotiating process are listed; their relevance will become clear as the story of the negotiations unfolds.

The issues of financial resource transfers and technology transfers, which had been so prominent in the UNCED forest negotiations, also became subjects of contention in the ITTA negotiations. Other substantive questions arising were the scope of the successor agreement (namely whether it should cover just tropical timber or all timber), the degree of commitment to conservation objectives, trade discrimination and the future of ITTO project work.

THE PREPARATORY PROCESS

The first Preparatory Committee meeting (PrepCom) was preceded by an informal working group meeting at the US State Department, Washington (September 1992).[1] The working group consisted of seven participants from producer countries (Brazil, Ghana, Indonesia and Malaysia) and eleven from consumer countries (European Community, Japan and the USA). No observers were present and no formal record was kept of the group's discussions. Participants expressed their views in a personal capacity, and not as representatives of governments or groups. The group considered comments submitted to the chairman from 13 countries; these had been collated into a working document by the ITTO secretariat. Following discussions the chairman of the group prepared an 'Issues and Options' paper to be introduced at the first PrepCom meeting.

The first PrepCom was held in parallel with the ITTC's thirteenth session in November 1992 at Yokohama. A significant development at this session was WWF's withdrawal from all ITTO national delegations. This decision had been taken the previous month[2] and was designed to apply pressure on national delegations to expand the scope of the new agreement to embrace all timbers. WWF also felt that by withdrawing from ITTO delegations it could highlight the absence from delegations of other interested NGOs.[3]

Delegations divided into the producers' caucus and the consumers' caucus, each of which met in closed session. Amha Buang of Malaysia was elected as the producers' spokesman while Milton Drucker of the USA was elected as spokesman for the consumer caucus. The 'Issues and Options' paper had noted several differing views between the producers and consumers, and these differences became prominent when discussions commenced at Yokohama. In the remainder of this section details are given of the positions of the four sets of actors involved in the negotiations at the end of the preparatory process: the producers' caucus, the consumers' caucus, the environmental NGOs and the timber trade organizations.

The central dispute concerned the scope of a renegotiated ITTA, with disagreement focusing on whether or not the agreement should cover all timbers. The consumers advocated a continuation of the existing tropical-timber-only format. The Malaysian delegation formally tabled a proposal that the agreement be expanded to include all timbers.[4] Following this proposal, the producers subsequently argued for an expanded international timber agreement for all the world's forests in which the producer/consumer distinction would be replaced by a developed country/developing country distinction. The former would undertake to provide the latter with financial resources to achieve sustainable forest management. The producers cited the UNCED Statement of Forest Principles to support their arguments; it was inconsistent, they argued, for the consumer countries to adopt this document, which relates to all the world's forests, and then to discriminate between tropical and non-tropical forests in the first post-Rio forest-related negotiations.

The producers received support from the environmental NGOs on the question of scope. A common concern of the producers and the NGOs was

Table 5.1   *Meetings, conferences and seminars of significance in the negotiation of a successor agreement to the International Tropical Timber Agreement, 1983*

Date	Place	Meeting
1992		
24–25 September	Washington DC	Informal Working Group meeting at the US State Department
11–24 November	Yokohama	1st session of PrepCom, 13th session of ITTC, 11th sessions of permanent committees
1993		
22–30 January	Quito	2nd session of PrepCom
30 January	Quito	1st special session of ITTC
13–16 April	Geneva	UN Conference on Negotiation of a Successor Agreement to ITTA 1983, Part 1
7–8 May	Atlanta	Symposium, *Forests and the Environment: A US Response to the Rio Earth Summit*
11–19 May	Kuala Lumpur	14th session of ITTC, 12th sessions of permanent committees
16–17 June	Helsinki	2nd Ministerial Conference on Protection of Forests in Europe
21–25 June	Geneva	UN Conference on Negotiation of a Successor Agreement to ITTA 1983, Part 2
11–13 August	Yokohama	Informal consultations (participants invited by president of UN Conference)
13–18 September	Kuala Lumpur	14th Commonwealth Forestry Conference
27 Sept–1 Oct	Montreal	Conference on Security and Cooperation in Europe (CSCE) Seminar of Experts on Sustainable Development of Temperate and Boreal Forests
4–15 October	Geneva	UN Conference on Negotiation of a Successor Agreement to ITTA 1983, Part 3
10–17 November	Yokohama	15th session of ITTC, 13th session of permanent committees
1994		
10–26 January	Geneva	UN Conference on Negotiation of a Successor Agreement to ITTA 1983, Part 4
26 January	Geneva	2nd special session of ITTC

that there should be no trade discrimination between tropical and non-tropical timbers. The NGOs also noted that many temperate timber species are substitutable for tropical timber, and pointed to the attention sections of the timber industry have been paying to Siberia's forests since the collapse of the Soviet Union.[5]

The commonality between the producers' and the NGOs' positions is worthy of further comment. It is noted in Chapter 2 that during the TFAP restructuring process the NGOs and the tropical forest countries were divided on the question of NGO participation in a consultative group, with the former pushing for genuine NGO participation, and the latter seeking to limit NGO involvement by advocating a traditional intergovernmental format within the FAO. There is a striking difference in the common ground established between the NGOs and the producers during the negotiations under consideration here, where the two sets of actors formed an alliance to push for an expansion of the scope of the new agreement. What explains this difference?

The answer lies in the distinction between an actor's overall strategy and the tactics adopted to pursue this strategy. During the ITTA renegotiations the producers and the NGOs agreed on tactics, while their strategies differed profoundly. Both favoured an expansion of scope, but for different reasons. While advocating expansion, the NGOs simultaneously argued for a contraction of the mandate of the new agreement, which should focus purely on trade and trade-related issues. Disillusioned with the poor conservation track record of the ITTO, the NGOs favoured a reduced role for projects dealt with by the ITTO's permanent committees on reforestation and forest management and on forest industry although, in arguing that a new organization should focus primarily on trade-related issues, they did see a continuing role for projects under the auspices of the Permanent Committee on Economic Information and Market Intelligence. In arguing this the NGOs favoured a fundamental revision of the ITTO's permanent committee structure. The NGOs considered that the ITTO did not have the competence to deal properly with forest conservation and that other mechanisms, such as the Forest Stewardship Council (Chapters 3 and 6), were better suited to dealing with environmental objectives. However, as tropical and non-tropical timbers are substitutable there was a need for a holistic view of the global trade in a new international timber trade organization. The producers also favoured an expansion of scope, but, in continuing to favour a high project profile for the new organization, they disagreed with the NGOs on its mandate. Despite this difference, the two groups of actors had sufficient common ground to offer support to each other throughout the negotiations. This is in marked contrast to the TFAP restructuring debate, where the tropical forest governments argued for a restriction of NGO participation in a consultative group so that, not surprisingly, no cooperation between the two sets of actors was possible.

The consumers and producers disagreed on some fundamental points. As noted earlier, the consumers favoured a continuation of the tropical-timber-only format; in an effort to break the unity of the producers' caucus they argued that an expansion of scope would marginalize the small tropical timber producing countries. Whereas the producers favoured a renewed emphasis on further processing in the new agreement, the consumers advocated that further processing should be limited solely to tropical timber from sustainable sources. The consumers advocated a greater emphasis on conservation objectives, with the ITTO guidelines receiving mention in the new agreement. There was a strong emphasis on national reporting in the consumers' position: national reporting requirements were to be strengthened and should include the progress made by the producers towards Target 2000. The consumers also favoured tightening the ITTO's purse strings: projects should meet the objectives of the new agreement if they are to qualify for funding; there should be a greater emphasis on post-project evaluation with project reports introduced; and a new and strong Permanent Committee on Finance and Administration should be introduced to oversee the ITTO's financial expenditure. The consumers advocated that the new agreement mention the importance of local communities and indigenous peoples. In this they differed slightly with the NGOs, many of which wished to see the new agreement recognize the ancestral land rights, and not merely the 'importance', of indigenous peoples. Finally, there was one important consumer objective that was not pursued in the negotiations; it will be argued below that although many consumer governments continued to favour a global forests convention (GFC), for various reasons they did not consider the ITTA negotiations to be an appropriate forum in which to pursue this objective.

While the producer countries received support from the NGO community, the consumers were backed by the timber trade organizations, which also favoured a remit covering only tropical timber. Prior to the first PrepCom, representatives from the timber traders held a series of consultations from which there emerged a firm trade view. This favoured a tropical-timber-only agreement containing commitments to the ITTO guidelines and Target 2000, including annual reports by members towards Target 2000. The traders advocated that the international tropical timber trade should be maintained 'at least at its present value and possibly with the prospects of increased revenue'.[6] The FAO, as it had during the UNCED process, supported the North's view, with its observer stating that the FAO believed that a new agreement should remain focused on tropical timber.[7]

Although there were no significant differences between the timber traders and the consumer countries, some differences may be discerned between the environmental NGOs and producer countries. The first concerns Target 2000. NGOs agreed that Target 2000 was not achievable at current levels of trade and they therefore argued for the promotion of a high-value/low-volume trade. Most NGOs wished to see Target 2000 written into an expanded agreement, although it should be noted that some NGOs did not, most noticeably Friends of the Earth (FoE) who felt that to insist on this point could break the producers/NGO alliance.[8] The producers appeared divided on this issue. Many opposed any mention of the target, while others stated that if it was to be included there should also be quantified statements of the resources needed to attain sustainability. The latter view was an attempt to strike a bargaining issue-linkage between Target 2000 and financial resource transfers.

Indeed, such a linkage helps explain the second difference between the NGOs and the producers, namely on projects. The producers remained, in line with their UNCED claims for financial transfers and technology transfers on a grant or concessional basis, keen on a high profile for projects, funded primarily by the developed countries. However, and as previously noted, consistent with their view that the mandate of the new organization should embrace trade-related issues only, NGOs favoured a reduced role for project activity.

Third, the NGOs argued that the new agreement should mention financial and non-financial incentives. Here they found some common ground with the consumers, although in line with their different views on scope and mandate the two sets of actors differed on the types of incentives for which they wished the new agreement to provide. The consumers favoured the promotion of incentives for the sustainable management of tropical forests, whereas the NGOs advocated the promotion of trade-related incentives for all timbers. The producers, who it will be recalled had blocked the introduction of incentives in the ITTO (Chapter 5), shared neither view and opposed any mention of incentives in the new agreement.

In addition, there were other issues that the NGOs wished included in a new agreement that the producers did not. For example, and as previously noted, the NGOs favoured strong provisos for consultation with local communities and indigenous peoples. The NGOs also wished to see clauses against the illegal timber trade written into the new agreement. They had raised this issue several times at ITTC sessions and, in November 1992, concerted NGO campaigning against the illegal trade occurred in Yokohama, where the first PrepCom for the conference was held alongside the ITTC's thirteenth session.[9] In 1991 leading figures from the Global Legislators' Organisation for a Balanced Environment (GLOBE), an NGO consisting of legislators from the European Community, the USA, Japan and Russia, had lobbied on this issue when they wrote to the ITTO's executive director stating that the illegal trade 'falls within the purview of ITTO's mission'.[10]

A further position where the NGOs and the consumers shared some common ground was the need for clear standardized reporting on all trade-related issues. In this the NGOs shared a position in common with the timber traders as well as the consumers, both of which advocated strengthened reporting requirements. The NGOs supported national reporting in part to expose the illegal trade. This was a position that the producers, sensitive to possible erosions of sovereignty, did not share.

However, the differences between the producers and the NGOs did not prove to be significant, and the producers/NGO alliance was firmly established by the end of the first PrepCom. Indeed, the Malaysian Minister for Primary Industries cited research by WWF–International showing that temperate deforestation was increasing,[11] and called upon NGOs to focus further on temperate forest destruction.[12] The alliance between the two sets of actors lasted until the end of the negotiation process.

The producers' and consumers' caucuses met in closed session during the PrepCom and did not exchange drafts. Following deadlock the ITTC voted to convene a second session of the PrepCom in Quito in January 1993.

Between the two PrepComs the fifth ministerial-level meeting of the Malaysian–Indonesian Joint Working Group on Forestry reiterated the line taken by the producer countries.[13] Meanwhile, the European Commission announced that it supported the inclusion of Target 2000 in a new tropical-timber-only agreement.[14]

An indication of how little consensus had been reached at Yokohama occurred when the second PrepCom in Quito continued to use the Washington 'Issues and Options' paper as the main working document. As at Yokohama, all caucus meetings were held in closed sessions. However, an NGO representative was permitted to make joint NGO statements to each caucus. At the end of the Quito PrepCom the first special session of the ITTC was convened. This recommended institutional arrangements for the conference which were then conveyed by the executive director of the ITTO to the UNCTAD Secretary-General.

Table 5.2 below shows the positions of the four main groups of actors by the end of the preparatory process.

FIRST PART OF CONFERENCE, GENEVA, 13–16 APRIL 1993

Following the two PrepCom sessions, which were open to ITTO members only, the process moved to the UNCTAD at Geneva where all UNCTAD members were eligible to partake. The institutional arrangements recommended by the first special session of the ITTC were adopted. An executive committee of the whole (or main committee) was established. Wisber Loeis of Indonesia was elected as the president of the conference in which capacity he chaired meetings of the executive committee of the whole. Two vice-presidents were elected, namely Yolanda Goedkoop van Opijnen (Netherlands) and Jorge Barba (Ecuador). In addition, two committees were established reporting to the executive committee of the whole, namely Committee 1, dealing with economic and technical clauses (chaired by

Table 5.2 Positions of delegations, environmental NGOs and timber traders at the preparatory committee meetings for the United Nations Conference for a Successor Agreement to the International Tropical Timber Agreement, 1983

	Producer Countries	Consumer Countries	Environmental NGOs	Timber Traders
SCOPE	An expanded agreement dealing with all timber. The producers/consumers distinction to be replaced with a developed/developing countries distinction.	A revised agreement dealing with tropical timber only. (Expansion to deal with all timber would marginalize the interests of small tropical timber producers.)	An expanded agreement dealing with all timber.	A revised agreement dealing with tropical timber only.
TARGET 2000	Two views: (1) Target 2000 should not be incorporated into an expanded agreement; (2) Any mention of Target 2000 should be accompanied by a quantified statement of the resources needed to attain the target.	Target 2000 to be written into a revised agreement.	Most NGOs favoured inclusion of a clearly-defined Target 2000 to be written into an expanded agreement. (Some NGOs did not lobby for this, arguing that to do so ran the risk of breaking the producers/ NGO alliance.)	Target 2000 to be written into a revised agreement.
PROJECTS	Projects to be an integral part of an expanded agreement. New and additional financial resources to be made available by developed countries to fund projects.	Projects should meet the objectives of a revised agreement to qualify for funding. A greater emphasis on post-project evaluation. Project progress reports to be introduced.	The ITTO does not have the competence to deal with a heavy project burden. Project work should focus on trade related issues, such as economic information and market intelligence.	No view offered on this subject by the timber traders.

Producer Countries	Consumer Countries	Environmental NGOs	Timber Traders
NATIONAL REPORTING			
No need for new provisions for national reporting.	National reporting requirements to be strengthened with producer members required to report on *inter alia* progress towards Target 2000.	Clear standardised reporting on all trade related issues is essential, in part to expose the illegal timber trade.	Producer members to report on *inter alia* progress towards Target 2000.
MANDATE			
Basic objectives to be retained, but to apply to all timbers. The special needs of developing countries to be provided for by financial transfers and technology transfers on a grant or concessional basis. Emphasis on further processing of tropical timber in developing countries.	Greater priority to be given to conservation. ITTO guidelines to receive mention in the new agreement. Further processing of timber from sustainable sources only. Incentives to promote sustainable tropical forest management.	Contraction of the mandate to deal solely with trade related issues. A holistic view of the global timber trade is required. Elimination of the illegal timber trade. The ancestral land rights of indigenous peoples to be respected. Incentives to promote a sustainable global timber trade.	A revised agreement should include articles providing for commitments to Target 2000 and ITTO guidelines.
OTHER NEGOTIATING POSITIONS			
There should be no unilateral bans or boycotts against tropical timber, and no trade discrimination between tropical and non-tropical timbers.	A new Permanent Committee on Finance and Administration to be established.	An emphasis on the promotion of a high-value/low-volume trade. The Permanent Committee structure requires fundamental revision.	International trade to be maintained at least at its present value, with prospects for increased revenue.

David Boulter of Canada), and Committee 2, dealing with financial and administrative clauses (chaired by Eugene Capito of Gabon).[15] The spokesmen for the two caucuses remained unchanged, namely Amha Buang of Malaysia and Milton Drucker of the USA. The first special session of the ITTC had also decided upon the arrangements for NGO participation at the conference. Originally there had been some confusion over this. With invitations to the conference issued by UNCTAD, and not by the ITTO, some NGOs had been concerned that only NGOs registered with the Economic and Social Council (ECOSOC) of the UN would be invited. However, these concerns were allayed when it was announced that all NGOs that had attended ITTC sessions would also be invited.

Two parts of the conference were originally scheduled. The first part (April 1993) opened with two main documents on the table; the ITTA 1983, and a comparative tabulation of the proposed texts of the producers and consumers. Both the proposed texts included a reference to the eighth session of the UNCTAD (UNCTAD VIII). Concluded in February 1992 this adopted *A New Partnership for Development: The Cartagena Commitment*, as well as a shorter declaration, *The Spirit of Cartagena*.[16]

The first part of the conference saw neither producers nor consumers move from the positions agreed at the PrepComs. Few significant developments occurred. The producers staked a claim for access to Western technology to help in forest conservation.[17] Agreement was reached that the new organization should be based in Yokohama, but there was no consensus concerning the scope of a new agreement. Representatives of Indonesia, Brazil and the Philippines made statements in favour of expansion, while the EC, USA, Japan and Canada favoured retention of the tropical-timber-only format.[18] The WWF accused the USA, Canada and Sweden of leading the consumers' opposition to expansion.[19] After four days the first part of the conference ended in deadlock.

ENVIRONMENTAL DIPLOMACY BETWEEN THE FIRST AND SECOND PARTS OF THE CONFERENCE

The fourteenth session of the International Tropical Timber Council

Between the first and second parts of the conference the ITTC's fourteenth session convened in Kuala Lumpur (May 1993) where the renegotiation featured in the discussions. Following informal consultations between the chairman of the ITTC and the president of the conference a reconciliation group was formed 'to explore areas of convergence' between the consumers' and producers' proposed texts.[20] The work of the group, which was of a non-binding nature, centred on streamlining a composite text based on the proposed texts of the two caucuses.[21] The outcome of the group's work was a series of minor textual amendments, and no substantive progress was made in reconciling the two caucuses.

The ITTC's fourteenth session was marred by a physical assault by the Sarawak director of forests on the executive director of WWF–Malaysia, which ended with the latter falling into the swimming pool of the hotel where the ITTC session was being staged. The assault occurred following an argument concerning the ITTO Mission to Sarawak. NGOs, including WWF, have continued to lobby the state government of Sarawak and the Malaysian government on the continuing deforestation of Sarawak since the ITTO Mission recommended that the annual cut of timber logged in the state be reduced by 30 per cent (Chapter 3). This recommendation has not been fully implemented. After the incident some NGOs considered boycotting the remainder of the session.[22] However, a boycott did not materialize, and the NGOs remained.

In an effort to overcome the polarized positions between the producers and the consumers Milton Drucker circulated a paper compiled on the basis of informal consultations seeking to deal with areas of producer concern. This proposed that timber traded should not face discrimination based on its geographical origin and that nothing in the renegotiated agreement 'would justify the restraint of trade'. A second point in the paper was that, while consumers would not wish a commitment to Target 2000 for non-tropical timbers to be written into the new agreement, there could be 'a commitment by consumer members of the ITTO, to consider their own shift towards producing supplies of non-tropical timber in international trade from sustainably managed sources'.[23] NGOs, most of whom wished the new agreement to include a commitment to Target 2000 for all timbers, subsequently accused the consumers of applying 'double standards' with respect to tropical and non-tropical timbers.[24]

The Helsinki Ministerial Conference on the Protection of Forests in Europe

In June 1993, one month after the Kuala Lumpur session, the Second Ministerial Conference on the Protection of Forests in Europe convened in Helsinki. Thirty-eight European countries attended, with three observer countries, namely the USA, Canada and Japan. The conference made an important contribution to the post-UNCED discussions on the criteria and indicators for sustainable forest management (Chapter 6). The conference also saw discussion on the ITTO's Target 2000. The delegation from the Netherlands, whose government had signed up to WWF's 1995 target (Chapter 3), proposed that a commitment by European governments similar to the ITTO's Target 2000 be written into the Helsinki resolutions. The proposal was blocked within the European Community and opposed by Sweden. However, the United States, although only having observer status, used the conference to announce that, 'The US is committed to the national goal of achieving sustainable management of its forests by the Year 2000.'[25]

The United States and Target 2000

The announcement of the US delegation at Helsinki embarrassed the Europeans. Two factors explain this policy shift. The first is the change in the US administration from Presidents Bush to Clinton in January 1993. Al Gore, Clinton's running mate in the 1992 presidential campaign, has strong environmentalist credentials: he led the US legislators' delegation to the UNCED in Rio; and at the time he announced his candidacy for vice-president of the USA he was president of GLOBE International. The change in US policy on the environment in general and towards Target 2000 in particular can, in large measure, be attributed to Gore's influence.[26]

The second factor concerns NGOs. There is evidence of considerable US NGO lobbying prior to the Helsinki meeting pressing the US government to agree to Target 2000 within an expanded ITTA. In March 1993 a joint paper by the National Wildlife Federation, the Sierra Club and FoE was sent to the US government providing a rationale for US support for an expanded agreement. Aware that the US remained committed to a GFC after UNCED, the paper urged the USA to support an expanded ITTA arguing that unless there was a resolution of the producer–consumer stalemate 'it will be impossible to begin a productive international process towards such an agreement'.[27] One month later the president of WWF–US wrote to an assistant to President Clinton on the ITTA negotiations urging the US delegation to support an expanded ITTO which should then 'make a more concerted effort to achieve its Target 2000'.[28]

Although this lobbying did not result in the US government shifting its position with respect to an expansion in scope, it was instrumental in causing it to reconsider its position on Target 2000. Shortly before the ITTC's fourteenth session a two-day symposium on *Forests and the Environment: A US Response to the Rio Earth Summit* was convened in May 1993 in Atlanta chaired by former US President Jimmy Carter. While the symposium was in progress it received a letter from Gore. Noting the ITTO's efforts 'to develop a policy to manage sustainably rainforests', the letter concludes that, 'The President and I strongly support that effort, and believe that all forests — tropical, temperate and boreal — should be sustainably managed.'[29]

Despite the fact that Gore's letter did not explicitly commit the US government to support for either Target 2000 or an expanded ITTA, it was welcomed by NGOs at the ITTC's fourteenth session,[30] which convened in Kuala Lumpur shortly after the Atlanta symposium. An interesting question therefore is how far Drucker was mandated by other consumer delegations at Kuala Lumpur to include reference to consumer support for the sustainable management of non-tropical forests outside the ITTO. The reluctance of most European countries to support a commitment to Target 2000 for their own forests at Helsinki suggests that Drucker — who would certainly have been aware of the letter from Gore to Carter, and possibly also aware that the USA was planning to make a statement in support of Target 2000 at Helsinki — was attempting to rally the consumers around a US-inspired position, and that his paper at Kuala Lumpur was not based on a genuine joint-consumer position. Nonetheless, Drucker's proposal was eventually to prove decisive in breaking the deadlock.

SECOND PART OF CONFERENCE, GENEVA, 21–25 JUNE 1993

Less than one week after the Helsinki conference the second part of the negotiating conference commenced in Geneva. Although events since the first part had seen divisions appear among the consumers' caucus with respect to Target 2000, they remained committed to an agreement covering tropical timber only. Once again no agreement was reached, with Target 2000 and scope remaining the overarching points of contention. The question of financial resources was a further contentious issue, with producer countries claiming that financial assistance from the consumers was necessary if sustainable forest management was to be achieved. Some consumer delegates sought to deflect attention on the issue of trade discrimination by claiming that the UN Commission on Sustainable Development, scheduled to deal with forests at its 1995 session, was the appropriate forum for dealing with this issue. NGO representatives responded that the CSD would have too heavy a schedule to deal thoroughly with trade discrimination.[31]

At the end of the second part of the conference a third part was scheduled. The conference president announced that he would convene informal discussions in Yokohama prior to the third part, and he issued a discussion paper in advance of this meeting.[32]

ENVIRONMENTAL DIPLOMACY BETWEEN THE SECOND AND THIRD PARTS OF THE CONFERENCE

Informal discussions in Yokohama

The meeting called by the president of the conference was held in Yokohama in August 1993 with 14 individuals in attendance. With the exception of the president and the producers' and consumers' spokesmen, all individuals attended in their personal capacities.[33] NGOs were not permitted to attend, although a statement was forwarded to the meeting by representatives from WWF–International and the National Wildlife Federation, USA. This noted that none of the suggestions in the president's paper addressed the central issue of scope.[34] The exact details of the informal meeting, which was held in closed session, are not known, although it is clear that the meeting was unable to reconcile the central points of contention. The president issued a revised discussion paper which was on the table at the third part of the conference in October 1993. One noteworthy point from this document concerned ITTO-collated statistics. It was noted that producer countries held the view that 'notwithstanding the outcome of the discussions on the scope of the Agreement' information on non-tropical timber 'should be provided in order to attain transparency in the market'. The president added his personal authority to this point, arguing that there was 'a compelling necessity of taking a global perspective'.[35]

The Fourteenth Commonwealth Forestry Conference in Kuala Lumpur

The Fourteenth Commonwealth Forestry Conference, with its theme of *People, The Environment and Forestry: Conflict or Harmony*, was held in Kuala Lumpur in September 1993. The conference was the first to be held since the UNCED, where all Commonwealth governments had adopted the Statement of Forest Principles. Consistent with the strong line taken by the Malaysian government on forests during the UNCED process and the ITTA negotiations, the Malaysian prime minister, Dr Mahathir Bin Mohamad, referred in his opening speech to many of the producers' concerns, including a claim that discrimination between tropical and non-tropical timbers was 'a glaring case of double standards and a clear contradiction to the decisions of UNCED'. Mahathir also staked a claim for 'new and additional resources' and 'access to environmentally-sound technologies on favourable terms'.[36] However, no reference to the negotiating process was made in the list of recommendations the conference issued.[37]

The CSCE Forests Seminar in Montreal

Shortly after the close of the Fourteenth Commonwealth Forestry Conference a seminar of experts on the sustainable development of temperate and boreal forests opened under the auspices of the Conference on Security and Cooperation in Europe (CSCE) in Montreal (27 September–1 October 1993). The seminar, hosted by Forestry Canada, was convened with the objective of helping to define the criteria and indicators for the sustainable management of forests in Europe and North America. Like the earlier Helsinki meeting (see above), the Montreal seminar made a significant contribution to post-UNCED discussions on this subject (Chapter 6).

No joint closing statement, declaration or list of recommendations was issued. In the absence of any formal statement from the seminar, the following account draws upon the recollections of Simon Counsell of FoE.[38]

For most of its duration the seminar divided into two workshops, one dealing with the social and economic criteria of sustainability, and the second dealing with environmental and ecological criteria. Many of the delegates in attendance were also ITTO representatives. Representatives of some ITTO producer countries attended as observers, including Cameroon, Ghana, Gabon and Papua New Guinea. The invited ITTO producer countries were viewed by the consumers as among the 'weaker' members of the ITTO producer caucus. Some of the CSCE delegates attempted to impress upon these delegates that the seminar was evidence that the ITTO consumers were dealing with the sustainable management of their own forests outside the ITTO, therefore an expansion of scope to ensure sustainable management of non-tropical forests was unnecessary.

THIRD PART OF CONFERENCE, GENEVA, 4–15 OCTOBER 1993

Most of the two weeks of the third part of the conference were spent in informal working groups as opposed to formal committee meetings. NGOs were initially excluded from these meetings, although after protest they were permitted to attend. Some of the consumers made significant moves towards the producers' position, while continuing to advocate a tropical-timber-only format. An early development was the commitment by Australia to 'the goal of achieving sustainable management of its own forests by the year 2000'.[39] This was followed by a statement from Canada that it was prepared to associate itself 'with sound internationally-agreed-upon measurements [including] association with Objective 2000 of the ITTO'.[40] (The third part of the conference saw a shift in nomenclature, with delegates referring to 'Objective 2000' rather than 'Target 2000'. Producer members, who remained as sensitive as ever on sovereignty, were not prepared to accept that the goal of sustainable forest management by the year 2000 was a legally-binding target, and the change of wording to Objective 2000 signified a softening of the original target.) The statement from Canada brought to three the number of consumer countries agreeing to observe Target/Objective 2000 outside the ITTO. Although no other consumer delegations made such a commitment, a draft non-paper was leaked from the consumer caucus, indicating that some consumers favoured a formal statement within the ITTO in favour of Target/Objective 2000.[41]

The EC moved to allay another producer concern, namely unilateral trade discrimination, when its spokesman stated that the EC did not support 'the use of trade restrictions of any type' and favoured language in the new agreement to the effect 'that unilateral trade measures cannot be taken'.[42] This statement moved the EC behind one of the points made by Drucker at the ITTC's fourteenth session.

The third part also saw developments on the financial resources issue. The consumers issued a proposal advocating the establishment of an 'Objective Year 2000 Fund'. This proposed the establishment of a third ITTO account to be used 'only for the implementation and management of approved projects and pre-projects relating to the attainment of the Objective Year 2000'.[43]

The producers also submitted a proposal which laid claim to new and additional financial resources provided by the developed member countries to enable developing member countries 'to meet the agreed full incremental costs to them of implementing measures which fulfil the Objective Year 2000'.[44] This proposal was more of a normative statement than a realistic negotiating claim by the producers, and its real significance was that it suggested, for the first time, that the producers would accept mention of Target/Objective 2000 in the new agreement, a move welcomed by the NGOs who criticized the consumers for not making a similar commitment.[45]

The only other significant event was a statement by a Canadian indigenous peoples' NGO, namely the Nuu-Chah-Nulth Tribal Council. Noting the deforestation taking place in large areas of British Colombia, the representative informed delegates that indigenous peoples have blockaded logging roads in an effort to protect their lands; these blockades are similar to those in Sarawak.[46] A statement by the Netherlands Committee of IUCN also noted concern at deforestation in Canada.[47] The two interventions, a source of irritation for the Canadian delegation, were part of an NGO tactic to support arguments in favour of an expanded agreement by drawing attention to the fact that forest dwellers live not just in tropical forests but also in temperate forests, where the threat of deforestation is often as acute as in the tropics.

Overall, the third part of the conference saw notable movement by the consumers towards the producers' position. However, in the draft articles submitted by the president at the conclusion of the meeting the word 'tropical' appeared throughout in square brackets, indicating that there was still no consensus on the scope of the new agreement.[48]

FOURTH PART OF CONFERENCE, GENEVA, 10–26 JANUARY 1994

Two days after the start of the fourth part of the conference the European Union (EU)[49] stated that it was 'fully committed to sustainable forest management ... notably in the framework of the Year 2000 Objective'. The EU statement also noted that this commitment should allow for the integration 'of the aims of the ITTO Year 2000 Objective into a legally binding instrument covering all types of forests',[50] in other words a GFC.

By now Japan had also stated that it intended to work towards the sustainable management of its forests by the year 2000,[51] and with the support of the USA, Canada and Australia previously pledged, the way was paved for a formal statement on this issue by the consumers. This was issued on 21 January 1994 and was based on the consumer non-paper 'leaked' at the third part of the conference. Endorsed by all members of the consumer caucus, the statement committed all consumers to 'the national objective of achieving sustainable management of their forests by the year 2000'.[52]

On the same day that the consumers issued their formal statement on Objective 2000, a text for the International Tropical Timber Agreement, 1994 (hereafter the ITTA 1994) was agreed (see Annex C). The consumers' statement does not form a part of the ITTA 1994 and is not legally-binding, although mention of the commitment appears in the Preamble to the Agreement.[53] The commitment was one of two concession made by the consumers in a compromise formula. The second concession was the agreement to establish the Bali Partnership Fund, thus meeting in part the producers' claim for new and additional resources. (The reader will recall from Chapter 3 that Bali is the venue where the ITTO's Permanent Committee on Forest Industry adopted the original Target 2000.) The clause on this fund in the ITTA 1994 draws heavily from the consumers' proposal for an 'Objective Year 2000 Fund' made at the third part of the conference.[54] The Bali Partnership Fund will form a third ITTO account for projects and pre-projects designed to achieve Objective 2000. In return for these two consumer concessions the producers dropped their insistence on expansion of scope, although they did succeed in inserting a clause that the scope of the agreement shall be reviewed by the Council after four years.[55] At the conclusion of the conference in Geneva reservations were tabled by the EU and China indicating that the text of the new agreement did not meet with the approval of all consumers. The reservations, and the reasons for them, will be considered below. The agreement may enter into force for an initial period of four years, with two options for three-year extensions.

By the conclusion of the conference 65 countries had participated in the negotiations. Of these 50 were ITTO members (see Box 3.2, Chapter 3) and 15 were non-members. Of the 15 countries which were not ITTO members, nine are listed in the ITTA 1994 as producers and six as consumers. Of the nine producers, seven had also taken part in the negotiations for the ITTA 1983 without subsequently depositing articles of ratification for the agreement. Of the six consumers, two had taken part in the negotiations for the ITTA 1983; see Box 5.1 below.

Box 5.1 *NON-ITTO MEMBERS THAT TOOK PART IN THE NEGOTIATION OF THE INTERNATIONAL TROPICAL TIMBER AGREEMENT, 1994*

Producing countries	Consuming countries
Costa Rica*	Afghanistan +
Dominican Republic*	Algeria +
El Salvador*	Bahrain +
Equatorial Guinea +	Bulgaria*
Mexico*	Chile*
Myanmar (Burma)*	Slovakia +
Paraguay +	
Tanzania*	
Venezuela*	
Total producers 9	*Total consumers 6*

Key: * Countries that took part in the negotiation of the International Tropical Timber Agreement, 1983 but did not ratify this agreement. + Countries taking part in the negotiation of an International Tropical Timber Agreement for the first time.

After the closing of the conference the second special session of the ITTC was convened. A minor, but interesting, procedural point here is that the second special session of the ITTC was held in the same venue as, and immediately after the close of, the fourth part of the conference, with the only changes being that the president of the negotiating conference left the chair to make way for the ITTC chairman, and non-ITTC members who had taken part in the conference could remain only as observers, not as participants. The second special session of the ITTC agreed that the ITTO would continue to operate under the auspices of the ITTA 1983 until the ITTA 1994 enters into legal effect.[56]

THE NORTH'S DEMAND FOR A GLOBAL FORESTS CONVENTION

At this juncture let us remind ourselves that during the UNCED process the North favoured inserting language on the negotiation of a post-UNCED global forests convention (GFC) into the Statement of Forest Principles. The G77 successfully opposed this (Chapter 4). These overall forest strategies of North and South played a major role in the negotiations, and help explain why the consumers were unwilling to agree to an expansion of scope. It could be argued that the consumers should have seized the opportunity to agree to an international timber agreement on the grounds that such an agreement would represent an important first step towards a GFC. However, there are three reasons why the consumers were not prepared to use the negotiations towards this end.

First, and most obvious, the PrepCom process for the negotiations began in November 1992, just five months after the North had suffered its GFC defeat at Rio. Rather than risk a further failure, the North kept the subject of a GFC off the agenda of the negotiations, while simultaneously making clear that it remained an overall policy objective.

Second, an expanded ITTA would not be a GFC (although it would undeniably contain features that a GFC may be expected to contain with respect to the timber trade). Here it is useful to consider what a GFC may look like. Two such drafts have so far appeared. One was introduced in Chapter 4, namely the FAO draft GFC. The second is a draft produced for GLOBE International by the Centre for International Environmental Law, London (now the Foundation for International Law and Development) with the assistance of the Dutch NGO AIDEnvironment, the IIED and the Gaia Foundation (see Annex D). This draft was circulated at UNCED PrepCom 3 as a basis for informal discussions[57] although it made no significant impact on the UNCED forests negotiations, with PrepCom 2 having previously decided that the negotiations would focus upon a non-legally binding Statement of Forest Principles (Chapter 4).

Despite the fact that the FAO and GLOBE International draft GFCs emanate from very different sources, a comparative analysis of the two documents reveals that they share several features in common. These are the role of forests in national development policy, including fuelwood and energy requirements, establishment of a national forest policy, regeneration of forest resources including reforestation, observation, monitoring and surveying of forest cover, the establishment of a national forest authority or service, the establishment and management of protected forest areas, indigenous forest peoples' rights, prevention of air and water pollution, biodiversity protection, mitigation of the effects of climate change, international cooperation, scientific research, education and public information and the importance of environmental impact assessments.[58]

To date no government from the North has produced a draft GFC. (Indeed, a possible reason why the North did not use the ITTO negotiations to push for a convention could be that no consumer government has moved beyond support for a GFC to a precise formulation of the features such an instrument would contain.) However, the fact that there is a high degree of similarity between the FAO and GLOBE International drafts, despite the fact that the two drafts emanate from non-governmental sources, provides some indication of the features that a finalized GFC may be expected to contain. Clearly, if the countries of the North had chosen to use the ITTA negotiations to promote a GFC, they would have had to agree not only to an expansion of scope, but also to have successfully advanced the case for a large-scale expansion of mandate. It is worth emphasizing that the ITTO is a commodity organization that deals not with all tropical forests, or even with all tropical timber, but only with internationally-traded tropical timber. This is not to say that many of the issues noted in the preceding paragraph could not be dealt with in an expanded ITTA; it is possible for a commodity organization such as the ITTO to expand its mandate over time, thus evolving into a fundamentally different type of organization. The International Whaling Commission, for example, was created primarily as a trading organization; over time it has become more of a conservation organization, although tensions between the claimed right to trade and conservation remain. However, most Northern governments are keen to secure a GFC at the earliest opportunity and would reject a strategy requiring several lengthy rounds of negotiations to complete. Rather than elect for an evolutionary track, the consumers have decided that other options for a GFC are more likely to be successful.

This leads on to our third point. The negotiations began shortly after the proposal for a World Commission on Forests and Sustainable Development (WCFSD) was made (Chapter 2). At this time some consumer governments saw the WCFSD as offering a more promising track for a GFC (Chapter 6). It is noteworthy that the organizing committee for the WCFSD was established by Ola Ullsten, one of the authors of the TFAP independent review which recommended a GFC in 1990. Furthermore, previous world commissions have established a good record for bridging North–South divisions and establishing a new consensus.

Why, therefore, did the producers support an expansion of scope? It was argued earlier that the producers supported an expanded scope in part because they wished all timbers to be dealt with by the same body to alleviate what they perceived as trade discrimination against tropical timber. But there is a further reason why the producers favoured an expansion in scope, namely to defuse any subsequent pressures for a GFC. An expanded ITTA would not have been a GFC. However, the existence of the former would have permitted the South to respond to any future Northern demands for the latter that, 'There is an international timber agreement, therefore why do you want a GFC?' The South therefore argued that the negotiations should not deal with the question of a GFC and should instead deal with timber and trade-related issues. Similarly, the consumers resisted expansion for the same reason. Any subsequent consumer arguments that a GFC was necessary would have been weaker if an international timber organization had been in place.

AN ANALYSIS OF THE INTERNATIONAL TROPICAL TIMBER AGREEMENT, 1994

The conference ended with the *status quo* prevailing regarding scope. The consumers/timber-trader alliance prevailed over the producers/NGO alliance. This section will compare the ITTA 1983 with the ITTA 1994 and will provide an assessment of the latter, bearing in mind the history to date of the ITTO considered in Chapter 3. A clause-by-clause comparative analysis of the ITTA 1994 compared with the ITTA 1983 reveals five new articles, and ten substantially reworded articles. Note that textual amendments considered by the author as 'minor' are not included in the latter category. A detailed breakdown of these changes appears in Boxes 5.2 and 5.3 below. An analysis of the significance of the main changes now follows.

The consumers succeeded in their overriding objective of maintaining the tropical timber scope of the agreement. However, in addition to the establishment of the Bali Partnership Fund and the clause providing for a subsequent review of scope by the International Tropical Timber Council (ITTC), the producers were able to obtain other concessions as part of the compromise formula. First, there is a clause on non-discrimination which notes that nothing in the agreement authorizes 'the use of measures to restrict or ban international trade in ... timber and timber products' (Article 36). The insertion of this clause follows Austria's timber-labelling case (Chapter 3), the statement made by Drucker at the ITTC's fourteenth session and the EC's statement at the third part of the conference

Second, among the new objectives of the agreement, there are clauses on the development of mechanisms for 'new and additional financial resources' and on the transfer of technology 'including on concessional and preferential terms and conditions, as mutually agreed' (Article 1). The ITTA 1994 is therefore the second intergovernmental forest document in which the South has succeeded in inserting clauses on these two bargaining claims, with the UNCED Statement of Forest Principles being the first (see Annex B).

Remit

Although the producers were unsuccessful in their efforts to expand the scope of the new agreement, they did succeed in broadening the remit of the ITTO with respect to information sharing and statistics collation. Here the support given by Wisber Loeis, the conference president, to the producers' position on this issue after the Yokohama informal consultation appears to have been influential. A comparison between the ITTAs of 1983 and 1994 reveals a change of emphasis in the articles on statistics and the annual report. Whereas the ITTA 1983 emphasized 'tropical timber', the ITTA

1994 lays stress on 'timber'. The ITTA 1994 also states that the ITTO will consider 'relevant information on non-tropical timber' (Article 29). The word 'tropical' has been deleted from some other clauses.

Box 5.2   *THE FIVE NEW ARTICLES IN THE INTERNATIONAL TROPICAL TIMBER AGREEMENT, 1994*

1. Article 21, The Bali Partnership Fund
- *a new ITTO account is established to help producing members attain Objective 2000.*

2. Article 24, policy work of the organization
- *the ITTO should aim to integrate policy work and project activities.*

3. Article 35, review
- *the scope of the agreement shall be reviewed four years after the agreement's entry into force.*

4. Article 36, non-discrimination
- *the ITTO does not authorize measures to restrict or ban international trade in timber and timber products.*

5. Article 48, supplementary and transitional provisions
- *the ITTA 1994 is the successor agreement to the ITTA 1983.*
- *acts by the ITTO under the ITTA 1983 shall remain in effect when the ITTA 1994 enters into force unless prohibited by the latter.*

Reservations

A proposed rewording by the producers of Article 1(l) — formerly Article 1(h) of the ITTA 1983 on the conservation of tropical forests and their genetic resources — caused concern for some NGOs and consumer delegations. First, the word 'tropical' was deleted from the original article. Second, a new proviso was added, namely 'in the context of tropical timber trade'. The amended article now reads: 'To encourage members to develop national policies aimed at sustainable utilization and conservation of timber producing forests and their genetic resources and at maintaining the ecological balance in the regions concerned, in the context of tropical timber trade.'[59]

The producers had originally proposed that the proviso read 'in the interests of the tropical timber trade'. NGOs and some consumer delegations argued forcefully for the substitution of 'context' for 'interests'. Nonetheless, certain members of the EU remained sufficiently concerned to table a reservation on the clause at the end of the conference, principally because it left unclear the degree to which the EU members would be accountable to the ITTO for the management of their forests.

Box 5.3 *THE TEN SUBSTANTIALLY-REWORDED ARTICLES IN THE INTERNATIONAL TROPICAL TIMBER AGREEMENT, 1994*

1. Preamble (Preamble, 1983)
- *mention of the New Partnership for Development and the Cartagena Commitment and the Spirit of Cartagena adopted at the eighth session of the UNCTAD;*
- *two mentions of the UNCED Statement of Forest Principles, including reference to principle 10 on new and additional financial resources;*
- *the need to apply guidelines for all types of timber producing forests;*
- *the need to improve transparency in the international timber market; and*
- *two mentions of Objective 2000, including one reference to the consumer countries' commitment to the objective.*

2. Article 1, Objectives *(Article 1, 1983)*
There are six new objectives:

- *to promote non-discriminatory timber trade practices;*
- *to contribute to the process of sustainable development;*
- *to enhance the capacity of members to achieve Objective 2000;*
- *to develop mechanisms for new and additional financial resources;*
- *to promote technology transfer and technical cooperation, including on concessional and preferential terms; and*
- *to encourage information sharing on the international timber market.*

Other changes to Article 1 are:
- *mention of principle 1(a) of the UNCED Statement of Forest Principles on the sovereignty of states over their natural resources;*
- *mention of the need to conserve and enhance other forest values; and*
- *mention of 'due regard for the interests of local communities'.*

3. Article 14, Cooperation and coordination with other organizations *(Article 14, 1983)*
- *mention of cooperation with the UN Commission on Sustainable Development and CITES.*

4. Article 25, Project activities of the Organization *(Article 23, 1983)*
- *a list of new criteria to be taken into account when judging projects;*

Box 5.3 (continued from p127)

- *a schedule and procedure to be established for submission, appraisal and prioritization of pre-projects and projects; and*
- *granting of powers to the executive director to suspend disbursement of funds being used contrary to the project plan.*

5. **Article 26, Establishment of Committees** *(Article 24, 1983)*
- *establishment of a Committee on Finance and Administration.*

6. **Article 27, Functions of the Committees** *(Article 25, 1983)*
- *to review information on the undocumented timber trade;*
- *the committees shall be responsible for effective appraisal, monitoring and evaluation of pre-projects and projects, including implementation, and regular review of the ITTO's action plan;*
- *transfer of technology to be encouraged; and*
- *the Committee on Finance and Administration shall make recommendations on the ITTO's budget and management operations, on actions needed to secure resources, and on modifications to the Rules of Procedure and the Financial Rules.*

7. **Article 29, Statistics, studies and information** *(Article 27, 1983)*
- *emphasis on non-tropical timber; and*
- *emphases on 'tropical timber' in the ITTA 1983 amended to read 'timber'.*

8. **Article 30, Annual report and review** *(Article 28, 1983)*
- *reference to 'members' and 'timber' (whereas the equivalent clause in the ITTA 1983 saw reference to 'producing members' and 'tropical timber').*

9. **Article 43, Withdrawal** *(Article 39, 1983)*
- *members' financial obligations shall not be terminated by withdrawal.*

10. **Article 46, Duration, extension and termination** *(Article 42, 1983)*
- *the Agreement is valid for an initial period of four years, with two options for extension of three years. (The ITTA 1983 was valid for an initial period of five years, with two options for extension of two years).*

Further EU reservations tabled concerned Article 1(g) on new and additional resources, which left unclear the amount of money that consumers would be expected to pay, and on the fact that no references to Objective 2000 appeared other than in the Preamble and Article 1. The opinion of some delegations in the EU was that language was manipulated to cover differences that should have been subject to further negotiation. It should be noted that the reservations tabled by the EU did not represent the common concerns of all EU members; rather the list is a collection of separate reservations by individual members. Meanwhile, China also tabled reservations. One concern centred on the producer/consumer distinction, with China considering that she fell into both categories. A second concern was that the Bali Partnership Fund will be used strictly for the disbursement of funds for ITTO producers, thus ruling China out of funding from this account.[60]

From Target 2000 to Objective 2000

Only three references to Objective 2000 appear in the new agreement, namely two in the 'Preamble' and one in Article 1, 'Objectives'. While the objectives of a legal agreement are, in a broad sense, 'legally-binding', they are unlikely to have a real impact unless accompanied by references in the main body of the text to legally-binding implementation, review or monitoring mechanisms. This is not the case with respect to Objective 2000 in the ITTA 1994. Here it should be borne in mind that the producers agreed only reluctantly to any mention of Objective 2000 in the text, and they worked hard to ensure that references to it were kept to a minimum.[61]

The confusion that has surrounded Target/Objective 2000 was not clarified in the ITTA 1994; indeed its meaning is now more ambiguous. The wording in the agreement is different with respect to producers and consumers. Article 1 refers to a strategy 'for achieving exports of tropical timber and timber products from sustainably-managed sources by the year 2000', whereas the Preamble refers to 'the commitment to maintain, or achieve by the year 2000, the sustainable management of their respective forests made by consuming members'. The commitment by the consumers is clearly broader than that of the producers and applies to forests as opposed to internationally-traded timber. A possible explanation here is that the consumers made a commitment, albeit not a legally-binding one, to a broad interpretation of Objective 2000 in order to use this as a future leverage to extract a similar commitment from the producers. Such an explanation would be consistent with the demand of many consumers for a GFC, and would also explain the statement by the EU on the integration of Objective 2000 into a legally-binding instrument covering all forests. There will certainly be further concentrated debate, both at future ITTC sessions and in other fora, on the meaning of Objective 2000 and how it is to be applied.

NGO lobbying

There is evidence that NGO lobbying both at the ITTO and at other fora dealing with forest-related issues had some impact on the negotiations, with three references in the ITTA 1994 to issues on which NGOs have a campaigning history. First, there is reference to the need to give 'due regard for the interests of local communities dependent on forest resources' (Article 1). Second, the need to 'conserve and enhance other forest values' receives mention (Article 1). This would appear to be derived in part from the work of the LEEC in the incentives debate. This issue is one on which many NGOs have lobbied and which many consumer delegations also support. Finally, there is a reference to the need for the permanent committees to include information on 'undocumented trade' (Article 27), a veiled reference to the illegal trade against which NGOs have long campaigned.

Monitoring

An increased emphasis on monitoring and evaluation is a feature of the new agreement. The establishment of the Committee on Finance and Administration as the result of a consumer proposal during the PrepComs is an indication that the consumers are not prepared to make 'blank cheque' donations to ITTO accounts, including the newly-established Bali Partnership Fund, without the establishment of strict criteria on project expenditure, guarantees of prudent financial management and a follow-up process to determine the effectiveness of money disbursed. The agreement also empowers the executive director to suspend disbursement of ITTO funds to projects and pre-projects 'if they are being used contrary to the project document or in cases of fraud, waste, neglect or mismanagement' (Article 25). These clauses were inserted at the insistence of the European Union, which was concerned at the emphasis on projects at the ITTO which some producer delegates see as a 'donor of the last resort', with project proposals seldom rejected outright.

With UNCED failing to produce any legally binding agreement on forests, the ITTA 1994, when it enters into effect, will be the only legal forest-related instrument operating at the global level. Since 1995 one legal instrument on forest conservation operates at a regional level, namely the *Protocol on the Sustainable Management of Forest Resources* to the Lomé IV Convention; we will return to this instrument in Chapter 6. There are also various natural resource and pollution control instruments operating at the global and regional levels, the provisions of which impinge upon forests. Box 5.4 below lists the principal international forest-related instruments.

CONCLUDING REMARKS

The tension between timber production and forest conservation, so evident in the work of the ITTO throughout the lifespan of the ITTA 1983, was not resolved in the successor agreement which, like its predecessor, has

Box 5.4 PRINCIPAL INTERNATIONAL FOREST-RELATED INSTRUMENTS

Forest-related instruments operating at the global level
International Tropical Timber Agreement, 1983 (Geneva) — to be superseded by the International Tropical Timber Agreement, 1994 (Geneva) when the latter enters into legal effect

Forest conservation instruments operating at a regional level
Protocol on the Sustainable Management of Forest Resources to the Lomé IV Convention (Brussels, 1995).

Natural resource instruments operating at the global level
Convention on Wetlands of International Importance Especially as Waterfowl Habitat (Ramsar, 1971)

Convention for the Protection of the World Cultural and Natural Heritage (Paris, 1972)

Convention on International Trade in Endangered Species of Wild Fauna and Flora (CITES) (Washington, 1973)

Convention on Biological Diversity (Rio de Janeiro, 1992)

Natural resource instruments operating at a regional level
Convention on Nature Protection and Wildlife Preservation in the Western Hemisphere (Washington, 1940)

African Convention for the Conservation of Nature and Natural Resources (Algiers, 1968)

Amazonian Cooperation Treaty (Brasilia, 1978)

Convention on European Wildlife and Natural Resources (Berne, 1979)

ASEAN Agreement on the Conservation of Nature and Natural Resources (Kuala Lumpur, 1985)

Pollution control instruments operating at the global level
Convention for the Protection of the Ozone Layer (Vienna, 1985), and subsequent protocol:

> *Protocol on Substances that Deplete the Ozone Layer (Montreal, 1987)*

Convention on Climate Change (Rio de Janeiro, 1992)

Box 5.4 (continued from p131)

Pollution control instruments operating at a regional level
Convention on Long-Range Transboundary Air Pollution (Geneva, 1979), and subsequent protocols:

Protocol on Long-term Financing of the Co-operative Programme for Monitoring and Evaluation of the Long-Range Transmission of Air Pollutants in Europe (Geneva, 1984)

Protocol on the Reduction of Sulphur Emissions or their Transboundary Fluxes by at least 30 per cent (Helsinki, 1985)

Protocol Concerning the Control and Emission of Nitrogen or their Transboundary Fluxes (Sofia, 1988)

Protocol Concerning the Control of Emission of Volatile Organic Compounds or their Transboundary Fluxes (Geneva, 1991)

Sources: FAO document COFO–90/3(a), 'Proposal for an International Convention on Conservation and Development of Forests', September 1990, p11. Struan Simpson, *The Times Guide to the Environment* (London: Times Books, 1990), p67. Updated by the present author.

been drafted to suggest that the two objectives are simultaneously attainable. Despite the fact that NGOs and producer governments formed an alliance during the negotiations, this is an area where the two sets of actors are almost diametrically opposed, with the NGOs pressing for forest conservation, and the producers asserting that forests are a national sovereign resource to be exploited in line with national development policy.

The ITTO's mandate overlaps with other international institutions (Chapter 3). There is evidence that ITTO delegations have taken advantage of the confusion this situation has engendered by attempting to move debate on an issue from one institution to another. Take, for example, Malaysia's argument at the March 1992 meeting of the CITES conference that the ITTO was the appropriate forum to deal with trade restrictions on endangered tropical tree species (Chapter 3). Given that the ITTO has not become involved in the endangered species debate there must be a suspicion that Malaysia, whose present government is opposed to any kind of timber trade restriction, was acting to block the proposal in CITES on the assumption — correct as subsequent events proved — that it would receive no attention at the ITTO.

The problem of institutional overload on the forest issue has since been compounded by the creation of the UN Commission on Sustainable Development (CSD). Consumer delegations to the second part of the conference attempted to take advantage of this by suggesting that producer concerns on trade discrimination be dealt with by the CSD at its 1995 session. Again this can be seen as an attempt to deflect unwelcome attention on an issue in one forum into another, presumably with the intention that the latter will not have the time or the mandate to deal meaningfully with the issue in question. With progressively more environmental international institutions being created, and with more institutions receiving environmental mandates, this sort of tactic may become more commonplace in international environmental diplomacy, at least by those delegations who do not wish to deal meaningfully with a particular issue.

The future of the ITTO depends on several factors. First, there is the question of whether the organization will take action on the labelling and incentives issues. Second, there is the question of the composition of the ITTO's membership. Following severe deforestation, Thailand introduced a complete logging ban in 1989,[62] while remaining an ITTO 'producer'. Current timber logging and consumption figures indicate that other ITTO producers are set to become net importers of tropical timber.[63] Should these trends result in a fall in the number of ITTO producers' countries, this would inevitably effect the future functioning of the ITTO. Third, the nature of the relationship between the ITTO and the newly-created World Trade Organization appears to require redefinition given that contradictions between the GATT and the ITTO were never satisfactorily resolved (Chapter 3). Fourth, there is the question of the ITTO's relationship with other international institutions with forest or forest-related mandates, such as the CITES and the Forest Stewardship Council. But from the viewpoint of those environmental NGOs that have lobbied the ITTO and its members, the most salient question is whether the ITTO that will operate during the lifespan of the ITTA 1994 will be more successful at promoting forest conservation than was the case throughout the validity of the ITTA 1983.

Chapter 6

The Global Politics of Forest
Conservation since the UNCED

The mutual suspicion between North and South dominated the UNCED forest negotiations and was a feature of the negotiations for the International Tropical Timber Agreement, 1994. Since then mistrust has slowly yielded to a new cooperative spirit. Following the failure of the UNCED to produce a global forests convention (GFC), several new international initiatives on forest conservation emerged. This led initially to fragmented dialogue and piecemeal decision-making. However, throughout 1994 the situation became more coherent, first because governmental actors from North and South undertook confidence-building measures, and second because the actors involved in the various processes focused their efforts for the approaching third session of the UN Commission on Sustainable Development (April 1995), which dealt *inter alia* with forests. Since the UNCED 11 major international political processes on forests have taken place which may be grouped into five categories as shown in Box 6.1.

The previous chapter dealt at length with (c) in Box 6.1, namely the negotiation of the International Tropical Timber Agreement, 1994 (ITTA 1994). The objective of this chapter is to disentangle the ten other processes and to analyse the significance of, and relationships between, them.

CONFIDENCE-BUILDING INITIATIVES

Negotiations for the UNCED Statement of Forest Principles and ITTA 1994 were characterized by mutual suspicion between North and South. Since then there has been a gradual move from confrontation to attempts at confidence-building involving a search for areas of agreement. Significantly, the two countries most strongly opposing a GFC during the UNCED process — Malaysia and India — each launched bridge-building exercises with countries that advocated a GFC prior to Rio, namely Canada and the UK respectively.

Box 6.1　POST-UNCED INTERNATIONAL FOREST POLITICAL PROCESSES

1. **Activities under the auspices of the UN and UN specialized agencies**
(a) *Meetings held under the auspices of the UN's Food and Agriculture Organization (FAO)*
(b) *Third session of the Commission on Sustainable Development, and the subsequent establishment of the Intergovernmental Panel on Forests*

2. **Negotiation of intergovernmental agreements**
(c) *Negotiation of the International Tropical Timber Agreement, 1994*
(d) *Negotiation of a 'Protocol on the Sustainable Management of Forest Resources' to the EC–ACP Lomé IV Convention*

3. ***Ad hoc* processes initiated by governments**
(e) *Intergovernmental Working Group on Forests (cosponsored by Canada and Malaysia)*
(f) *Workshop 'Towards Sustainable Forestry: Preparing for CSD 1995' (cosponsored by India and the UK)*

4. **Processes to determine criteria and indicators for sustainable forest management**
(g) *Second Ministerial Conference on the Protection of Forests in Europe and follow-up meetings (Helsinki process)*
(h) *Working Group on Criteria and Indicators for the Conservation and Sustainable Development of Temperate and Boreal Forests (Montreal process)*
(i) *Amazonian Cooperation Treaty Workshop to Define Criteria and Indicators of Sustainability in the Amazon (Amazonian process)*

5. **Processes initiated by actors in international civil society**
(j) *Forest Stewardship Council*
(k) *World Commission on Forests and Sustainable Development*

Intergovernmental Working Group on Forests

At the Fourteenth Commonwealth Forestry Conference (Kuala Lumpur, September 1993) the Malaysian prime minister suggested that an 'intergovernmental task force on forestry' be established under the aegis of the

CSD to undertake consultations for the CSD's third session in 1995.[1] However, the CSD did not authorize the establishment of such an organ prior to its 1995 session. Following discussions between Malaysian and Canadian forestry officials, the Intergovernmental Working Group on Forests (IWGF)[2] was subsequently established, cosponsored by Malaysia and Canada. Goree notes that 'initiation of this process by these two unlikely partners represents a bilateral accommodation ... on a series of premises', including the recognition that states have the sovereign right to sustainably develop their forests which are an important resource for economic development and basic human needs, and that forests fill important environmental roles at the local, national, regional and global levels.[3] The IWGF was formed not as an intergovernmental negotiating forum but as a confidence-building process, with its main objective being to facilitate dialogue. Its first meeting, attended by 15 countries in Kuala Lumpur in April 1994, had an agenda of seven issues; see Box 6.2 below.

Box 6.2 *AGENDA OF THE INTERGOVERNMENTAL WORKING GROUP ON FORESTS*

A   *Management, conservation, sustainable development and enhancement of all types of forests to meet basic human needs*
B   *Criteria and indicators for sustainable forest management*
C   *Trade and the management, conservation and sustainable development of all types of forests*
D   *Approaches to mobilizing additional financial resources and environmentally sound technologies*
E   *Institutional linkages*
F   *Participation and transparency in forest management*
G   *Comprehensive cross-sectoral integration including land use and planning and management and the influence of policies external to the traditional forest sector.*

Source: 'Report: First Meeting of the Intergovernmental Working Group on Global Forests, Kuala Lumpur, Malaysia, 18–21 April 1994.'

The second meeting took place in Ottawa in October 1994 with 32 countries, more than double the number at the first meeting, in attendance. Syntheses of the meeting's discussions were prepared for each issue; these were intended to be a crystallization of the main points raised by discussants, including possible policy options, rather than consensus statements.[4] The syntheses were forwarded to the CSD's 1995 session where, it will be seen below, they had an impact upon the debate.

Indian–UK Initiative

The second confidence-building exercise between countries from North and South was cosponsored by the governments of India and the UK. A hindrance to efficient data compilation by intergovernmental organizations is often the absence of a standardized reporting format, with different countries frequently submitting data using different reporting styles. The purpose of the Indian–UK process, which culminated with a workshop in New Delhi in July 1994, was to provide a framework to guide and facilitate country reporting on forests to the CSD's 1995 session.[5] The workshop, attended by representatives of 39 countries, reached agreement on a reporting format designed to reflect the activities outlined in the Statement of Forest Principles, and Agenda 21 Chapter 11, 'Combating Deforestation'. After the New Delhi workshop the format was forwarded to the CSD secretariat, which transmitted it to other countries.

The Indian–UK process was unsuccessful in its efforts to encourage governments to adopt a standardized format. A study of the country reports submitted to the CSD's 1995 session reveals that while some countries adopted the format, such as Norway, many others did not, such as France, Germany and Ireland. In fact only 34 countries completed the surveys,[6] and the main significance of the Indian–UK process was that it saw North–South cooperation on forests led by two countries that during the UNCED forests debate had occupied diametrically-opposed positions. Like the IWGF meetings, the Indian–UK process was a positive indication of the emergence of a more cooperative atmosphere in post-Rio international forest politics.

THE CRITERIA AND INDICATORS PROCESSES

It was noted above that one of the issues considered by the IWGF was criteria and indicators for sustainable forest management. It is to this subject that attention now turns. Paragraph 8(d) of the Statement of Forest Principles states that, 'Sustainable forest management and use should be carried out in accordance with national development policies and priorities and on the basis of environmentally sound national guidelines. In the formulation of such guidelines, account should be taken, as appropriate and if applicable, of relevant internationally-agreed methodologies and criteria.'[7]

The importance of 'formulating scientifically sound criteria and guidelines for the management, conservation and sustainable development of ... forests' also receives mention in Chapter 11 of Agenda 21.[8] The recognition that criteria and indicators are necessary to determine whether sustainable forest management is being achieved led to three separate initiatives in the period 1993–5, namely the Helsinki process,[9] the Montreal process and the Amazonian process. Starting from the *a priori* assumption that sustainable forest management is possible, the basic premise of each process is that there are various features, or criteria, of sustainable forest management and that, in order to demonstrate sustainability, these criteria should be measured by indicators.

Similar definitions of 'criterion' and 'indicator' have been provided by the Helsinki and Montreal processes. In the case of the former, a criterion describes an aspect of sustainability on a conceptual level and is 'a distinguishable element or set of conditions or processes by which a forest characteristic or management is defined'. An indicator shows 'changes over time for each criterion and demonstrates how well each criterion reaches the objectives set for it'.[10] Indicators may be a quantitative measure of change or they may be qualitative. These definitions correspond closely with those of the Montreal process, which defines a criterion as 'a category of conditions or processes by which sustainable forest management may be assessed [and] is characterized by a set of related indicators which are monitored periodically to assess change'. An indicator is a measurement of an aspect of the criterion and is a 'quantitative or qualitative variable which can be measured or described and which when observed periodically demonstrates trends'.[11]

Helsinki process

The first criteria and indicators (C&I) process to be launched was the Helsinki process. In June 1993, 38 European countries participated in the Second Ministerial Conference on the Protection of Forests in Europe at Helsinki. (The First Ministerial Conference on the Protection of Forests in Europe, cosponsored by France and Finland, took place in Strasbourg, December 1990, and was attended by 31 countries, compared to the 38 that attended the Helsinki conference. This increase is due to the creation of new countries following the break-up of the Soviet Union, Czechoslovakia and Yugoslavia.) It will be recalled from Chapter 5 that the European governments attracted NGO criticism at Helsinki for not adopting the target date of the year 2000 for sustainable forest management. The Europeans were embarrassed when the US delegation, attending the conference as observers, used the conference as the forum to announce their own commitment to Target 2000.

The conference finalized the drafting of a general declaration and of four resolutions which were opened for signature.[12] The Dutch delegation 'seriously considered the possibility of not signing the resolutions' in protest at the lack of any mention of Target 2000.[13] After the conference, work began on devising C&I for measuring sustainable forest management.

The organization of the Helsinki follow-up process is as follows. First, each country has appointed a national coordinator. Second, there is a

Liaison Unit, which at present is directed by the Finnish national coordinator and based in Helsinki. Third, there is a General Coordinating Committee (GCC), which at present consists of the national coordinators and experts from Finland, Portugal, Austria and Poland. This meets approximately three times a year. Fourth, various informal pan-European round tables have been convened to discuss the Helsinki process C&I. Fifth, there is a Scientific Advisory Group consisting of some GCC members and experts from other countries; its main task is 'to advise the GCC and to give scientific support to the Liaison Unit'.[14]

The Scientific Advisory Group has participated, along with other invited participants, in the informal pan-European round tables as well as in two expert level meetings. The first round table (Brussels, March 1994) agreed on a provisional set of criteria. The first expert level meeting (Geneva, June 1994) subsequently adopted six criteria and 20 quantitative indicators for the sustainable management of European forests; see column 1 of Table 6.1 below. While at present the Helsinki process is using only quantitative indicators, the head of the Liaison Unit has stated that the defining of indicators is 'a continuous process' and new indicators will be added as and when scientific evidence develops and political consensus is reached.[15] Following a further round table meeting (Brussels, December 1994), the second expert level meeting (Antalya, January 1995) reviewed progress in those European countries using the C&I.[16]

Montreal process

The second C&I process is the Working Group on Criteria and Indicators for the Conservation and Sustainable Development of Temperate and Boreal Forests, also known as the Montreal process. The origins of this process can be traced to a seminar on the sustainable development of temperate and boreal forests hosted by Canada in Montreal in September 1993 and held under the auspices of the Conference on Security and Cooperation in Europe. The seminar's executive director considered the CSCE to be an appropriate forum to deal with forest conservation in the Northern hemisphere due to its 'pan-European and trans-Atlantic membership as well as its comprehensive concept of security ... linking military, political, humanitarian, economic and environmental issues'.[17] The governments of the USA and Canada shared this view and wished the Helsinki and Montreal processes to be brought together. However, this proposal was effectively vetoed by France, Germany and Britain, which preferred the development of European C&I to take place solely within the Helsinki process. Effectively, the Canadian government, which had initiated the CSCE meeting, then had two choices — to abandon, or to restructure, the process. Keen to maintain a leadership role in forest politics, it chose the latter option. After three informal meetings in Kuala Lumpur, Geneva and New Delhi in mid-1994, the process was relaunched in September 1994 in Olympia, Washington with ten countries involved. These are Australia, Canada, Chile, China, Japan, South Korea, Mexico, New Zealand, Russian Federation and the USA. Of these, only the Russian Federation is a participant in the Helsinki process. In Tokyo in November 1994 agreement was reached on the first drafts of seven criteria.[18] In February 1995 in Santiago, Chile, seven criteria and 67 indicators for sustainable forest management were approved; see column 2 of Table 6.1 below.

Amazonian process

The third set of C&I deals with the sustainable management of Amazonian forests and emerged from a workshop held in Tarapoto, Peru in February 1995. Representatives from Bolivia, Brazil, Colombia, Peru, Surinam and Venezuela attended. Observers included the European Union, FAO and the World Resources Institute. The meeting, hosted by the Ministry of Foreign Affairs of Peru in its capacity as *pro tempore* secretariat of the Amazonian Cooperation Treaty, reached agreement on 12 criteria and 76 indicators for the sustainable management of Amazonian forests; see column 3 of Table 6.1 below.

A comparative analysis of the three criteria and indicators processes

The criteria of the Helsinki and Montreal processes are designed to assess sustainability only at the national level, whereas with the Amazonian initiative seven of the twelve criteria deal with sustainability at the national level, four at the forest management unit level, while the twelfth criterion deals with services performed by Amazonian forests at the global level. The fact that in the post-UNCED era the Amazonian countries are prepared to acknowledge the global functions of forests, whereas to date the developed countries are not, is somewhat ironic given that during the UNCED negotiations the developed countries stressed the global dimension when advancing their case for a GFC, while the developing countries continually asserted that forests are a sovereign natural resource.

Comparison between the Helsinki and Montreal processes on the one hand, and the Amazonian process on the other, is problematic for two reasons. First, the two former processes deal with temperate and boreal forests, while the Amazonian one deals with tropical forests. Second, and as noted above, the two former processes deal solely with sustainability at the national level, whereas the Amazonian one has a broader focus. Nonetheless, a comparison should be attempted, not least because an important international policy debate is whether there should be 'convergence' or 'harmonization' between the three processes to yield a uniform global set of

Table 6.1   *The criteria for sustainable forest management of the Helsinki,
Montreal and Amazonian processes*

Helsinki Process 6 criteria/20 indicators (all quantitative)	Montreal Process 7 criteria/67 indicators (quantitative and qualitative)	Amazonian Process 12 criteria/76 indicators (quantitative and qualitative)
1. Maintenance and appropriate enhancement of forest resources and their contribution to global carbon cycles (3)	5. Maintenance of forest contribution to global carbon cycles (3)	
2. Maintenance of forest ecosystem health and vitality (4)	3. Maintenance of forest ecosystem health and vitality (3)	10. Conservation of ecosystems [management unit level] (6)
3. Maintenance and encouragement of the productive functions of forests (wood and non-wood) (3)	2. Maintenance of productive capacity of forest ecosystems (5)	3. Sustainable forest production [national level] (5)
		9. Sustainable forest production [management unit level] (5)
4. Maintenance, conservation and appropriate enhancement of biological diversity (5)	1. Conservation of biological diversity (9)	4. Conservation of forest cover and biological diversity [national level] (8)
5. Maintenance and appropriate enhancement of protective functions in forest management (notably soil and water) (2)	4. Conservation and maintenance of soil and water resources (8)	5. Conservation and integrated management of water and soil resources [national level] (4)
6. Maintenance of other socioeconomic functions and conditions (3)	6. Maintenance and enhancement of long-term multiple socioeconomic benefits to meet the needs of societies (19)	1. Socioeconomic benefits [national level] (16)
		11. Local socioeconomic benefits [management unit level] (8)
	7. Legal, institutional and economic framework for forest conservation and sustainable management (20)	2. Policies and legal institutional framework for sustainable development of the forests [national level] (4)
		7. Institutional capacity to promote sustainable development in Amazonia [national level] (4)
		8. Legal and institutional framework [management unit level] (3)
		6. Science and technology for sustainable development of the forests [national level] (6)
		12. Economic, social and environmental services performed by Amazonian forest [global level] (7)

Sources: Ministerial Conference on the Protection of Forests in Europe, *European List of Criteria and Most Suitable Quantitative Indicators*, (Helsinki: Ministry of Agriculture and Forestry, 1994); Montreal Process, *Criteria and Indicators for the Conservation and Sustainable Management of Temperate and Boreal Forests*, (Quebec: Canadian Forest Service, 1995); United Nations document E/CN.17/1995/34, 'Regional Workshop on the Definition of Criteria and Indicators for Sustainability of Amazonian Forests, Final Document, Tarapoto, Peru, February 25, 1995', 10 April 1995.

C&I. Table 6.1 has been constructed to facilitate comparison across the three sets of criteria which are not necessarily listed in their original order. The number of indicators for each criterion is shown in brackets. The following points emerge from a study of Table 6.1.

Comparison between the Helsinki and Montreal processes reveals that each has arrived at the same or very similar criteria in six cases, with the exception being a qualitative criterion. This is the legal, institutional and economic framework, namely the degree to which the law and institutions in a given country facilitate sustainable forest management, which only the Montreal process includes.

Across the three processes there are very similar criteria in five respects: ecosystem health/conservation; the productive functions of forests; biological diversity conservation; soil and water resource conservation; and the maintenance of the socioeconomic functions of forests. Of this latter category it is noteworthy that the Montreal and Amazonian processes include, using almost identical wording, the indicator of 'area and percentage of forest lands, in relation to total forest lands area, managed to protect cultural, social and spiritual needs and values'.[19] However, the Helsinki process indicators contain no mention of the importance of the human dimension of sustainable forest management. Furthermore, the Montreal and Amazonian processes include the qualitative criterion of the legal and institutional framework which is not a criterion for the Helsinki process, although once again it is noted that the present Helsinki criteria are only those that can be measured by quantitative indicators.

At first glance it appears that the Amazonian process does not take into account the role of forests in global carbon regulation. However, one of the indicators for criterion 12, which is an amalgam of economic, social and environmental services performed at the global level, is the contribution of Amazonian forests to the global carbon balance. Finally, it can be seen that only in the Amazonian process is science and technology included as a criterion of sustainable forest management. Among the indicators for this criterion are the 'quantity and quality of adequate technology for forest management and sustainable production' and 'investment in research, education and technology transfer'.[20]

THE THIRD SESSION OF THE CSD: CREATION OF THE INTERGOVERNMENTAL PANEL ON FORESTS

Forests dominated proceedings at the CSD's third session held in New York in April 1995. This was the first time since Rio that forests had been debated at such a high-level UN organ. The session was tasked to deal with national level follow-up of the Statement of Forest Principles and of Agenda

---

Box 6.3　PROGRAMME OF WORK OF THE
INTERGOVERNMENTAL PANEL ON FORESTS

I.　*Implementation of UNCED decisions on forests through an 'open, transparent and participatory process', including considering 'ways and means for the effective protection and use of traditional forest-related knowledge'.*

II.　*Forestry aid, including financial assistance and technology transfer.*

III.　*Promotion of scientific research on forest functions, encouragement of national implementation of criteria and indicators for sustainable forest management and promotion of methodologies for valuing the multiple benefits derived from forests and their inclusion into national accounting systems.*

IV.　*Trade and the environment, including the issue of voluntary certification and labelling schemes.*

V.　*The role of international institutions and legal instruments, including the identification of 'any gaps and areas requiring enhancement, as well as any areas of duplication'.*

Source: UN document E/CN.17/1995/36, 'Commission on Sustainable Development, Report of the Third Session (11–28 April 1995), Economic and Social Council Official Records, 1995, Supplement No 12', pp50–2.

21 chapters dealing with land resources, including Chapter 11, 'Combating Deforestation'. However, forest discussions were dominated by institutional questions, with national reporting and C&I assuming only a secondary place in the deliberations. Prior to the meeting the CSD's Ad Hoc Working Group on Sectoral Issues (New York, February 1995) had recommended that the CSD 'consider the establishment of an open-ended intergovernmental panel on forests, under the aegis of the Commission'.[21] This idea was welcomed by the Ministerial Meeting on Forestry hosted by FAO (Rome, March 1995). At its third session the CSD agreed to create the Intergovernmental Panel on Forests (IPF). The IPF will report to the CSD's fifth session in 1997, which will undertake a review of all Agenda 21 issues in the five years since the Rio conference and will also decide on the future, if any, of the IPF after 1997. The five areas of work for the panel, as agreed by the CSD's third session, are summarized in Box 6.3 above.

A comparison of the IWGF's seven issues (Box 6.2 above) with the IPF's mandate reveals that the former helped to define the latter; IWGF syntheses B, C, D and E correspond closely with IPF mandates III, IV, II and V respectively. Furthermore, IWGF syntheses F on participation and transparency fed into IPF mandate I, while a suggested policy option from IWGF synthesis G, namely that the CSD could commission research on the development of 'methodologies for comprehensive assessment of the costs and benefits of forest services and goods' and their inclusion into national accounting systems,[22] found expression in IPF mandate III. The significance of the IWGF is not just that it played a role in confidence-building, but that it has also helped to set the international forestry agenda for the two years between the CSD's 1995 and 1997 sessions.

The creation of the IPF would not have been possible without the new spirit of trust and cooperation that emerged prior to the CSD's third session. The recognition that the various post-Rio processes lacked a unifying focus was also a factor. The decision to call the new institution a 'panel' reflects the hopes of some governments that the IPF will fill the same consensus-building role for forests that the Intergovernmental Panel on Climate Change did for global warming. The panel will commission research from relevant bodies and will draw on the resources and expertise of organizations such as the FAO, ITTO and the United Nations Environment Programme.

THE WORLD COMMISSION ON FORESTS AND SUSTAINABLE DEVELOPMENT

The findings from a further international process, the World Commission on Forests and Sustainable Development (WCFSD), will feed into the work of the IPF.[23] The idea of establishing a WCFSD originated at a seminar at the Woods Hole Research Center in Massachusetts in October 1991. In May 1992 a statement in support of the idea was made by the InterAction Council of Former Heads of State and Government. In July 1992, shortly after the UNCED, the first meeting of the organizing committee for a WCFSD was held in Rome. After a second meeting in Ottawa (November 1992) the proposal that a WCFSD be established was presented to an international gathering of foresters, the Global Forests Conference in Bandung (February 1993), where it found favour. The third and final meeting of the organizing committee was held in New Delhi in April 1993. Thereafter, there was a delay of over two years before the first meeting of the WCFSD was convened in Geneva in June 1995. Two principal reasons explain this delay.

First, the organizing committee was unable to garner sufficient support for the establishment of a commission along the lines of the World Commission on Sustainable Development where the UN Secretary General appointed the chairperson and vice chairperson, the General Assembly welcomed the establishment of the commission, and the commission presented its findings to the General Assembly. This was the model that the WCFSD

organizing committee envisaged.[24] Although some actors expressed support
for a WCFSD, most notably the Forestry Advisers Group (Chapter 2),
support for a WCFSD established within the UN system was not forthcoming. This was despite a meeting between the organizing committee and
UN Secretary General Boutros Boutros-Ghali in November 1992. There was
no debate in the General Assembly on whether support should be given to a
commission established within the UN system. However, it seems likely that
the proposal was the victim of North–South divisions. Some developed
countries viewed the WCFSD as a track that could establish consensus for
the negotiation of a post-Rio GFC,[25] and it appears that developing countries opposed it for the same reason. The organizing committee was chaired
by Ola Ullsten, the former Swedish prime minister who coauthored the
1990 independent review of the TFAP, which recommended the negotiation
of a GFC (Chapters 2 and 4). With their emphasis on the sovereignty of
states over their natural resources, most developing countries prefer to limit
international debate on forests to intergovernmental fora, hence the WCFSD
proposal received little support from the G77. Certainly those countries
opposed to a GFC would not welcome the establishment of a commission in
which a strong GFC advocate, namely Ullsten, was likely to play a
prominent role.

The second reason for the delay in the establishment of the WCFSD was
that it was overtaken by developments within the UN system, namely by
moves to create the IPF. By delaying the first meeting of the WCFSD until
June 1995 its members were able to take into account the outcome of the
1995 CSD session. With the IPF established within the UN system and the
WCFSD established as an *ad hoc* forum outside the UN, the report from the
June 1995 meeting of the WCFSD emphasized that its activities will
'complement and support' the activities of the IPF and CSD.[26]

In June 1994 a plenary session of the InterAction Council of Former
Heads of States and Government meeting in Dresden, chaired by former
West German Chancellor Helmut Schmidt, formally gave its support to the
creation of the WCFSD. This meeting also established the commission's
mandate, namely to raise understanding on the dual role of forests in
environment and sustainable socioeconomic development, to establish consensus on scientific data on forest conservation and management, and to
build confidence between North and South. The InterAction Council invited
Ullsten and a former Indonesian Minister of State for Population and
Environment, Professor Emil Salim, to co-chair the commission.

The first meeting of the commission in June 1995 agreed to establish
three working panels dealing with sustainable and equitable use and
management of forest resources; trade and the environment; and financial
mechanisms, international agreements and international institutions. There
is thus an overlap between the mandates of the IPF and the WCFSD. Two

institutions will report to the WCFSD. The first is the Science Council, which will assemble scientific evidence on the roles of forests and fill scientific knowledge gaps that are 'a constraint to policy formulation'. Second, a Policy Advisory Group will be established to facilitate the work of the three working panels and to prepare the commission's report.

One of the objectives of the WCFSD is to 'ensure that the views of *all main stakeholders* are fairly represented and adequately taken into account' (emphasis in original).[27] In pursuit of this objective a series of public hearings will be held. Such a format means that the commission could serve as a channel through which views from international civil society can be aired outside the UN system and later fed into the CSD, either directly or via the IPF. The WCFSD's final report will be presented to the CSD's fifth session in 1997, with any interim reports being copied to the CSD and the IPF.

PROSPECTS FOR THE FUTURE

For the first two years after the UNCED the diversity and range of international forest conservation initiatives confused both policy makers and policy analysts alike. After the 1995 session of the CSD the situation became much clearer. However, important unanswered questions remain in four important areas, namely the future of the C&I processes, timber certification, the future role of the FAO, and the contentious question of whether negotiations should begin for a GFC. These four areas are briefly considered in this section.

The future of the criteria and indicators processes

The first area relates to the C&I processes. The concept of sustainable forest management has attracted popular support among many forest stakeholders, including governments, forest industry, forest managers and some NGOs. All three C&I processes assumed, without the need for prior inquiry, that sustainable forest management is attainable. This assumption, however, is not shared by all actors. The question of whether or not sustainable forest management can be achieved is a vast one, and differences of opinion have arisen both between and within NGOs on this question. Insufficient space exists here to consider this question in detail. What is necessary is that the question be noted.

At the heart of this debate, and one of the most politically contentious issues in global forest politics, is the role of modern science in forest management. According to what may be termed a postmodernist interpretation, that is a view that rejects the Enlightenment notions of rationality and science,[28] environmental degradation is an intrinsic feature, rather than an accidental condition, of modernity.[29] This view has been extended to scientific forestry management which, according to its critics, displaces the knowledge systems of indigenous peoples and is thus 'a form of intellectual and ideological imperialism'.[30] According to this view, to cast forest management as a 'technical problem' privileges those who lay claim to technical expertise and marginalizes those who do not. 'Technical knowledge' is treated as of greater worth than the knowledge of indigenous peoples with the result that the latter are peripheralized in decision-making processes. A postmodernist view would argue that the forests can only be saved by returning power to local community level, and that despite the fact that the C&I processes aim to minimize conflict between forest stakeholders, the insistence that sustainable forest management is technically possible will serve in the long term to heighten conflict between industrial interests and local communities. This viewpoint will continue to be advanced by those NGO campaigners for whom it is salient. However, it is not a view that forestry advisers involved with the three C&I processes have paid heed to, nor is it likely to find any sympathy with government delegates and scientific advisers to the IPF. Unlike the postmodernists, mainstream international society considers scientific knowledge and indigenous peoples' knowledge to be complementary, rather than mutually exclusive, knowledge systems. Hence, and as noted above, the roles of 'traditional forest-related knowledge' and of science each fall within the IPF's remit.

There are currently three mainstream policy questions on C&I. First, there is the question of whether C&I should be developed only at the national level or whether separate C&I should be developed for the forest management unit and global levels. Second, there is the question as to whether, and if so how, countries not participating in the three processes, principally the tropical forest countries of Africa and Asia, should be brought in to the C&I debate.

Third, is the question of whether 'harmonization', or 'convergence', of the various initiatives should occur. In February 1995 a joint ITTO/FAO meeting in Rome endorsed the idea of global criteria with specific indicators developed for national or management unit levels.[31] At the twelfth session of FAO's Committee on Forestry (Rome, March 1995) opinion among delegates was evenly divided. Two members of the Montreal process adopted different positions; Canada favoured C&I convergence, while the USA was opposed.[32] One of the suggested policy options to emerge from the second meeting of the IWGF was that the CSD should 'encourage the harmonization' of the three processes and consider the preparation of criteria and indicators for adoption at the global level.[33] However, the view that harmonization is desirable or even possible is not shared by all actors.

Timber certification and the Forest Stewardship Council

As noted above, 'voluntary certification and labelling schemes' fall within the IPF's remit. Timber labelling has a troubled history in international forest politics. As we saw in Chapter 3, in 1988 the WWF criticized the ITTO for failing to promote forest conservation and stated that if this state of affairs were to continue the conservation organizations would seek other mechanisms. One year later a proposal by Friends of the Earth for the labelling of sustainably-produced timber was blocked in the ITTO by Malaysia. Since then the ITTO has taken no action on labelling, although the issue has arisen in debate.

To satisfy consumer demand for timber labelling, WWF, applying its earlier threat to seek alternative mechanisms, played the leading role in establishing the Forest Stewardship Council (FSC), a voluntary, private-sector labelling scheme for tropical and non-tropical timbers. The FSC brings together conservation NGOs and private companies from the timber industry. Its founding assembly, held in Toronto in October 1993, agreed upon an innovative institutional format, namely a general assembly comprising two chambers: the first, holding 75 per cent of voting rights, consists of representatives from social, environmental and indigenous peoples' groups, while the second, with 25 per cent, consists of representatives from the timber industry.

The goal of the FSC is 'to promote environmentally-responsible, socially-beneficial and economically-viable management of the world's forests by establishing a worldwide standard of recognized and respected Principles of Forest Management'.[34] The FSC aims to end the worldwide proliferation of timber labelling schemes by working towards a globally harmonized labelling system. It will authorize national certifying authorities to award the FSC label to timber from forest concessions that adhere to the FSC's principles for 'well-managed forests'. (The FSC deliberately avoids the term 'sustainably-managed forests' because of the ambiguity surrounding it.) The label will be awarded to owners or managers at the forest management unit level. There will be a legal chain of custody whereby timber certified as well-managed passes from owner to owner, from tree-felling, through intermediate stages such as saw-milling and transport, to the point of sale. The rationale behind the FSC is that while the scheme is voluntary, it will attract sufficient support and authority over time to become the sole recognized timber labelling scheme worldwide. For this to happen, two things are necessary in the medium to long term. First, a majority of timber traders must be prepared to cooperate with the scheme and, second, the scheme must attract sufficient support among a majority of the world's consumers who will be prepared to purchase only timber carrying the FSC label.

In June 1994 the FSC, meeting in Oaxaca, Mexico, agreed on nine principles and associated criteria for well-managed forests; these are shown

---

**Box 6.4   THE FOREST STEWARDSHIP COUNCIL'S NINE
PRINCIPLES FOR FOREST MANAGEMENT**

1. *Forest management shall respect all applicable laws of the country
   in which they occur, and international treaties and agreements to
   which the country is a signatory, and comply with all FSC principles and criteria.*
2. *Long-term tenure and use rights to the land and forest resources
   shall be clearly defined, documented and legally-established.*
3. *The legal and customary rights of indigenous peoples to own, use
   and manage their lands, territories and resources shall be recognized and respected.*
4. *Forest management operations shall maintain or enhance the long-
   term social and economic well-being of forest workers and local
   communities.*
5. *Forest management operations shall encourage the efficient use of
   the forest's multiple products and services to ensure economic
   viability and a wide range of environmental social benefits.*
6. *Forest management shall conserve biological diversity and its
   associated values, water resources, soils and unique and fragile
   ecosystems and landscapes, and, by so doing, maintain the ecological functions and the integrity of the forest.*
7. *A management plan — appropriate to the scale and intensity of the
   operations — shall be written, implemented, and kept up to date.
   The long-term objectives of management, and the means of achieving them, shall be clearly stated.*
8. *Monitoring shall be conducted — appropriate to the scale and
   intensity of forest management — to assess the condition of the
   forest, yield of forest products, chain of custody, management
   activities and their social and environmental impacts.*
9. *Primary forests, well-developed secondary forests and sites of major
   environmental, social or cultural significance shall be conserved.
   Such areas shall not be replaced by tree plantations or other land uses.*

Source: Forest Stewardship Council, 'Forest Stewardship Principles and Criteria for
Natural Forest Management, June 1994, Oaxaca, Mexico', pp2–6.

in Box 6.4 above. A principle is defined as 'an essential rule or element
… of forest management', while a criterion is 'a means of judging whether
or not a Principle (of Forest Management) has been fulfilled'.[35] The FSC's
principles and criteria should not be confused with the criteria and

indicators of the Helsinki, Montreal and Amazonian processes. First, and as noted above, the FSC refers to 'well-managed' rather than 'sustainably-managed' forests. Second, the objective of the C&I processes is to provide a system to gauge whether or not sustainable forest management has been achieved. In the case of the FSC, the principles and criteria are, first and foremost, prescriptions for action. By following the FSC principles, it is claimed, well-managed forests will ensue. However, the FSC system can also be used to evaluate whether an area of forest has been well-managed; providing the criteria demonstrate that the principles have been properly fulfilled, the area of forest in question is considered to be 'well-managed' and may thus qualify for the FSC label. It is worth noting that principles for sustainable forest management have not been stipulated by the Helsinki, Montreal and Amazonian processes. Indeed, only the Helsinki process has arrived at an agreed definition of 'sustainable forest management'.[36]

Two main policy questions are likely to arise with respect to timber labelling and the C&I schemes. (For the sake of the following discussion the objections of the postmodernist, who eschews the notion of management of ecosystems, are set aside.) First, although the FSC has stated that it 'intends to complement, not supplant, other initiatives that support responsible forest management worldwide',[37] if the FSC is successfully to promote a global labelling scheme, consideration must be given eventually to its relationship with those C&I processes that seek to evaluate sustainable forest management at the forest management unit level. (At present only the Amazonian process seeks to do this.) Second, those involved with the three C&I schemes may wish to consider how sustainable forest management can be successfully evaluated if the concept is not first defined, and if the principles by which it can be realized have not been clearly stipulated. The IPF, and possibly the WCFSD, may wish to consider the nature of the relationship between principles for, and measurements of, sustainably-managed/well-managed forests. In doing so, consideration may be given to whether existing and future C&I schemes may contribute to certification and labelling, and if so what their relationship with the FSC will be.

The role of the FAO

Prior to 1990 FAO was the undisputed leading international agency on forestry issues. However, in 1990 it attracted serious criticism over its handling of the Tropical Forestry Action Programme and has since struggled to regain its previously eminent role in international forestry (Chapter 2). In 1995 the FAO Council responded to this erosion of its authority by investigating how it can strengthen its normative role in forest management, by offering to provide a forum for dialogue on C&I and by calling for a greater share of the FAO's budget to be spent on forestry programmes.[38]

In an effort to recapture some support from the international community, FAO convened the 1995 Ministerial Meeting on Forestry, as well as separate meetings with the private sector and NGOs which took place prior to the twelfth session of FAO's Committee on Forestry (COFO) in March 1995. According to the *Earth Negotiations Bulletin,* FAO hoped that the COFO and the ministerial meeting would provide endorsements that: 'FAO is the appropriate forum for policy and strategy decisions on global forestry issues; that within the UN system it is the competent agency on all matters related to forestry; and that COFO is the intergovernmental body to deal with technical forestry issues at all levels.'[39]

However, neither the COFO nor the ministerial meeting provided such endorsements. The COFO requested 'FAO to concentrate on those areas in which it felt that FAO had a comparative advantage' such as 'the collection, analysis and dissemination of data and information, policy advice and ... technical assistance in its fields of concentration'.[40] The Rome Statement on Forestry confined itself to recommending that FAO 'muster its technical expertise to advise and cooperate with member countries in developing their capacity in the management, conservation and sustainable development of forests' while explicitly recognizing the CSD as 'the political body mandated to review and promote the implementation of the UNCED's decisions in the field of forests in their entirety'.[41] Attempts by some countries to insert stronger language on FAO's role were deleted from earlier drafts of the statement.[42] Political authority has ebbed from FAO to the CSD, and the IPF is the institution that will generate political recommendations, at least until the 1997 CSD session. While FAO is certain to be requested to make an input to the work of the IPF, it is unlikely that FAO will be able to recapture the leadership role on forests that it enjoyed prior to 1990.

Towards a global forests convention?

The question of whether a GFC should be negotiated has been on the agenda of several fora since Rio. In September 1992 a German–Japanese Expert Meeting on Tropical Forests held in Berlin agreed on the 'desirability' of a GFC.[43] The overall objective of pursuing a post-Rio GFC defined the position taken by the consumer caucus at the negotiations for the ITTA 1994 (Chapter 5). During these negotiations Switzerland and Canada[44] directly signalled their support for a GFC. Resolution 1 of the Helsinki conference noted that signatories would work towards 'the preparation of a global convention on the management, conservation and sustainable development of ... forests'.[45] Further Northern support for a GFC has come from the Global Legislators Organization for a Balanced Environment (GLOBE), an NGO consisting of legislators from the European Parliament, the US Congress, the Japanese Diet and the Russian Duma. In September

1993 the GLOBE International General Assembly called for its members to urge their respective governments 'to reinitiate the debate' on a GFC.[46]

While there remains significant support for a GFC in the North, many governments from the South continue to oppose the idea. Despite this a degree of North–South unity on this question is discernible in three respects. First, consensus was reached between the European Community and the African, Caribbean and Pacific (ACP) countries in February 1992 when the ACP–EC Joint Assembly, meeting in the Dominican Republic, passed a resolution stressing how a GFC might contribute to the protection and sustainable use of forests in the regions and worldwide.[47] As the Dutch NGO AIDEnvironment notes, this is 'the first occasion that a formal South–North forum positively propagates further discussion about a Forest Convention'.[48]

A second indication of an emerging North–South consensus on a GFC came with the conclusion in June 1995 of a protocol to the Lomé IV Convention on the sustainable management of forest resources. The protocol, negotiated under the auspices of Directorate General VIII, was one of the outcomes of the mid-term review of the Lomé IV Convention. Most of the protocol consists of declarations of intentions made already in the Statement of Forest Principles, such as the need for institution building, reforestation, technology transfer and action plans. However, three clauses in the protocol do not appear in the UNCED document: first, signatories agree to ensure, by the year 2000, the sustainable management of forests destined for the production of timber; second, signatories agree to support 'the definition and development of certification systems' for timber; and third, it is agreed to take into account 'internationally harmonized criteria and indicators on sustainable management of forests'.[49] The protocol is a brief document consisting of just five paragraphs, and its real significance lies in the fact that it is the first ever international legal agreement on sustainable forest management between governments from North and South. The International Tropical Timber Agreements of 1983 and 1994 are first and foremost commodity agreements rather than forest management agreements, and while both contain a conservation clause, this relates only to tropical forests from which timber is harvested for international trade. Note also that the CITES, while it has since 1992 listed timber species (Chapter 3), is a trade monitoring, rather than an ecosystem management, agreement.

The third indication that a consensus on a GFC is emerging comes from the semantics adopted in post-Rio ministerial and intergovernmental statements compared with the wording in the Statement of Forest Principles. While the word 'convention' remains anathema to the G77, some post-Rio ministerial and intergovernmental statements contain reference to a 'legally-binding instrument on forests'. This formulation could be taken to mean a GFC, thus serving an interpretation favoured by many Northern govern-

ments, although it may also be read as a protocol to another legal instrument, such as the conventions on climate change or biodiversity. It is
significant that four international meetings on forestry have adopted very
similar language with respect to a legal instrument, namely the FAO European Commission on Forestry (Antalya, January 1995), the twelfth session
of the FAO's Committee on Forestry (Rome, March 1995), the FAO
Ministerial Meeting on Forestry (Rome, March 1995) and the Commission
on Sustainable Development (New York, April 1995), namely that, 'concerning the discussion on the controversial idea of a legally binding
instrument on forests, the way forward should be based on consensus-
building in a step-by-step process'.[50] Apart from the fact that such
similarities demonstrate how language from one international meeting may
be fed into another, thus reinforcing a particular consensus, this is of
relevance in that the South has not blocked, as it did at Rio, any mention of
a possible legal instrument on forests. Overall, it is clear that some
governments from the South support, or at least are not prepared to veto the
negotiation of, a GFC. Indeed, at the twelfth session of FAO's Committee
on Forestry two developing countries, namely South Korea and Senegal,
stated their support for a legal instrument on forests.[51] There appears to be
some latitude in the South, and the hard line of the G77 at Rio is clearly no
longer shared by all its members.

Chapter 7

Conclusions

In the latter part of this chapter we will return to the forest conservation problematic outlined in Chapter 1 to assess whether the actors involved in the international processes considered in this book have addressed its three dimensions, namely the causal, institutional and proprietorial dimensions. First, however, we will examine the reasons for the persistent North–South disagreement at the intergovernmental level that has dominated the history of global forest conservation politics. Although a measure of trust was established between governments from North and South in the period prior to the 1995 session of the Commission on Sustainable Development (CSD), substantial differences remain. In considering the reasons for this we will draw upon the work of theorists of international relations, particularly those specializing in regime theory.

THE CONTINUING DEMAND FROM THE NORTH FOR A GLOBAL FORESTS CONVENTION

In international relations literature, regimes, according to the most oft-cited definition, are 'sets of implicit or explicit principles, norms, rules and decision-making procedures around which actors' expectations converge in a given area of international relations'.[1] In the pages that follow, it will be seen that an application of the analyses of regime theorists helps to explain why a global forests convention (GFC) has yet to be concluded. An international legal agreement should not be conflated with a regime, although the former may provide a framework within which the latter may emerge. Hence, Keohane distinguishes between *agreements*, defined 'in purely formal terms (explicit rules agreed by more than one state)' and *regimes*, which arise 'when states recognize these agreements as having continuing validity'.[2] To Rittberger 'a regime is said not to have come into existence if the pertinent norms and rules are disregarded by states at their discretion'.[3] A GFC would not therefore qualify as a global forests conservation regime although, following Keohane, if the provisions of a GFC were to be observed by its parties, then a GFC could provide the international legal framework for the development of such a regime.

The only politicians to demand a GFC prior to the UNCED, through proposals by the G7 and the European Parliament, were from the North (Chapter 4). Since Rio Northern governments have continued to express their interest in a GFC (Chapter 6). The question therefore arises of why Northern, but not Southern, governments have demanded a GFC. Traditional interest-based explanations for regime creation emphasize the realization of mutual gains by actors. An actor may demand a regime if it is anticipated that the gains, or payoffs, from regime participation will exceed the costs. Regimes should be mutually beneficial, although inevitably the arrangements will satisfy some actors more than others. Such a line of reasoning does not explain the Northern governments' demand for a GFC, for such a policy would not be pursued by governments motivated solely by the desire to realize short- to medium-term payoffs. Hence an alternative explanation must be sought.

The demand for a GFC should be seen as arising from a collective shift in state–society relations in the North. Ruggie sees the postwar economic order as a compromise between two demands. The first was the demand of the postwar hegemon, namely the USA, to establish a liberal, multilateral free-trading system. The second was the demand from Northern societies for domestic social and economic stability after the Second World War. This was expressed by a new collective balance in state–society relations in the industrialized world. Ruggie refers to the compromise between these two demands as embedded liberalism.[4]

A shift in state–society relations since the Second World War, and especially over the last 20 years, explains in large part the negotiation of international environmental agreements, as well as the demand from the North for a GFC. The role of social groups and movements in global environmental politics is central to this shift. As Litfin notes, 'Rarely has a nation taken the lead in preserving the global environment without substantial pressure from social movements and other non-state actors',[5] while Keohane, Haas and Levy argue that '*If there is one key variable accounting for policy change, it is the degree of domestic environmentalist pressure in major industrialized democracies*' (emphasis in original).[6] As I have argued elsewhere, the environmental concerns of Northern societies challenge Ruggie's notion of embedded liberalism as a set of economic relations that is compatible with the requirements of domestic social and economic stability. A new emphasis has emerged, one that stresses the need for economic relations that are compatible with global and domestic ecological stability.[7]

However, a similar shift has not occurred in the South, which remains pre-occupied with domestic economic instability, and in particular with perceived underdevelopment.

Having argued that the demand for international regimes has come principally from the North, it should be asked why the South has agreed to partake in many international environmental regimes, but has yet to agree to a GFC. The rational state will not join a regime if to do so is counter to its perceived interests or if it will compromise its power position relative to other actors. However, it will agree to join a regime where it may achieve gains. One of the arguments that will be developed in the following section is that the South considers that it potentially has much to lose by agreeing to a GFC on the terms outlined by the North during the UNCED process. Furthermore, the South believes it can extract greater concessions from the North than the latter has so far been prepared to offer.

EXPLAINING NORTH–SOUTH DISAGREEMENT IN GLOBAL FOREST POLITICS

The power of the veto coalition

Porter and Brown note that veto coalitions may arise during international negotiations. A veto coalition is a group of government delegations whose cooperation is necessary for agreement on a particular issue, but which has the power to block regime creation if it so chooses.[8] Institutional arrangements such as consensual decision-making procedures may empower groups of states to block change. The veto coalition concept involves interest-based factors; the rational state will refrain from joining a regime if the anticipated costs exceed the anticipated payoffs, and it may seek to enlist the support of other states for a veto coalition if it is perceived that regime creation will threaten their collective interests.

The veto coalition concept may also introduce a power-based factor if the possession of material capabilities empowers states to block the demands of other states. The South possesses important material capabilities in its tropical rainforests from which it produces goods demanded by other actors in international society; such goods are produced either directly (tropical timber) or indirectly (carbon sinks). The North also possesses relevant material capabilities; international forest conservation politics has a history of the South seeking to create bargaining issue-linkages between tropical forest conservation and other issues. Initially such linkages consisted of demands for increased aid flows, but since the start of the UNCED forest negotiations they have consisted of demands that were first expressed in the South's programme for a New International Economic Order (NIEO) in the 1970s. These linkages are pertinent to any proper enquiry on veto coalitions. For there to be agreement between North and South on forest conservation, it is necessary first that no veto coalition against a GFC is formed, and second that no veto coalition emerges against issues linked by important forest states to forest conservation. In the following paragraphs it will be argued that two mutually-reinforcing veto coalitions have arisen in global forest politics.

In the TFAP, the ITTO and the UNCED process decisions were made by consensus, and in all three cases such procedures empowered veto coalitions. In the case of the TFAP the approval of all actors was necessary for institutional reform, and consequently the Forestry Forum for Developing Countries (FFDC), fearful of a possible erosion of sovereignty, was able to prevent the establishment of an independent consultative group. In short, the South has partaken in, yet also blocked initiatives designed to strengthen, the TFAP (Chapter 2). Similarly, the ITTO's consensual decision-making procedures have prevented the introduction of policy reforms designed to achieve a sustainable tropical timber trade. While the issues of incentives and labelling have been on the ITTO's agenda since 1989, as a result of producer opposition no ITTC decision has been passed on these issues. Indeed, the ITTO has passed no decision that could be interpreted as a trade restriction. Like the TFAP, actors from the South have been involved in the ITTO while also acting as a veto coalition against reform. When it became clear that the ITTO would take no action on labelling, the WWF took steps to create the Forest Stewardship Council, a non-governmental labelling mechanism (Chapters 3 and 6). In the UNCED process a strong veto coalition in the G77, led by India and Malaysia, blocked the negotiation of a GFC. India and Malaysia exercised structural leadership against a regime arising from their status as important tropical forest countries and from the leading role they assumed in the forests negotiations for the G77.

Governments from the North have also acted as a veto coalition against the South's forest-linked bargaining demands. By funding a limited number of NFAPs and ITTO projects, Northern aid agencies have met in part the claims of the South for increased aid flows. However, they have not been prepared to meet in full the South's claims. For example, claims by the ITTO producers for increased funding to meet Target 2000 have not been met. In the UNCED forests debate bargaining was complex and issue-dense, with a wide range of NIEO-related demands linked by the South to forest conservation. However, the North was able to block these demands as it has control over international markets and a comparative advantage in international trade. While some commitment was made to increase aid flows at Rio, this fell far short of the US$ 125 billion UNCED Secretary-General Maurice Strong had advocated. As Stanley Johnson suggests, had the North been more willing to negotiate on financial resources and technology transfer prior to Rio, the G77 may have been unable to maintain the unity of its opposition to a GFC.[9] Hence two mutually-reinforcing veto coalitions emerged during the UNCED process, with the intransigence of the North informing that of the South and vice versa.

The quest for bargaining power

Young and Osherenko distinguish between two types of bargaining in international negotiations, namely integrative and distributive bargaining. Distributive bargaining occurs where actors seek to demarcate boundaries within which authority may be exercised. In so doing, an actor seeks to attain the best possible outcome for itself. However, actors engaged in integrative bargaining seek to develop 'new opportunities for mutually beneficial relationships'[10] so that 'a search for mutually beneficial solutions assumes a prominent place in the [bargaining] process'.[11] In other words, distributive bargaining involves an actor seeking to improve his or her position relative to other actors (relative gains), whereas integrative bargaining involves a search for absolute gains for all actors. With distributive bargaining, actors frequently disagree on which issues are at stake, whereas agreement on the issues at stake is a necessary feature of integrative bargaining. Young and Osherenko conclude that the chances of regime formation are enhanced if actors engage in predominantly integrative bargaining. It should be noted that bargaining is a 'mixed-motive activity' with actors seeking to attain relative gains even while searching for mutual gains;[12] bargaining is 'seldom wholly integrative or distributive, but rather constitutes a hybrid involving both types'.[13]

The evidence gathered in this study indicates that, while integrative bargaining has been discernible, overall the bargaining between states on the forests issue has been predominantly distributive in nature during the period covered in this book. For example, while there are indications that at times bargaining in the ITTO has been driven by a search for mutually beneficial solutions, with the adoption of Target 2000 as the most obvious example, it is also the case that agreement among actors on the issues at stake, an important precondition of integrative bargaining, is not in evidence. The producers have been reluctant to debate meaningfully the labelling and incentives issues that were placed on the ITTO agenda by the consumers. Further lack of agreement concerns the producers' claims for extra resources to meet the full incremental costs for Target 2000, an issue which the consumers have yet to recognize (see below). There has been no sustained search for effective and durable forest conservation policies by all actors within the ITTO, with distributive bargaining more in evidence than integrative bargaining.

During the UNCED forests negotiations there was no agreement on the issues at stake and consequently the focus of the two sides differed. The North and South each pursued positions of perceived self-interest, with each side seeking to shift the responsibilities of the other. The North attempted to shift the jurisdictional boundaries of forests conservation by invoking 'stewardship', a concept which governments from the South, fearing it could restrict the use of forests within their geographical domain, refused to recognize. Instead, the South sought to redefine the responsibilities of the North with respect to debt relief, reversal of net South-to-North financial flows and other NIEO-related claims. By linking forests to these issues the South, correctly perceiving that forest conservation was a highly salient issue to the North, raised the stakes.

The South negotiated from a position of increased strength in the UNCED forests negotiations in comparison with some of the previous NIEO negotiations when, as Renninger notes, it had to cope with a 'lack of significant bargaining power' because, in exchange for what was demanded, the South had relatively little to offer.[14] However, the rise of forest conservation as an international issue has enhanced the value of the South's tropical forests which, in turn, has increased its bargaining leverage. Realizing this, the South has attempted to translate the North's concerns about tropical forests into the hard currency of economic and political gain.

On a theoretical level this indicates that in international negotiations the mere possession of economic capabilities by an actor is no guarantee of bargaining strength. A cognitive factor enters the equation, for an actor must first appreciate the full value other actors assign to those capabilities if it is to maximize its bargaining position. Indeed, it may be the case that the South has consciously taken the decision to delay signing a GFC in order to enhance further its position. As Young and Osherenko note, 'Parties anticipating gains in relative bargaining strength over time are apt to be reluctant to reach a negotiated settlement today, while parties expecting losses in relative bargaining strength will be anxious to conclude negotiations quickly.'[15] No evidence exists to suggest that this is the G77's strategy on forests, although it is certainly the case that, having appreciated the value the North attaches to tropical forest conservation, the G77 realizes that it has gained an important source of bargaining leverage.

The different priorities of North and South

Some theorists of international studies have noted that the wider international context in which regimes are planned or created is a key factor in determining whether states decide to join a regime. To Hurrell, states 'are often deterred from entering into cooperative arrangements if these entail negative implications for their relative power position'.[16] Keohane considers the wider context when arguing that if regimes existed in isolation they 'would be abandoned when governments calculated that the opportunity costs of belonging to a regime were higher than those of some feasible alternative course of action'; for the rational government to defect from a regime 'the net benefits of doing so must outweigh the net costs of the effects of this action on other international regimes'.[17] Zacher notes that if international regulations have a negative effect on 'the competitive positions of major states and coalitions, then strong regimes are not likely to develop'.[18] Contextual factors inevitably enter into the considerations and calculations the rational state makes with respect to regime creation. In short, when a government considers whether to join a regime, it does not solely evaluate the costs and gains that may arise from regime membership, but also takes into account those that may arise outside the regime.

How does such a view explain the behaviour of states from the North and South during the UNCED forests debate? Here it will be argued that states from North and South calculated that if they were to agree to a GFC on the terms of the other, the costs would have exceeded the benefits. The North resisted meeting the South's demands in order to protect its under-lying economic power position: technology transfer on preferential and concessional terms would have resulted in lower returns on research and development costs; substantial financial transfers could have increased the taxation burden in Northern countries; debt relief would have harmed Northern-based transnational banks; and reversal of the South's declining terms of trade would have compromised the North's competitive position in international trade. The overall result for the North would have been a decline over the long term of comparative advantages in international trade and finance. Similarly, governments from the South feared that to accept a GFC on the North's terms would be to lose full and sovereign control of a capability providing them with a source of leverage over the North, namely their tropical rainforests. The North feared that meeting the South's demands would erode its relative power position, with any benefits gained being exceeded by the substantial economic costs that would accumulate. Forest conservation was a salient issue to the North, owing to the shift in state–society relations described earlier, but of greater importance was the maintenance of a competitive role for its business and industry. Meanwhile, realizing the demands of the NIEO was more salient to the South's perceived interests than forest conservation. In short, North and South anticipated that to meet the other's terms would compromise their long-term interests.

The problem of veto coalitions is a real one for those advocating a GFC. During the UNCED process, a clumsy and unsuccessful attempt was made by some EC governments to break the G77 veto coalition against a GFC. When African governments from the G77 expressed a desire for a desertifi-ation convention, some EC governments then sought to link this issue with the negotiation of a post-UNCED GFC. The intention of this tactic was to

split the G77's unified stance against a GFC by luring the African govern-
ents into the pro-GFC camp with promises of EC support for a desertifi-
ation convention. However, the tactic was not endorsed by all EC members,
and when the USA voiced support for a desertification convention the EC
then dropped its insistence that the desertification and forests issues be
linked.[19]

Noting that the CITES and the London Convention on the Dumping of
Waste at Sea were negotiated outside the UN, Stanley Johnson has proposed
that a similar course of action be pursued by the North for a post-UNCED
GFC. To Johnson, this would avoid 'the tyranny of consensus rule as we
saw it practised in Rio'.[20] To date, no proposal for GFC negotiations outside the UN has been made. If such negotiations were conducted on
majority voting this could, in theory, circumnavigate a small veto coalition
from the South. However, the cohesive front presented by the ITTO producers in the negotiation of the International Tropical Timber Agreement,
1994 suggests that the South's unity on the forests issue will be difficult to
erode.

Until confidence-building measures were initiated in 1994 a high degree
of mutual suspicion existed between North and South. Consider the
suggestion from the Malaysian delegate, Ting Wen Lian, during the UNCED
process that the FAO was being utilized by some governments in the North
to promote a GFC, and of the US negotiator, Curtis Bohlen, that countries
from the South were trying to receive guarantees of financial assistance
before agreeing to make conservation commitments (Chapter 4). It was also
clear from sentiments expressed at governmental level that leading politicians were pursuing positions of self-interest: the Malaysian prime minister
urged the South to speak together with one voice, while President Bush
commented prior to Rio that he would not allow the USA's way of life to be
threatened. Although the UNCED was principally concerned with the common survival of humanity, there is no indication that key actors were truly
motivated by the notion of intergenerational equity, and neither North nor
South were prepared to make sacrifices in order to realize long-term gains
for all humanity. Furthermore, North and South were informed by different
vulnerabilities; the North felt vulnerable to global environmental degradation, while the South felt more vulnerable to underdevelopment, to
poverty and to local environmental problems.

The role of metaphors in the building of trust

The confidence-building measures initiated in 1994 helped to establish
trust between governments from the North and the South, a factor that subsequently led to the establishment of the Intergovernmental Panel on Forests
(Chapter 6). North and South have set aside their suspicions of each other

and engaged in a search for areas of mutual self interest. There is a recip-rocal willingness to discuss each other's concerns, informed by a recognition that neither North nor South can deal with their concerns about forest con-servation and related issues by acting alone. The emphasis has shifted, for the time being at least, to a search for absolute gains for all. The Inter-governmental Working Group on Forests played an important role in reach-ing consensus on the issues that the Intergovernmental Panel on Forests will consider; as noted earlier, Young and Osherenko believe that agreement on the issue at stake is a prerequisite of integrative bargaining. If a durable trust can be established, and if future bargaining on forests is more integrative than distributive in nature, then a global agreement on forest conservation may ensue. However, if there is to be a GFC it is clear that the developing countries will expect economic concessions from the developed countries. At the insistence of the South, clauses on the need for new and additional financial resources and for technology transfer on favourable or con-cessional terms were written into the Statement of Forest Principles and the ITTA 1994.[21] In September 1993 the Delhi Declaration of the Forestry Forum for Developing Countries also stated that these two issues are important if tropical countries are to achieve sustainable forest manage-ment.[22] And, despite the recent emphasis on confidence-building, the two issues were considered by the Intergovernmental Working Group on Forests at Malaysia's insistence and are also agenda items for the Intergovernmental Panel on Forests.

Some regime theorists have noted the importance of metaphors in regime formation. To Jönsson, 'one should not underestimate the role of a formula consisting of a systematic metaphor which makes the actors view their com-mon problem in a new light'.[23] Young and Osherenko note that successful regime creation may depend on 'the ability of those formulating proposals to draft simple formulas that are intuitively appealing'.[24] So far, the states from the North and the South have yet to agree upon a metaphor or formula acceptable to all. Some agreement on core concepts would appear to be necessary, especially with regard to aid assistance. The South's demands for finance and technology cannot be considered in isolation from the concept of 'burden-sharing'. The principle first appeared in global forest politics as part of the sovereignty/stewardship/burden-sharing formula out-lined in the FAO draft GFC of 1990. Following FAO's failure to secure a mandate to host GFC negotiations from the international community, the formula reappeared in the UNCED forests debate as sovereignty/stewardship/common-responsibility. In response to the North's claim that all countries have a 'common responsibility' to cure deforestation, the South responded that as the North bears the greatest responsibility for defores-tation, it was necessary to talk of 'common but differentiated respon-sibilities' (Chapter 4).

Aid transfers and burden-sharing

The principle of 'burden-sharing' appeared in the UNCTAD negotiations for the ITTA 1994, but on this occasion it was raised by the South, and without linkage to stewardship.[25] The principle was not written into the ITTA 1994. At the fourteenth session of the International Tropical Timber Council the concept of 'common but differentiated responsibilities' arose in a slightly modified guise when the president of the UNCTAD conference, Wisber Loeis of Indonesia, argued that the new ITTA should provide for new and additional resources 'characterized by shared and differentiated responsibilities'.[26] However, the consumers failed to agree to the introduction of the concept into the ITTA 1994, preferring instead that this and other contentious questions should be deferred until the South agrees to negotiations for a GFC.

The relationship between aid assistance from the North and forest conservation in the South will remain central to global forest politics. There is evidence that the ITTO producers have felt they are expected to carry alone the burden of forest conservation. For example, the Indonesian Ministry of Forestry has estimated that Indonesia requires a total of US$ 20.2 billion to achieve Target 2000, and a further US$ 637 million to implement the Indonesian NFAP until the year 1999.[27] An ITTO expert panel estimated that Brazil will need to spend US$ 6.5 billion in order to attain sustainable forest management by the year 2000.[28] The panel based its findings on a consultancy report which noted that many tropical forest countries fail to capture sufficient revenues from log sales and that if fees were increased 'a considerable part of the additional resources needed for these programs could be borne by the producer countries', although the difference (a monetary or percentage figure is not given) would have to be borne by the consumers.[29] Substantial extra aid allocations to assist tropical forest governments achieve sustainable forest management have so far not been forthcoming from the North, either inside or outside the ITTO.

If agreement is to be reached on the core concepts guiding aid assistance from North to South, two conceptual questions require addressing. First, a precise formulation of a principle acceptable to all actors must be agreed upon. It is significant that the principles of 'burden-sharing' and 'common but differentiated responsibilities' each have a precedent in international environmental law. 'Burden-sharing' appears in the Convention on Biological Diversity (Articles 20.2 and 21.1). The principle of 'common but differentiated responsibilities' appears in the Rio Declaration on Environment and Development (Principle 7) and in the Convention on Climate Change (Preamble, Articles 3.1 and 4.1) where it asserts that the polluting North has a special obligation to assist developing countries deal with global warming.[30] It is significant that at the third session of the Commission on Sustainable Development in 1995 the developing countries reasserted the principle of 'common but differentiated responsibilities'.[31]

Second, there is the question of whether 'burden-sharing' (or some formulation based on this concept) will be linked to other principles. While the South has so far resisted the notion of 'stewardship', the principle has, in the case of the Svalbard Archipelago, provided a point of convergence for actors in an environmental regime. A group of islands in the polar North, the Svalbard Archipelago is legally the territory of Norway, which has agreed to accept the responsibility of acting as a 'steward' for the archipelago while also allowing other states access to the islands' natural resources. Young and Osherenko note that the success of this regime 'stands as a monument to the proposition that state sovereignty need not constitute a barrier to effective international cooperation when individual states ... are willing to accept explicit restrictions on the exercise of their sovereign authority'.[32] There are of course substantial differences between the Svalbard Archipelago regime and a possible global forests conservation regime; in the former only one country has renounced sovereignty, and then only for a small part of its territory, while in the latter the principle would require widespread acceptance among major forest states. Nonetheless, it is significant that the principle is established as a precedent.

While Northern governments that support the principle of stewardship presumably are prepared to renounce some degree of sovereignty over their forests as one of the terms of a GFC, tropical forest governments have so far shown no such inclination. Fully cognizant of the value that Northern governments attach to tropical forests, the South is determined to use its forests as a bargaining chip to extract economic concessions from the North. It is clear that if the developing countries are to agree to a GFC, the developed countries will be expected to pay a price. Here it is worth recounting an anecdote that Bernardo Zentilli, the Senior Programme Officer (Forests) of the UNCED secretariat, shared with me shortly after the UNCED. Zentilli referred to the North's interest in a GFC during the UNCED process, and the South's subsequent attempts to link this demand to economic concessions from the North, as the 'oriental bazaar syndrome'. If a customer seeking to buy a rug in an oriental bazaar intentionally or unintentionally indicates which rug he most wants, the owner raises its price. Similarly, when the North expressed a strong interest in a GFC, the South put up its price.[33] At the UNCED the North was unwilling to pay, and since Rio has shown no indication of being prepared to raise its offer; nor has the South been willing to lower its price. Despite the more cooperative atmosphere currently prevailing in the international relations of forest conservation, the financial assistance and technology transfer issues will prove crucial; the South will not accept a GFC unless agreement is reached on these two bargaining issue-linkages. Agreement on debt relief may also be necessary.

However, the North cannot conceivably meet in full the South's bargaining claims. The costs that would be borne by tax payers and Northern-based TNCs to meet the South's claims on, say, financial and technology transfers and debt relief would be prohibitively high, and no democratically-elected government could meet these claims in full and confidently expect to be returned in future elections. Nonetheless, it is clear that the governments from the South expect their Northern counterparts to make further economic concessions. For an acceptable *quid pro quo* to be agreed upon it appears necessary that, first, the North be prepared to move further towards meeting in part the bargaining claims of the South and that, second, the South be prepared to lower its demands; in short, there is an agreement to meet half way. However, if the North is to make substantial concessions, it will expect the South to make, and to adhere to, binding commitments on forest conservation. This inevitably touches upon the South's fears of erosions of sovereignty. These are highly contentious issues, and if they are to be fully and frankly broached the newly-found cooperative spirit between North and South may well come under severe strain.

THE FOREST CONSERVATION PROBLEMATIC

The author's formulation of the forest conservation problematic was introduced in Chapter 1, where it was argued that a solution of this problematic would increase the chances of effective global forests conservation. In this section we consider how, if at all, the international political processes examined in this work have sought to deal with the three dimensions of the problematic. We first consider the causal dimension.

The causal dimension

The case studies yield no evidence that the transnational causes of deforestation are being arrested, and give no clear conception of what policies are necessary to deal with this problem. The TFAP and ITTO are constrained by the intergovernmental system within which they operate. National action cannot address global causes; it can at best tackle only those that arise at country level. However, an NFAP or ambitious ITTO project cannot in practice address national causes, for attempts to do so are likely to draw accusations of interference in national sovereign affairs or of 'eco-colonialism' from tropical forest governments. Hence, despite the emphasis on the need to address the causes of tropical deforestation during the TFAP restructuring process, no serious consideration was given to dealing with causes that have their loci in the global political economy. The ITTO has rarely debated causes, focusing almost entirely on trade-related issues and projects. The question of causes did not arise in the negotiations for the ITTA 1994.

The UNCED process presented a real opportunity to address the structural causes of deforestation meaningfully. This opportunity was not taken, and the UNCED forests debate became the arena for a complicated bargaining issue-linkages battle between North and South. The UNCED was a standard United Nations conference, involving the negotiation of agreements between governments, and no attempt was made to obtain binding agreements from non-state actors contributing to environmental destruction. Despite NGO protests, one area of international activity not covered by the UNCED was the role of TNCs. One NGO campaigner notes that negotiations resulted in 'repeated deletion of Agenda 21 references to TNCs during PrepCom 4' resulting in 'a text which gives [TNCs] free reign to pursue their activities without any accountability'.[34] Various business and industrial interests were represented at the UNCED with NGO status; this irritated many traditional NGOs, which were neither pacified nor impressed by the creation of the Business Council for Sustainable Development (BCSD) by some leading industrialists. The BCSD's view that environmental conservation should take place within the free market was heavily criticized by NGOs. WWF's view is typical of many in the NGO community: 'Opportunity for substantial discussion within UNCED on the role of business and industry was in effect sidetracked by the creation of the Business Council for Sustainable Development which . . . made no suggestions for concrete action plans.'[35] After the UNCED the WWF called for governments and international institutions to adopt by 1995 a regulatory framework establishing global environmental standards for TNCs, including environmental cost accounting. However, no such process was initiated. While some TNCs have been prepared to adopt unilateral guidelines, such as in the oil industry (Chapter 1), these are no guarantee that TNCs will cease engaging in activities that degrade the environment. Such guidelines have at best served only to integrate environmental concerns into TNC activities, rather than to constrain and restrict such activities by a genuine conservation imperative.

A prerequisite to effective global forests conservation is the arrest of the structural causes of deforestation, yet the activities of those transnational actors who contribute to deforestation remain beyond independent scrutiny. There remains a need for the development of mechanisms to monitor and regulate the activities of TNCs and of other transnational actors whose activities impact upon forests, such as regional development banks, the UN and its specialized agencies, and even NGOs.

There are two reasons why a new regulatory framework is necessary. The first is the failure of traditional institutions such as governments and UN organs to deal successfully with the complex global dynamics that give rise to environmental problems such as deforestation. International economic activity crosscuts the state system, while traditional international diplomacy ignores those global economic dynamics that gave rise to environmental degradation. The failure to contain deforestation is symptomatic of this failure. A problem that is generated at the global level has to be tackled by collective global action. (At the same time it should not be forgotten that many causes of environmental degradation are to be found at the national and local levels.)

Second, to tackle TNCs' activities would be to challenge a great many powerful vested political, economic and financial interests, principally in the North. Clearly, if some TNCs were to ensure that conservation came before profit, while others ignored conservationist norms, the latter would gain a comparative advantage over the former. A new regulatory framework should ensure that all such interests abide by the same standards, and that they are accountable to the wider international community.

The institutional dimension

The second dimension of the forest conservation problematic is the institutional dimension. The development of new regulatory mechanisms for TNCs and other transnational actors will clearly require new institutions. As argued in Chapter 1, the development of new institutions is also necessary to give all forest stakeholders, especially those at the local level, a voice in policy-making. Is there any evidence of a new type of polity, in which forest stakeholders from all levels of international society are represented?

Following the 1990 TFAP legitimacy crisis there was an initiative to create an innovative type of institution, namely the consultative group, in which a diversity of state and non-state actors would have been represented. Had such a group been created in line with the 1991 Geneva recommendations, with a strong emphasis on NGO and local community representation, it could have broken the mould of international forest politics. As outlined in Chapter 2, this attempt failed due to a significant anti-NGO lobby among tropical forest governments.

The ITTO has probably the most open arrangement offered by any intergovernmental organization for NGO access. Nonetheless, and as in all other intergovernmental fora, only government delegations are granted voting rights. The ITTO Mission to Sarawak did provide an opportunity for a new type of polity to be created. The mission, led by the Earl of Cranbrook, investigated the sustainability of forest management practices in Sarawak between November 1989 and March 1990.[36] Its membership was composed of ten forestry experts from nine countries. Although the mission visited some local communities, the views of local people did not receive extensive consideration. The mission investigated only the technical aspects of forest sustainability and thus ignored broader social concerns. No visits were made to the settlements most affected by logging, although the mission did receive written invitations to do so.[37] Cranbrook, when asked by the author to comment why the mission did not pay greater attention to the complaints of local communities, replied that the mission adhered to its terms of reference which were worded so as to be acceptable to the Malaysian government 'without whose approval the mission would not have proceeded'.[38] Once again the clash between genuine NGO participation and the recourse of governments to national sovereignty becomes apparent.

With respect to the UNCED process the attendance at the PrepComs and at Rio of what at the time was an unprecedented number of NGOs (including business and industry), was a recognition that the environmental crisis cannot be solved by governments acting alone. However, at Rio the venues for the NGO Global Forum and the main conference were several miles apart, and in effect there were two separate debates. Despite the high rate of NGO attendance, NGOs' views did not impact on the later stages of the UNCED forests debate in any significant way. At present there is no agreement on a formal mechanism by which NGO views can be fed into intergovernmental negotiations other than the traditional methods of lobbying and pressure group activity.

Despite the fact that governmental actors prevail in international decision-making, a consensus is emerging that the voices of local communities and indigenous peoples should be heard. The 1991 Geneva recommendations on the TFAP emphasized that the revamped TFAP should ensure the participation of rural communities.[39] Although the needs of local communities did not receive mention in the ITTA 1983, indigenous peoples and local NGOs have attended ITTC sessions and have on occasion participated in ITTO projects. Local participation receives mention in the UNCED Statement of Forest Principles and in the ITTA 1994. It is also noteworthy that the newly-created Intergovernmental Panel on Forests aims to consider the ways in which traditional forest-related knowledge can be protected and used, while the World Commission on Forests and Sustainable Development aims to ensure that the views of all main forest stakeholders are fairly represented and taken into account (Chapter 6). However, at present such views are heard in existing institutions dominated by governments, rather than receiving equal treatment in new institutions. Here it is worth returning to the distinction made in Chapter 1 between participation and consultation; while there is a growing consensus that NGOs and local and indigenous peoples should be consulted, genuine participation remains a concession from the state, rather than a right.

At the time of writing no new polity or political process has been developed that successfully integrates the views of actors from all levels of international society. There remains a need, at least with respect to forests, for the development of institutions that, first, integrate the views and concerns of all actors with a stake in forest use, especially those at the local level, and that provide them with a fair and equitable voice in policy-making and, second, that monitor and hold accountable those actors from outside the forests whose activities may result in forest degradation or forest loss.

The proprietorial dimension

There has been no attempt in the TFAP or the ITTO to deal with the proprietorial dimension, namely to reconcile the three conflicting claims to forests of global common, national resource and local common. Tropical forest government delegates continually asserted in the TFAP Ad Hoc Group that forests are a national resource. The International Tropical Timber Agreements of 1983 and 1994 each recognize the sovereignty of producer members over their forest resources. Language from Principle 21 of the Stockholm Declaration on the Human Environment is reproduced in the Statement of Forest Principles and the Rio Declaration on Environment and Development. However, and as mentioned above, consultation of local communities is slowly becoming an accepted practice in forest politics.

With respect to the UNCED, had the North and South adopted the sovereignty/stewardship/common-responsibility formula this could have helped bridge two of the competing claims to the world's forests, namely those of global common and of national resource. However, the third claim, namely that forests are a local common, would not have been dealt with.

As noted in Chapter 4, in the period before the UNCED, statements referring to forests as a common heritage of mankind were made in Abidjan (in the African Common Position on Environment and Development) and in Caracas (by Venezuelan President Carlos Andres Perez at the second summit-level meeting of the G15). These statements suggest that sovereignty was not as overwhelmingly important an issue for all Southern governments as debates at the UNCED PrepComs might suggest. They represent the first, and so far as is known the only, references from politicians in the South that forests may be seen as a global common. This suggests that some Southern governments could be prepared to negotiate over sovereignty, and indicates that there is a certain latitude in the South's position. It implies that a solution may be found to the proprietorial dimension if the concerns of the South, such as the structural causes of deforestation and North–South inequities, are addressed by the North.

CONCLUSIONS

In Chapter 1 it was argued that the three dimensions of the forest conservation problematic are interlinked and must be dealt with together. The evidence gathered here reinforces this view. It is clear that the causal dimension and the proprietorial dimension are interlinked. The North cannot realistically expect the countries of the South to yield a fraction of their sovereignty over forests if their concerns, including those on the global causes of deforestation, are not addressed. Similarly, the North is unlikely to make such concessions unless it first receives genuine guarantees from the South of a positive commitment to conservationist norms. The South cannot expect the North to make such concessions if it continues to assert that forests are a national resource to be exploited in line with national policy.

In turn, the causal dimension is linked to the institutional dimension in two ways. First, the forests can only be saved with the active help of local communities. Frequently, they are destroyed when local communities lose control over their forests to outside interests. It is necessary that the voices of local communities who live in and rely on the forests are heard on a just and equal basis with those of other actors. Second, and as previously argued, a new form of regulatory mechanism is necessary to arrest the transnational causes of deforestation. As G77 delegations maintained during the UNCED forest negotiations, external indebtedness, unsustainable patterns of production and consumption and net South-to-North financial flows can be seen as causes of deforestation. In fact there is an incongruence, if not hypocrisy, in the position of those Southern governments that contend on the one hand that deforestation is caused by global dynamics, and assert on the other hand that the state has sovereignty over forest use. As David Potter argues, sovereignty is a legal fiction which has always been compromised by transnational economic and social forces, with assertions of sovereignty serving to insulate the state from the international environmental effects of its policies.[40]

Here the institutional and proprietorial dimensions become linked. The dominant principle that forests are a natural resource of the state serves only to empower the governments of the day and those economic interests with which they enter into a collaborative relationship. The recourse of developing countries to sovereignty has impeded effective international environmental cooperation, thwarted the attempt to develop an independent TFAP consultative group and constrained the independence of the ITTO Mission to Sarawak. As long as governments continue to assert state sovereignty, there is no realistic chance of developing new institutions in which non-state actors may genuinely participate. If local communities are to have an equitable say in such institutions, it is first necessary to recognize their oft-ignored claims that forests serve as local commons to local peoples. Deforestation is often a symptom of the displacement of traditional forms of land control and the concomitant disempowerment of local communities at the hands of powerful outside interests. As such, a recognition of the concerns and rights of local peoples is not only an environmental imperative, it is also one of social justice.

Notes

NOTES TO CHAPTER 1

1. UN document A/CONF.151/PC/27, 'Conservation and Development of Forests: Progress Report by the Secretary-General of the Conference', 5 February 1991, para 6, p8.
2. Norman Myers, *Deforestation Rates in Tropical Forests and their Climatic Implications* (London: Friends of the Earth, December 1989), p6.
3. Anthony L Hall, *Developing Amazonia: Deforestation and social conflict in Brazil's Carajás programme* (Manchester: Manchester University Press, 1989).
4. David Price, *Before the Bulldozer: The Nambiquara Indians and The World Bank* (Washington DC: Seven Locks Press, 1989); and Alexander Shankland, 'Brazil's BR-364 Highway: A Road to Nowhere?', *The Ecologist*, vol 23, no 4, July/August 1993, pp141–7.
5. Barbara J Cummings, *Dam the Rivers, Damn the People: Development and Resistance in Amazonian Brazil* (London: Earthscan, 1990); and Bernard Eccleston, 'Lessons from Environmental NGO Campaigns around the Bakun Dam development project in Sarawak', paper presented to the EUROSEAS Conference, Leiden, June 1995.
6. Kenneth Piddington, 'The Role of the World Bank', in Andrew Hurrell and Benedict Kingsbury (eds) *The International Politics of the Environment: Actors, Interests, and Institutions* (Oxford: Clarendon Press, 1992), esp pp216–18.
7. World Bank, *The Forest Sector: A World Bank Policy Paper* (Washington DC: World Bank, 1991), p65. See also, Chapter 5, 'Forest Activities' of World Bank, *The World Bank and the Environment: A Progress Report, Fiscal 1991* (Washington DC: World Bank, 1991), pp80–92.
8. Friends of the Earth, *Special Briefing, The World Bank's 'Forestry and Environment' Project for Gabon: A Test Case for the World Bank's New Forest Policy* (London: Friends of the Earth, September 1992).
9. See, for example, Bruce Rich, *Mortgaging the Earth: The World Bank, Environmental Impoverishment and the Crisis of Development* (London: Earthscan, 1994).
10. United Nations Centre on Transnational Corporations, *Transnational Banks and the International Debt Crisis (Report ST/CTC/96)* (New York: United Nations, 1991).
11. World Commission on Environment and Development, *Our Common Future* (Oxford: Oxford University Press, 1987), p68.
12. Simon Rietbergen, 'Africa', in Duncan Poore et al, *No Timber Without Trees: Sustainability in the Tropical Forest* (London: Earthscan, 1989), p64.
13. Considerable confusion may arise over net South to North financial flows. Countries from, or IGOs representing, the South, frequently refer to net South to North financial transfers, but usually desist from citing a figure. For example, the Tenth Meeting of the Non-Aligned Movement referred only to 'crippling net negative transfers': see 'Tenth Conference of Heads of State or Government of Non-Aligned Countries, Jakarta, 1–6 September 1992: Final Documents', p76. The phenomenon of net South to North transfers was introduced by the South in the forest negotiations of the UNCED process (see Chapter 4), although the national delegations which referred to the concept, usually those from the Group of 77, did not cite a figure. The confusion arises from definitional problems, such as what

constitutes 'North' and 'South' and, more importantly, what types of resource flows should be included in the net figure. Calculations taking into account only debt repayments from the South less aid donated to the South show persistent net South to North flows. However, calculations taking into account Northern public and private investments in Southern countries may show net North to South flows. For example, the World Bank reports that '[i]n 1992, aggregate net resource flows — that is, net flows on long-term debt, grants excluding technical assistance, and net flows on equity investments (foreign direct investments and portfolio equity investment) — to all developing countries ... [reached] an all-time high of US$ 158 billion... Aggregate net resource transfers (that is, net resource flows, less interest on debt and profits on equity investment) rose sharply too, reaching US$ 80 billion in 1992'. See, World Bank, *World Debt Tables 1993–94: External Finance for Developing Countries, Volume I, Analysis and Summary Tables* (Washington DC: World Bank, 1994), p9, and Tables 1.1 and 1.2, pp10–11.

14. Andrew Hurrell, 'Brazil and the International Politics of Amazonian Deforestation', in Hurrell and Kingsbury (eds) op cit, p425.
15. Susanna Hecht and Alexander Cockburn, *The Fate of the Forest: Developers, Destroyers and Defenders of the Amazon* (London: Penguin, 1989), p109.
16. Antonio Carlos Diegues, *The Social Dynamics of Deforestation in the Brazilian Amazon: An Overview*, United Nations Research Institute for Social Development, Discussion Paper 36 (Geneva: UNRISD, July 1992), pp8–9.
17. Susan George, *A Fate Worse than Debt* (London: Penguin, 1988), p156.
18. Susan George, *The Debt Boomerang: How Third World Debt Harms Us All* (London: Pluto Press, 1992), pxix.
19. Patricia Adams, *Odious Debts: Loose Lending, Corruption and the Third World's Environmental Legacy* (London: Earthscan, 1991), especially Chapter 5, 'The Debt Crisis' Silver Lining', pp49–57.
20. As noted in World Wide Fund for Nature, *ITTO: Tropical Forest Conservation and the International Tropical Timber Organization, WWF Position Paper 1* (Gland, Switzerland: WWF–International, June 1988), p6.
21. For example, Barbara J Bramble, 'The Debt Crisis: The Opportunities', *The Ecologist*, vol 17, no 4/5, July/November 1987, pp192–9.
22. Rhona Mahony, 'Debt-for-Nature Swaps: Who Really Benefits?', *The Ecologist*, vol 22, no 3, May/June 1992, pp97–103.
23. World Wide Fund for Nature, *Tropical Forest Conservation, WWF Position Paper 7* (Gland, Switzerland: WWF–International, September 1991), p11.
24. Friends of the Earth, *Tropical Rainforests and Third World Debt, Briefing Sheet* (London: Friends of the Earth, September 1991).
25. Fred Pearce, 'France swaps debt for rights to tropical timber', *New Scientist*, 29 January 1994, p7.
26. Dominic Hogg, *The SAP in the Forest: The Environmental and Social Impacts of Structural Adjustment Programmes in the Philippines, Ghana and Guyana* (London: Friends of the Earth, September 1993).
27. George (1992), op cit, p3.
28. Gérald Berthoud, 'Market', in Wolfgang Sachs (ed) *The Development Dictionary: A Guide to Knowledge as Power* (London: Zed Books, 1992), p73.
29. Hecht and Cockburn, op cit, p151 and pp202–3.
30. See, for example: James D Nation and Daniel I Kormer, 'Rainforests and the Hamburger Society', *The Ecologist*, vol 17, no 4/5, July/November 1987, pp161–7; Peter Utting, *The Social Origins and Impact of Deforestation in Central America*, United Nations Research Institute for Social Development, Discussion Paper 24 (Geneva: UNRISD, May 1991), p6.
31. World Rainforest Movement, *Rainforest Destruction: Causes, Effects and False Solutions* (Penang: World Rainforest Movement, 1990), p46.
32. Susanna Hecht, 'Landlessness, Land Speculation and Pasture-led Deforestation in Brazil', in Marcus Colchester and Larry Lohmann (eds) *The Struggle for Land and the Fate of the Forests* (London: Zed Books, 1993), pp164–78. See also Hecht and Cockburn, op cit, pp105–9.

33. Paul Harrison, The Third Revolution: Population, Environment and a Sustainable World (London: Penguin, 1992), pp96–9.

34 John Vidal, 'British Gas flees the forest ire', *Guardian*, 1 May 1992, p26.

35. See, for example: Friends of the Earth, *British Gas Oil Operation in Ecuador, Special Briefing* (London: Friends of the Earth, August 1990); and Sarita Kendall, 'Ecuador's oil imperative and the Amazon', *Financial Times*, 11 October 1991.

36. Francis Sullivan, 'Oil, Ecuador and British Gas: an unhappy trio', *WWF News*, October 1991, pp6–7.

37. International Union for the Conservation of Nature and Natural Resources, *Oil Exploration in the Tropics: Guidelines for Environmental Protection* (Gland, Switzerland: IUCN, September 1991).

38. E & P Forum, 'Oil Industry Operating Guideline for Tropical Rainforests', Report no 2.49/170, April 1991, p2. The Oil Industry Exploration and Production Forum was formed in 1974 to represent its members interests at the International Maritime Organization and other international agencies.

39. Shell International Petroleum Company and World Wide Fund for Nature, *Tree Plantation Review: Guidelines* (London/Godalming: Shell International Petroleum Company/World Wide Fund for Nature, 1993), pii.

40. See Norman Myers, *The Primary Source: Tropical Forests and Our Future* (London: WW Norton & Company, 1992), pp104–5.

41. Chris Philipsborn, 'Roads threaten tropical forest in Honduras', *Independent*, 16 November 1991, p14; and Oliver Tickell, 'Honduran chop logic', *Guardian*, 14 February 1992, p30.

42. The Report of the Commission of Inquiry into Aspects of the Timber Industry in Papua New Guinea was led by Judge Thomas Barnett. Attempts were made to suppress the findings of the report, which was not published in its entirety. However, a summary of the report was published by the Asia Pacific Action Group (PO Box 693, Sandy Bay, Tasmania, Australia). See Asia Pacific Action Group, *The Barnett Report: A Summary of the Report of the Commission of Inquiry into Aspects of the Timber Industry in Papua New Guinea* (Hobart, Tasmania: Asia Pacific Action Group, November 1990). This publication, edited by George Marshall, has been endorsed by Barnett as 'in general terms, an accurate reflection' of the Commission's findings (p8).

43. Barnett Commission of Inquiry Final Report, vol. 1, p7, cited in ibid, pp15–16.

44. George Marshall, 'The Political Economy of Logging: The Barnett Inquiry into Corruption in the Papua New Guinea Timber Industry', *The Ecologist*, vol 20, no 5, September/October 1990, p174.

45. Asia Pacific Action Group, op cit, p17, and ibid, p176.

46. Debra J Callister, *Illegal Tropical Timber Trade: Asia Pacific – A Traffic Network Report* (Cambridge: TRAFFIC International, 1992), pp6, 73. 'TRAFFIC' stands for Trade Records Analysis of Fauna and Flora In Commerce. TRAFFIC International is an international NGO which is financially supported by WWF and IUCN. It has a close cooperative relationship with the secretariat of the Convention on International Trade in Endangered Species of Wild Fauna and Flora (CITES).

47. Openg Onn, 'Sabah loses millions in timber fraud', *New Straits Times* (Malaysia), 15 May 1993, p6.

48. Germelino M Bautista, 'The Forestry Crisis in the Philippines: Nature, Causes, and Issues', *The Developing Economies*, vol xxviii, no 1, March 1990, p70.

49. Friends of the Earth, *Plunder in Ghana's Rainforest for Illegal Profit: An exposé of corruption, fraud and other malpractices in the international timber trade, Volume 1, Main Findings* (London: Friends of the Earth, March 1992); and George Monbiot, *Mahogany is Murder: Mahogany Extraction from Indian Reserves in Brazil* (London: Friends of the Earth, August 1992).

50. Scott B MacDonald, *Mountain High, White Avalanche: Cocaine and Power in the Andean States and Panama (The Washington Papers/137)* (New York/Washington DC: Praeger/The Center for Strategic and International Studies, 1989), pp65–6.

51. International Narcotics Control Board, *Report of the International Narcotics Control Board for 1990* (New York: United Nations, 1990), para 190, p36.

52. Noll Scott and Chris Taylor, 'Peruvian president says cocaine blighting Amazon', *Guardian*, 7 February 1991, p12.
53. This was noted in the annual report of General Rosso Serrano, commander of Colombia's Police Narcotics Brigade, as reported by Timothy Ross, 'Colombian drug barons open new smuggling route', *Independent*, 3 January 1992. See also Timothy Ross, 'Why the drug barons are talking green', *Independent*, 30 March 1992, p20.
54. International Narcotics Control Board, *Report of the International Narcotics Control Board for 1992* (New York, United Nations, 1992), para 327, p45.
55. International Narcotics Control Board (1990), op cit, para 167, pp33–4.
56. Anon (Our Special Correspondent in Guatemala), 'Cutting down the Mayan world', *The Economist*, 25 September 1993, p73.
57. George (1992), op cit, pp34–62.
58. Myers (1989), op cit, p170.
59. See the special edition of *The Ecologist* on 'Indonesia's Transmigration Programme: A Special Report in collaboration with Survival International and Tapol', vol 16, no 2/3, 1986.
60. Jack Westoby, *Introduction to World Forestry: People and Their Trees* (Oxford: Basil Blackwell, 1989), cited in Myers (1989), op cit, p7.
61. Norman Myers 'Tropical Forests', in Jeremy Leggett (ed) *Global Warming: The Greenpeace Report* (Oxford: Oxford University Press, 1990), p374.
62. Rod Harbinson, 'Burma's Forests Fall Victim to War'. *The Ecologist*, vol 22, no 2, March/April 1992, pp72–3.
63. Nicholas Hopkinson, *Fighting the Drug Problem, Wilton Park Papers 15*, Conference Report based on Wilton Park Conference 331, 23–27 January 1989 (London: HMSO, January 1989), p11.
64. Victor Mallet, 'Cambodia powerless to prevent plundering of its forests', *Financial Times*, 20 August 1992. On 28 April 1993, four British MPs placed a motion before the House of Commons calling for an increased United Nations presence in Cambodia, the prosecution of the Khmer Rouge leadership and for 'trade sanctions against Thailand to be considered if that country continues to encroach on Cambodian sovereignty and to facilitate trade in timber across the Thai/Cambodia border'. See, *House of Commons, Notice of Motions, No 163, 28th April 1993*, Motion 1881, p8026.
65. Peter Utting, *Trees, People and Power* (London: Earthscan, 1993), p167.
66. Bill Weinberg, *War on the Land: Ecology and Politics in Central America* (London: Zed Books, 1991), p41.
67. Heffa Schücking and Patrick Anderson, 'Voices Unheard and Unheeded', in Vandana Shiva et al, *Biodiversity: Social and Ecological Perspectives* (London: Zed Books, 1991), p20.
68. Hecht and Cockburn, op cit, p108.
69. Val Plumwood and Richard Routley, 'World Rainforest Destruction: The Social Factors', *The Ecologist*, vol 12, no 1, 1982. p11.
70. Friends of the Earth, *Poverty, Population and the Planet* (London: Friends of the Earth, December 1992), p4.
71. United Nations Population Fund (UNFPA), *The State of the World Population 1992* (New York: United Nations, 1992).
72. United Nations Development Programme, *Human Development Report 1992* (New York: United Nations, 1992).
73. Paul R Ehrlich and Anne H Ehrlich, *The Population Explosion* (London: Hutchinson, 1990), p116.
74. Dobson argues that ecologism should be seen as an ideology. See Andrew Dobson, *Green Political Thought* (London: Harper Collins Academic, 1990).
75. Peter M Haas, 'Obtaining International Environmental Protection through Epistemic Consensus', *Millennium: Journal of International Studies*, vol 19, no 3, Winter 1990, p351.
76. Wolfgang Sachs (ed) 'Global Ecology and the Shadow of "Development"', in Wolfgang Sachs (ed) *Global Ecology: A New Arena of Political Conflict* (London: Zed Books, 1993), p11.

77. David Humphreys, 'Hegemonic Ideology and the International Tropical Timber Organization', in John Vogler and Mark Imber (eds) *The Environment and International Relations* (London: Routledge, 1996).
78. Harrison refers not to 'ecologism' but to the 'radical school'. The two are considered the same in this study.
79. Myers in Leggett (ed), op cit, p390.
80. Harrison, op cit, pp126–7.
81. James Mayers of IIED, personal communication (letter), 7 March 1996.
82. Gerald Leach and Robert Mearns, *Beyond the Woodfuel Crisis: People, Land and Trees in Africa* (London: Earthscan, 1988), p1.
83. Barry Munslow et al, *The Fuelwood Trap: A Study of the SADCC Region* (London: Earthscan, 1988), p5. There are nine countries in the Southern African Development Coordination Conference, namely Angola, Botswana, Lesotho, Malawi, Mozambique, Swaziland, Tanzania, Zambia and Zimbabwe.
84. José Serra-Vega, 'Andean settlers rush for Amazonia', *Earthwatch*, no 39, third quarter, 1990, p9. The view that population planning is necessary to arrest deforestation is shared by others associated with the IPPF; see the articles by Don Hinrichsen and Jacqueline Sawyer in the same volume. See also the edition on population and the environment of *Populi: Journal of the United Nations Population Fund*, vol 18, no 3, September 1991.
85. 'Harare Declaration on Family Planning for Life', Harare, Zimbabwe, 6 October 1989. Those attending the conference included representatives of African family planning associations and the World Bank.
86. Harrison, op cit, p128.
87. Peter Dorner and William C Thiesenhusen, *Land Tenure and Deforestation: Interactions and Environmental Implications, United Nations Research Institute for Social Development Discussion Paper 34* (Geneva: UNRISD, April 1992).
88. José A Lutzenberger, 'Who is destroying the Amazon Rainforest?', *The Ecologist*, vol 17, no 4/5, July/November 1987, pp155–60.
89. Marcus Colchester, 'Colonizing the Rainforests: The Agents and Causes of Deforestation', in Colchester and Lohmann (eds), op cit, pp11–12.
90. Larry Lohmann, 'Against the Myths', in ibid, p16.
91. Friends of the Earth (December 1992), op cit, p9.
92. Colchester, in Colchester and Lohmann (eds), op cit, p10.
93. J Eric Smith, 'The Role of Special Purpose and Nongovernmental Organizations in the Environmental Crisis', *International Organization*, vol 26, no 2, Spring 1972, pp302–26.
94. Bolin Warrick and Döös Jäger (eds) *SCOPE 29: The Greenhouse Effect, Climatic Change and Ecosystems* (Chichester: John Wiley and Sons, 1986), pp136, 365.
95. World Meteorological Organization/United Nations Environment Programme, *Climate Change — Intergovernmental Panel on Climate Change: The IPCC Response Strategies*, 1990, pxxx.
96. World Meteorological Organization/United Nations Environment Programme, *Scientific Assessment of Climate Change: The Policymakers' Summary of the Report of Working Group I to the Intergovernmental Panel on Climate Change* (Geneva: WMO/UNEP, July 1990), p24.
97. 'Conference Statement', in J Jäger and H L Ferguson (eds) *Climate Change: Science, Impacts and Policy, Proceedings of the Second World Climate Conference* (Cambridge: Cambridge University Press, 1991), p497.
98. Jeffrey A McNeely et al, *Conserving the World's Biological Diversity* (Gland, Switzerland and Washington DC: IUCN/WRI/Conservation International/WWF–US/World Bank, 1990).
99. World Resources Institute, IUCN and UNEP in consultation with FAO and UNESCO, *Global Biodiversity Strategy: Guidelines for Action to Save, Study and Use Earth's Biotic Wealth Sustainably and Equitably* (Washington DC: WRI/IUCN/UNEP, 1992).
100. Christopher Joyce, 'Prospectors for tropical medicines', *New Scientist*, 19 October 1991, pp36–40.

101. M L Parry, A R Magalhães and N H Ninh, *Earthwatch Global Environment Monitoring System: The Potential Socio-Economic Effects of Climate Change — A summary of three regional assessments* (Nairobi: UNEP, 1991), p12.

102. Revkin reports that 7603 Amazonian forest fires were photographed in one day by the NOAA–9 satellite in 1987. See Andrew Revkin, *The Burning Season: The Murder of Chico Mendes and the Fight for the Amazon Rain Forest* (London: Collins, 1990), p231.

103. Jan Rocha, 'Smoke alarm in the Amazon', *Guardian*, 4 September 1992, p25.

104. Adrian Cowell, *The Decade of Destruction* (Sevenoaks: Hodder and Stoughton/ Channel 4, 1990), pp14–15. This book accompanied a television documentary of the same name.

105. See 'Chico' Mendes, *Fight for the Forest* (London: Latin America Bureau, 1989).

106. Aubrey Meyer, 'The campaign continues', *Geographical Magazine*, February 1990, p14. On the day the petition was handed to the UN Secretary-General in New York, the World Rainforest Movement released a declaration agreed upon by representatives from several local forest groups in April 1989. Entitled 'An Emergency Call to Action for The Forests and Their Peoples', this is reproduced in World Rainforest Movement, op cit, pp8–13.

107. The 'Summary of the Final Report of the Forests Resources Assessment 1990 for the Tropical World' were presented to the eleventh session of FAO's Committee on Forestry, Rome 8–12 March 1993.

108. 'The Forest Resources of the Tropical Zone by Main Ecological Regions', by Forest Resources Assessment 1990 Project, FAO, Rome, Italy, June 1992.

109. Barbara J Bramble and Gareth Porter, 'Non-Governmental Organizations and the Making of US International Environmental Policy', in Hurrell and Kingsbury (eds), op cit, pp313–25.

110. International Union for the Conservation of Nature and Natural Resources/ UNEP/WWF, *World Conservation Strategy: Living Resource Conservation for Sustainable Development* (Gland: Switzerland: IUCN/UNEP/WWF, 1980). An updated version of the World Conservation Strategy was published in 1991 in International Union for the Conservation of Nature and Natural Resources/UNEP/WWF, *Caring for the Earth: A Strategy for Sustainable Living* (London: Earthscan, 1991).

111. It is wrong to say that this report *introduced* the concept of sustainable development. The former executive director of UNEP wrote in 1979 of the importance of 'eco-development' and 'the constraints of a sustainable environmental resource base'. See Mostafa K Tolba, 'Foreword' in, Barbara Ward, *Progress for a Small Planet* (Middlesex: Penguin Books, 1979), pix. It is very likely that earlier references to concepts similar in content to 'eco-development' exist. Certainly, the argument that the present generation has a responsibility to future generations is not a new one. See, for example, Richard I Sikora and Brian Barry (eds) *Obligations to Future Generations* (Philadelphia: Temple University Press, 1978).

112. World Commission on Environment and Development, op cit, p43.

113. See, for example, two special issues of *Third World Resurgence*; issue no 20, April 1992, and issue no 40, December 1993; and the special issue of *New Internationalist*, August 1993.

114. Karen Litfin, 'Eco-regimes: Playing Tug of War with the Nation-State', in Ronnie D Lipschutz and Ken Conca (eds) *The State and Social Power in Global Environmental Change* (New York: Columbia University Press, 1993), p110.

115. Andrew Blowers and Pieter Glasbergen, 'The search for sustainable development', in Pieter Glasbergen and Andrew Blowers (eds) *Environmental Policy in an International Context: Perspectives on Environmental Problems* (London: Arnold, 1995), p174; and Pieter Glasbergen, *Managing Environmental Disputes: Network Management as an Alternative* (Dordrecht: Kluwer, 1995), pp6–8.

116. World Commission on Environment and Development, op cit, p310.

117. Principle 21 of the Stockholm Declaration of the United Nations Conference on the Human Environment, 1972.

118. Makumi Mwagiru, 'The Legal Milieu of the Environment: An Overview', *Paradigms: The Kent Journal of International Relations*, vol 7, no 1, Summer 1993, p5.

119. Namely resolution 2996 (XXVII), 15 December 1972. See Mark Imber, 'UNEP's Catalytic and Coordinating Mandate: The Labours of Hercules and Sisyphus', paper presented to

the Annual Conference of the British International Studies Association, University of Warwick, 16–18 December 1991, p25.

120. 'The Hague Recommendations on International Environmental Law, Peace Palace, The Hague, 16 August 1991', Preamble.

121. World Commission on Environment and Development, op cit, Chapter 11, 'Managing the Commons', pp261–89.

122. Lynton Keith Caldwell, *International Environmental Policy: Emergence and Dimensions* (2nd edn) (Durham: Duke University Press, 1990), Chapter 8, 'International Commons: Atmosphere, Outer Space, Oceans, Antarctica', pp257–302.

123. John Vogler, 'Regimes and the Global Commons: Space, Atmosphere and Oceans', in Anthony G McGrew et al, *Global Politics* (Cambridge: Polity Press, 1992, p120.

124. Gareth Porter and Janet Welsh Brown, *Global Environmental Politics* (Boulder, Colorado: Westview Press, 1991), p92.

125. Maurice F Strong, 'ECO '92: Critical Challenges and Global Solutions', *Journal of International Affairs*, vol 44, no 2, Winter 1991, pp297–8.

126. J Romm, 'Exploring Institutional Options for Global Forest Management', in David Howlett and Caroline Sargent (eds) *Technical Workshop to Explore Options for Global Forestry Management, Bangkok, 1991, Proceedings* (London: International Institute for Environment and Development), p187.

127. 'Act so that you treat humanity, whether in your own person or in that of another, always as an end and never as a means only', Immanuel Kant, *Foundations of the Metaphysics of Morals*, translated by Lewis White Beck (Indianapolis: Bobbs-Merrill Library of Liberal Arts, 1959), p47, cited in Rachel M McCleary, 'The International Community's Claim to Rights in Brazilian Amazonia', *Political Studies*, vol xxxix, 1991, p701.

128. Ibid, p706.

129. Hurrell, in Hurrell and Kingsbury (eds), op cit, p402.

130. Garrett Hardin, 'The Tragedy of the Commons', *Science*, 1968, cited by Hecht and Cockburn, op cit, pp104–5.

131. George Monbiot, 'The real tragedy of the commons', *Guardian*, 6 August 1993, p19.

132. Harrison, op cit, p264.

133. Dharam Ghai, *Conservation, Livelihood and Democracy: Social Dynamics of Environmental Change in Africa*, United Nations Research Institute for Social Development, Discussion Paper 33 (Geneva: UNRISD, March 1992), p11.

134. 'Charter of the Indigenous–Tribal Peoples of the Tropical Forests: Statement of the International Alliance of the Indigenous–Tribal Peoples of the Tropical Forests', Penang, Malaysia, 15 February 1992, pp2, 11.

135. Article 11 of International Labour Organization Convention No 107.

136. Article 7 of International Labour Organization Convention No 169.

137. Note that McCleary does not explicitly refer to international trade and external debt as transnational causes of deforestation.

NOTES TO CHAPTER 2

1. FAO document FO:TFAP/92/2, 'Second Meeting of the Ad Hoc Group on the Tropical Forests Action Programme, Secretariat Note, March 1992', para 3, p1.

2. FAO document FO:FDT/83/Rep, 'Committee on Forest Development in the Tropics, Report of the Sixth Session, Rome, 18–21 October 1983', para7, p1 and para 22, p4. The FAO Committee on Forest Development in the Tropics first met in October 1967 and has met at approximately two-year intervals since.

3. FAO document FO:FDT/85/3, 'Draft Proposals for Action Programmes in Tropical Forestry, April 1985'.

4. FAO document FO:FDT/85/Rep, 'Committee on Forest Development in the Tropics, Report of the Seventh Session, Rome, 10–12 June 1985', para 18, p3. Since the Third World Forestry Congress held in 1948 the Congress has been organized by a host country

with FAO acting as a cosponsor. The Ninth World Forestry Congress was cosponsored by the FAO and the government of Mexico.

5. The Committee on Forest Development in the Tropics of the FAO of the UN, *Tropical Forestry Action Plan* (Rome: FAO, October 1985).

6. FAO document FO:TFAP/92/2, op cit, para 7, p1.

7. FAO document COFO–86/REP, 'Report of the Eighth Session of the Committee on Forestry, Rome, 21–25 April 1986', paras 60–1, p9. The FAO Committee on Forestry first met in 1972 and has met at approximately two-year intervals since, convening in the years when the CFDT does not meet.

8. James Gustav Speth, 'Foreword', in World Resources Institute, World Bank and United Nations Development Programme, *Tropical Forests: A Call for Action* (Washington DC: World Resources Institute, October 1985), Part 1, pv.

9. Much of the TFAP literature refers to the FAO, World Bank, UNDP and WRI as the four TFAP 'cosponsors'. However, the WRI has stated that it 'never agreed to be listed as a "cosponsor" of TFAP, though it certainly was a "cofounder"'. See FAO document FO:TFAP/92/2(a), 'Second Meeting of the Ad Hoc Group on the Tropical Forests Action Programme: comments of the co-sponsors, April 1992', p2.

10. FAO, World Bank, WRI and UNDP, *The Tropical Forestry Action Plan* (Rome: FAO, June 1987), p3.

11. See Peter T Hazelwood, *Expanding the Role of Non-Governmental Organizations in National Forestry Programs* (Washington DC: World Resources Institute, 1987); Owen J Lynch, *Whither the People? Demographic, Tenurial and Agricultural Aspects of the Tropical Forestry Action Plan* (Washington DC: World Resources Institute, September 1990); Cheryl Cort, *Voices From the Margin: Non-Governmental Organization in the Tropical Forestry Action Plan* (Washington DC: World Resources Institute, January 1991).

12. 'A Statement by Non-Governmental Organizations to the Bellagio Strategy Meeting on Tropical Forests', July 1987, p3.

13. 'A Global Research Strategy for Tropical Forestry: Report of an International Task Force on Forestry Research' (Sponsored by the Rockefeller Foundation, UNDP, World Bank, and FAO, September 1988).

14. Recommendation number 1 of The Hague meeting, cited by Winterbottom, op cit, p42.

15. Recommendation number 3 of The Hague meeting, cited by Winterbottom, ibid.

16. Tapani Oksanen et al, 'A Study on Coordination in Sustainable Forestry Development, Prepared for the Forestry Advisers' Group', June 1993, Appendix 7.5, 'Present Role and Mandate of TFAP Forestry Advisers' Group', pp70–1.

17. Antonio Carillo et al, *Project for the Implementation of the German Contribution to the Tropical Forestry Action Plan (TFAP): Critical Assessment of the Current Status of the TFAP, PN 88.2283.5–03.100* (Eschborn, Germany: Deutsche Gesellschaft für Technische Zusammenarbeit [GTZ] GmbH, October 1990), p69.

18. 'Statutory authority' refers to the authority of the FAO constitution as contained in FAO, *Basic Texts of the Food and Agriculture Organization of the United Nations, Volumes I and II — 1980 edition* (Rome: FAO, 1980).

19. FAO document FO:TFAP/92/2, op cit, para 14, p3.

20. FAO, *Guidelines for Implementation of the TFAP at Country Level* (Rome: FAO, 1989). In 1991 these guidelines were superseded by FAO, *Tropical Forestry Action Programme: Operational Principles* (Rome: FAO, November 1991).

21. The brief account of the NFAP process that follows has been trawled and amalgamated from five sources: FAO (1989), ibid; FAO document FO:TFAP/92/2, op cit; Winterbottom, op cit; Herman Savenije, *BOS-DOCUMENT 11: Tropical Forestry Action Plan — Recent Developments and Netherlands Involvement* (Wageningen, The Netherlands: Ministry of Agriculture, Nature Conservation and Fisheries, January 1990); and Marcus Colchester and Larry Lohmann, *The Tropical Forestry Action Plan: What Progress* (Penang, Malaysia/Sturminster Newton, Dorset: World Rainforest Movement/The Ecologist, 1990). The FAO documents have been used as the basic frame of reference, while the secondary sources have provided some of the finer descriptive detail.

22. *TFAPulse*, Number 18, December 1992, p1.

23. WRI comments in FAO document FO:TFAP/92/2, op cit, p2.

24. James Mayers of IIED, personal communication (telephone conversation), 19 March 1996.
25. The countries involved in the Caribbean Community (CARICOM) Plan, which has been formulated in a similar way to an NFAP, are Antigua and Barbuda, Barbados, Dominica, Grenada, Montserrat, St Christopher and Nevis, St Lucia, St Vincent and the Grenadines, and Trinidad and Tobago.
26. For example, The Editors, 'Tropical Forests: A Plan for Action', *The Ecologist*, vol 17, No 4/5, 1987, pp129–33, and Vandana Shiva, *Forestry Crisis and Forestry Myths: A critical review of 'Tropical Forests: A Call for Action'* (Penang: World Rainforest Movement, 1987).
27. Francis Sullivan, WWF–UK, personal communication (interview), Godalming, 13 March 1992.
28. 'Minutes from the Non Governmental Organization Retreat on the Tropical Forestry Action Plan hosted by the World Wildlife Fund, 2 March 1990', para II, p1.
29. George Marshall, *Tropical Forestry Action Plan: Campaign Dossier* (Sturminster Newton, Dorset: The Ecologist/World Rainforest Movement, October 1990), p26.
30. Ola Ullsten et al, *Tropical Forestry Action Plan: Report of the Independent Review* (Kuala Lumpur: FAO, May 1990), p1.
31. Winterbottom, op cit, p24.
32. Ullsten et al, op cit, p12.
33. Colchester and Lohmann, op cit, p83.
34. For example, Aubrey Meyer, 'Save the Forests, Save the Planet: The Campaign Continues', *Geographical Magazine*, February 1990, pp14–15; Maria Elena Hurtado, 'Root versus Branch', *Panascope*, no 22, January 1991, pp3–4; and Sierra Club leaflet, 'Protect Tropical Forests: Stop the Tropical Forestry Action Plan', Sierra Club, 730 Polk Street, San Francisco, CA 94109, June 1990.
35. Ullsten et al, op cit, p13.
36. Winterbottom, op cit, p5.
37. Elizabeth A. Halpin, *Indigenous Peoples and the Tropical Forestry Action Plan* (Washington DC: World Resources Institute, June 1990), p33.
38. Colchester and Lohmann, op cit, pp1, 86. Other actors had advocated a funding moratorium prior to the publication of the WRM review, including Friends of the Earth–USA: 'FOE–USA Position: Tropical Forestry Action Plan' (unpublished), February 20, 1990. This paper was circulated at the NGO retreat on TFAP hosted by WWF–US, Washington DC, 2 March 1990.
39. Ullsten et al, op cit, p31.
40. Colchester and Lohmann, op cit, p2.
41. Winterbottom, op cit, p21.
42. Ibid, p29.
43. FAO documents and other sources originally refer to a 'consultative forum'. Subsequently a minor change in nomenclature occurred, with reference made instead to a 'consultative group'. To avoid confusion, reference is made throughout to the consultative group.
44. Ullsten et al, op cit, p39.
45. Colchester and Lohmann, op cit, p3, and letter from Martin Khor Kok Peng, WRM International Secretariat, Penang, to Dr Caroline Sargent, International Institute of Environment and Development, London, 4 September 1990.
46. Winterbottom, op cit, p15.
47. Ullsten et al, op cit, p39.
48. Colchester and Lohmann, op cit, p84.
49. Letter from the chairman of the United States Senate Committee on Foreign Relations to The Honourable Barber Conable, president of the International Bank for Reconstruction and Development, 2 May 1990.
50. 'The Houston Declaration of the Group of Seven Industrialized Countries, July 1990', Article 66.
51. Clive Wicks of WWF–UK, cited in Hurtado, op cit, and WWF–UK press release, 'WWF Declares UN Tropical Forestry Plan a Failure, 3 October 1990'.
52. Sara Parkin, 'Rooting for the rainforests: West and East European Greens Unite', *Econews: Newspaper of the Green Party*, February/March 1990, no 49, p1. It seems safe to assume

that this resolution was primarily the work of West European green parties. There is no evidence of East European environmentalists campaigning for tropical forest conservation or against the TFAP prior to the fall of the Berlin Wall. The suppression of civil society in many of these countries gave such groups limited political space in which to operate. Nonetheless, in the late 1980s there did emerge groups concerned about environmental degradation in their own country.

53. Letter to the president of the World Bank from Friends of the Earth–USA (signed by representatives from 19 NGOs from the North and South) dated 6 September 1990. This letter urged the World Bank to suspend TFAP funding. The WRM had previously written to the president of the World Bank on 14 August 1990 urging a restructuring of TFAP.

54. FAO document (not numbered), 'Committee on Forestry, Tenth Session, Rome, 24–28 September 1990: Report', para 87, p10.

55. The criticism that the TFAP had 'no formal objectives, except in the very broadest sense' was also made by the IIED in a report distributed at the Tenth Session of the Committee of Forestry; Caroline Sargent, *Defining the Issues: Some thoughts and recommendations on recent critical comments on TFAP* (London: IIED, 1990) p1.

56. FAO document CL 99/22, 'Outcome of Meeting of Ad Hoc Group of Experts on TFAP', Appendix B, para 2, pB1.

57. Ibid, para 19, p4.

58. Ibid, Appendix A, para 3 (e), p2. This quote also appears in a memorandum prepared for internal discussions within the World Bank, 'Recommendations for Revamping the Tropical Forestry Action Plan', p 3, circulated on 29 May 1991.

59. The author is grateful to Dr Marcus Colchester of the World Rainforest Movement for explaining this vision to him: personal communication (telephone conversation), 22 September 1994.

60. 'Options for institutional arrangements for a revamped TFAP', para 8, p2. This anonymous paper was drafted by a participant at the meeting of the TFAP cofounders at UNDP, New York, 15–16 April 1991.

61. The tenth World Forestry Congress was cosponsored by FAO and the government of France. See note 4 above.

62. Namely 'functions of the proposed new mechanism, composition and selection of members, establishment and organization, national-level arrangements, and technical support'; FAO document CL100/2, 'Developments regarding the TFAP', October 1991, Appendix A, pA1.

63. For the wording of the functions agreed at Paris, see FAO document FO:TFAP/91/2, 'First Meeting of the Ad Hoc Group on the Tropical Forests Action Programme: Institutional Arrangements for the TFAP, Secretariat Note', November 1991, Annex 1, p6.

64. According to Marcus Colchester, the Malaysian delegation was the sole one insisting that the consultative group respect national sovereignty at the Paris Consultation; personal communication (telephone conversation), 22 September 1994. Consequently, the revised definition of the group's functions contained a proviso to this effect: FAO document FO:TFAP/91/2, ibid. This insistence effectively ensured that the consultative group would not, in line with the Geneva recommendations, be a fully independent polity that could promote genuine local and indigenous peoples' participation.

65. 'World Resources Institute Statement on the Future of the Tropical Forestry Action Plan, November 25, 1991', p2.

66. FAO document FO:TFAP/91/Rep, 'Report of the First Session of the Ad Hoc Group on the Tropical Forests Action Programme, 9 December 1991', para 1, p1 and para 21, p3. No NGOs attended the first session. One NGO, namely IUCN, attended the second session, while four (IUCN, IIED, WRI and FoE) attended the third session.

67. 'Summary Report of the Thirteenth Meeting of the TFAP Forestry Advisers on Harmonizing International Forestry Development Cooperation, 2–6 December 1991, Rome', Annex 6, 'Statement of the Tropical Forestry Action Programme Advisers' Group, Rome, December 1991', section 2.

68. FAO document FO:FDT/91/Rep, 'Committee on Forest Development in the Tropics, Tenth Session, Rome, 10–13 December 1991', para 13, p3. In accordance with a resolution

passed at its ninth session, the tenth session of the CFDT invited the chairman of the FAG to attend and present the FAG's views on the TFAP.

69. As noted by Carlos Marx R Carneiro, 'Information Paper of the TFAP Coordinating Unit', Meeting of the TFAP National Coordinators in the Caribbean, Port of Spain, 13–16 September 1992, p8.

70. In a FAO internal memorandum, the assistant director-general of the Forestry Department noted that 'the gauntlet had been thrown down for FAO' which must 'face the challenge of providing technical advice to the [consultative group] promptly and without bias': FAO internal document, report on the 'Tropical Forestry Action Plan meeting, UNDP Headquarters, New York, 15–16 April 1991', memo by C H Murray, ADG Forestry Department, dated 23 April 1991.

71. As noted in FAO document FO:TFAP/92/Rep, 'Report of the Second Session of the Ad Hoc Group on the Tropical Forests Action Programme, 5 May 1992', para 14, p2. (Note that two FAO documents were given the same number, namely FO:TFAP/92/Rep; these were the reports of the two meetings of the ad hoc group held in 1992, namely in May and September.) It is not clear why the FAO suggested the FFDC as a possible consultative group when the countries that comprised this group were opposed to the FFDC assuming such a role.

72. FAO document FO:TFAP/92/Rep, 'Report of the Third Session of the Ad Hoc Group on the Tropical Forests Action Programme, 4 September 1992', pp17–18. (Note once again that two FAO documents were given the number FO:TFAP/92/Rep).

73. FAO document FO:TFAP/91/Rep, op cit, para 13, p2.

74. FAO document FO:TFAP/92/Rep (Third Session), op cit, para 14, p17.

75. Bundesministerium für Ernährung, Landwirtschaft und Forsten, *Schutz und Bewirtschaftung der Tropenwälder: Tropenwäldbericht der Bundesregierung, 3 Bericht (Bericht–Nr. B 245/93)* (Bonn, March 1993), p42. (This document, the Third German Government Report on Tropical Forests, was obtained for the author by Susanne Staab BA MA; excerpts were translated by Beate Münstermann BA MA, 31 January 1994.) Brazil, Malaysia and Indonesia participated in the Ad Hoc Group, while Brazil and Malaysia (but not Indonesia) participated in the sub-group of the Ad Hoc Group.

76. As reported in 'Summary Report on the Fifteenth Meeting of the TFAP Forestry Advisers' Group on Harmonizing International Forestry Development Cooperation, 30 November–4 December 1992, San José, Costa Rica', section 3, p4.

77. FAO document FO:TFAP/92/Rep (Third Session), op cit, pp2–3

78. FAO internal memorandum, dated 24 September 1992.

79. FAO document CL 102/21, 'Establishment of a Consultative Group on the Tropical Forests Action Programme, November 1992', paras 8, 9 and 19, pp3–5.

80. FAO document CL 102/Rep, 'Report of the Council of FAO, Hundred and Second Session, Rome, 9–20 November 1992', para 141, pp24–5.

81. References appear in some FAO documents to 'the OECD group'. The OECD is not institutionally involved in the TFAP, and the acronym was clearly employed here as a shorthand for delegates from the developed world.

82. FAO document CL 103/19, 'Consultative Group on the Tropical Forests Action Programme: Proposals of the Independent Chairman of the FAO Council, May 1993', p2–5.

83. Ibid, paras 26–33, pp5–6.

84. Jean Clément, NFAP Support Unit, personal communication (e-mail), 29 March 1995.

85. Ralph Roberts, 'TFAP, An Evolving Process: Role of the TFAP Forestry Advisers' Group', in *10th World Forestry Congress, Paris, 1991: Proceedings* (Nancy: Ministère de l'Agriculture et de la Forêt, 1992), vol 8, p327.

86. 'Summary Report on the Fifteenth Meeting of the [FAG]', op cit, section 3, p5.

87. 'Summary Report on the Sixteenth Meeting of the Forestry Advisers' Group on Harmonizing International Forestry Development Cooperation, 25–28 May 1993, UNDP, New York', Annex 3, 'Letter from Ralph W Roberts, Chairperson, Forestry Advisers Group to Members of the FAO Council, dated 27 May 1993'.

88. Ibid.

89. 'Summary Report on the Sixteenth Meeting of the [FAG]', op cit, 'Appendix 8, The New York Proclamation: Purpose, Objectives and Activities of the Forestry Advisers' Group, 28 May 1993'.
90. Ibid, p11.
91. FAO, *Formulation, Execution and Revision of National Forestry Programmes: Basic Principles and Operational Guidelines* (Rome: FAO, 1996).
92. These are (i) the CARICOM countries (see note 25 above); (ii) Central America (Belize, Costa Rica, El Salvador, Guatemala, Honduras, Nicaragua and Panama); (iii) Treaty for Amazonian Cooperation (aka Amazonian Pact: Bolivia, Brazil, Colombia, Ecuador, Guyana, Peru, Surinam and Venezuela); (iv) Inter-governmental Committee for Drought Control in the Sahel (CILSS: Burkino Faso, Cape Verde, Chad, Gambia, Guinea Bissau, Mali, Niger and Senegal); (v) Inter-governmental Authority on Drought and Development in Eastern Africa (IGADD: Djibouti, Ethiopia, Kenya, Somalia, Sudan, Uganda); and (vi) Southern African Development Coordination Conference (SADCC: Angola, Botswana, Lesotho, Malawi, Mozambique, Namibia, Swaziland, Tanzania, Zambia and Zimbabwe). This information is obtained from *TFAP Update*, no 29, June 1993.
93. Dyaa Ajkbache, NFAP Support Unit, personal communication (e-mail), 20 December 1995.
94. World Wide Fund for Nature, *Reforming the Tropical Forestry Action Plan: A WWF Position* (Gland, Switzerland: WWF–International, September 1990), p5.
95· 'Summary Report on the Twelfth Meeting of the TFAP Forestry Advisers on Harmonizing International Forestry Development Cooperation, 10–14 June 1991, Hull, Canada', section 12 'TFAP Implementation, What happens after the Round Table III? Financing the Projects', pp13–14.
96. Carneiro, op cit, p1.
97. Oksanen et al, op cit, p11.
98. Comments by Luis Gomez-Echeverry to the FAG; 'Summary Report on the Sixteenth Meeting of the [FAG]', op cit, p15.
99. United Nations, *Earth Summit Agenda 21: The United Nations' Programme of Action from Rio* (New York: United Nations, 1993), p89.
100. Jean Clément, NFAP Support Unit, personal communication (telephone conversation), 9 January 1996.

NOTES TO CHAPTER 3

1. For the text of the resolution, see United Nations, *Proceedings of the United Nations Conference on Trade and Development, Fourth Session, Nairobi, 5–31 May 1976, Volume I, Report and Annexes* (New York: United Nations, 1977), pp6–9.
2. Marcus Colchester, 'The International Tropical Timber Organization: Kill or Cure for the Rainforests?', *The Ecologist*, vol 20, no 5, September/October 1990, p166.
3. UN document TD/TIMBER.2/3, 'Background, status and operation of the International Tropical Timber Agreement, 1983, and recent developments of relevance to the negotiation of a successor agreement', 26 February 1993, para 4, p1.
4. Terence Hpay, *The International Tropical Timber Agreement: Its prospects for tropical timber trade, development and forest management (IUCN/IIED Tropical Forest Policy Paper No 3)* (London: IIED–Earthscan, 1986), p8.
5. UN documents TD/B/IPC/TIMBER/33, 'Report of the Intergovernmental Group of Experts on Research and Development for Tropical Timber, Geneva, 16 to 20 December 1981'; TD/B/IPC/TIMBER/34, 'Report of the Intergovernmental Group of Experts on Improvement of Market Intelligence on Tropical Timber, Geneva, 23 to 27 November 1981'; TD/B/IPC/TIMBER/36, 'Reforestation and forest management of tropical timber within the Integrated Programme for Commodities: Report by the Secretariats of UNCTAD and FAO', 2 April 1982; and TD/B/IPC/TIMBER/37, 'Prospects for the expansion of timber processing activities in developing countries: Report by the Secretariats of FAO and UNCTAD', 5 April 1982.

6. UN document TD/B/IPC/TIMBER/39, 'Report of the Sixth Preparatory Meeting on Tropical Timber, Geneva, 1 to 11 June 1982', para 73, p17.
7. UN document TD/TIMBER/3, 'Report of the Meeting on Tropical Timber, 29 November to 3 December 1982', para 27, p7.
8. Comment by Tatsuro Kunugi (Japan, chairman of the conference) in *UNCTAD Bulletin*, no 192, April 1983, p19.
9. Some UN and ITTO sources state that 69 countries took part in the conference. Belgium and Luxembourg, which have a joint customs union, send a joint delegation to the ITTO. Some sources treat the two countries as one. They are treated as two in this work.
10. UNEP Governing Council decision 12/12, section III. See United Nations Environment Programme, *Annual Report of the Executive Director, 1984* (Nairobi: UNEP, May 1985), para 164, p65.
11. Prior to exercising its catalytic mandate with respect to the TFAP and the ITTA 1983, the Governing Council of UNEP passed seven decisions between 1973 and 1982 with respect to forests. See UN document TD/TIMBER/4/Add.1, 'International activities in the field of tropical forestry, Addendum', 17 March 1983, para 37, pp2–3.
12. UN document TD/TIMBER/11/Rev.1, 'International Tropical Timber Agreement, 1983', p8.
13. Ibid, Article 37, para 2, p17.
14. Ibid.
15. Hpay, op cit, p11.
16. Colchester, op cit, p167.
17. Lachlan Hunter of the ITTO secretariat, personal communication, interview in Yokohama with Dr Peter Willetts (of City University, London), 23 March 1992.
18. Responsibility within Whitehall for the ITTO passed from the DTI to the ODA on 1 October 1990: House of Commons Environment Committee, Session 1990/1, *Third Report, Climatological and Environmental Effects of Rainforest Destruction*, HC 24 (London: HMSO, 1991), para 106, p xxxii.
19. Barbara J Bramble and Gareth Porter, 'Non-Governmental Organizations and the Making of US International Environmental Policy', in Andrew Hurrell and Benedict Kingsbury (eds) *The International Politics of the Environment: Actors, Interests, and Institutions* (Oxford: Clarendon Press, 1992), p349.
20. See ITTO document ITTC(XI)/4, 'Relations with the Common Fund for Commodities: Report by the Secretariat', 30 October 1991. The idea of a Common Fund for Commodities was incorporated into the Integrated Programme for Commodities adopted at UNCTAD IV in 1976. The negotiations establishing the Common Fund were concluded in 1980 (UN document TD/IPC/CF/CONF/25). The agreement entered into force in 1988. The first account of the Common Fund for Commodities deals with the financing of buffer stocks. The second account became operational in 1991 and is designed to increase further processing of, and market opportunities for, commodities so that producing countries may retain more of their products' end use value. It is not a separate operational agency, but funds projects through designated international commodity bodies. By March 1993 the Common Fund had established relations with 21 ICBs, including the ITTO.
21. Henrik Skouenborg, chief operations officer of the Common Fund for Commodities, personal communication (letter), 24 March 1993.
22. Friends of the Earth and World Rainforest Movement, *The International Tropical Timber Agreement: Conserving the Forests or Chainsaw Charter?*, A critical review of the first five years of the International Tropical Timber Organization (London: Friends of the Earth, November 1992), p11.
23. UN document TD/TIMBER.2/3, op cit, para 48, p10.
24. 'NGO Statement to the ITTO Council, November 16, 1988', section III, para C, p3.
25. ITTO document ITTC(V)/D.1, 'Draft Report of the International Tropical Timber Council, Fifth Session, Yokohama, 14–16 December 1988', para 20, p5. See also ITTO document ITTC(V)/4/Rev.5, 'International Tropical Timber Organization, Project Cycle'.
26. Evidence given by the Minister of Overseas Development to the Chairman of the House of Commons Environment Committee. See, House of Commons Environment Committee, Session 1990/1, op cit, Appendix 14, p195.

27. ITTO document ITTC(X)/15, 'Decision 2(X), Continuation of the Expert Panel for Technical Appraisal of Project Proposals . . . ', Annex 2, 'Adjustments to the Project Cycle', paras 1–2, p4.
28. ITTO document ITTC(XII)/16, 'Decision 5(XII), Further Improvements of the Project Cycle', 14 May 1992, Annex, para 1, p2.
29. Friends of the Earth and World Rainforest Movement, op cit, p19.
30. ITTO Draft Project Document, Serial Number PD 170/91(F), 'Modernization and Development of Egyptian Forest Nurseries', and ITTO Draft Project Document, Serial Number PD 184/91(F), 'Multipurpose Tree Planting in Egypt'. Both project proposals were submitted by the government of Egypt. Although Egypt is a tropical timber consumer it is eligible for the funding of projects meeting the objectives of the ITTA 1983.
31. ITTO document ITTC(XI)/11, 'Report of the Panel of Experts for Technical Appraisal of Project Proposals: Report of Second Meeting, 14–19 October 1991, Yokohama, Japan', Appendix III.
32. ITTO document PCF(IX)/18, 'Draft Report to the International Tropical Timber Council: Ninth Session of the Permanent Committee on Reforestation and Forest Management', cited in Friends of the Earth and World Rainforest Movement, op cit, p21.
33. Ibid, pp21–2.
34. Duncan Poore et al, *No Timber Without Trees: Sustainability in the Tropical Forest* (London: Earthscan, 1989), p xiv. (Originally presented to the ITTO as pre-project report PPR 11/88 (F), 'Natural Forest Management for Sustainable Timber Production', International Institute for Environment and Development, London, October 1988.)
35. Colchester, op cit, p167.
36. Chee Yoke Ling of Sahabat Alam Malaysia, cited by Fred Pearce, *Green Warriors: The People and the Politics Behind the Environmental Revolution* (London: Bodley Head, 1991), p189.
37. Poore et al, op cit, p 6.
38. Ibid, p5. Poore provided a similar definition to a seminar held alongside the fifth ITTC session. See, Duncan Poore, 'ITTO Seminar on Sustainable Utilization and Conservation of Tropical Forest, Yokohama, 12 November 1988: Sustainable Management', p2.
39. Peter Utting, *Trees, People and Power* (London: Earthscan, 1993), p116.
40. ITTO document ITTC(XI)/6, 'Report of the Expert Panel on possible methods of defining general criteria for and measurement of sustainable tropical forest management', 3 October 1991.
41. ITTO document ITTC(XI)/20 'Decision 6(XI), Sustainable Forest Management I', 4 December 1991, Annex, p 2. The annex to this document is reproduced as: International Tropical Timber Organization, *ITTO Policy Development Series No 3: Criteria for the Measurement of Sustainable Tropical Forest Management* (Yokohama: ITTO, 1992), p2.
42. FAO document FO:FDT/91/5, 'Committee on Forest Development in the Tropics, Tenth Session, Sustainable Management of Tropical Forests, Secretariat Note', October 1991, para 15, p3.
43. Christopher Elliott, *Sustainable Tropical Forest Management by 1995* (Godalming: WWF–UK, November 1990), p1.
44. Chris Elliott, *ITTO in the 1990s – Urgency or Complacency?, A WWF Discussion Paper* (Gland, Switzerland: WWF–International, January 1991), p3.
45. International Tropical Timber Organization, *ITTO Action Plan: Criteria and Priority Areas for Programme Development and Project Work* (Yokohama: ITTO, November 1990), p7.
46. WWF press release, 'WWF proposal targets 1995 for sustainable tropical forestry', 19 November 1990.
47. ITTO document ITTC(X)/16, 'Decision 3(X), Sustainable Tropical Forest Management and Trade in Tropical Timber Products', 6 June 1991, para 1.b, p1.
48. See Francis Sullivan, 'Are Forests a Renewable and Permanent Source of Supply?', *Journal of the Institute of Wood Science*, vol 12, no 5, 1992, pp263–5.
49. Government of the Netherlands, *The Dutch Government's Policy Paper on Tropical Rainforests* (The Hague, December 1992), p43.

50. Klaas Kuperus, 'Dutch Debate: Call for Clarity', *Timber Trades Journal* (Supplement, 'Timber and the Environment'), 29 January 1994, pp14–15.
51. Francis Sullivan of WWF–UK, personal communication (interview), Godalming, 15 March 1994.
52. ITTO document ITTC (VIII)/15/Rev.1, 'Decision 4(VIII)', 23 May 1990, p1.
53. 'Foreword' by B C Y Freezailah, executive director of ITTO in, International Tropical Timber Organization, *ITTO Technical Series 5: Guidelines for the Sustainable Management of Natural Tropical Forests* (Yokohama: ITTO, December 1990).
54. Ibid, p9.
55. International Tropical Timber Organization, *ITTO Policy Development Series 4: ITTO Guidelines for the Establishment and Sustainable Management of Planted Tropical Forests* (Yokohama: ITTO, 1993).
56. ITTO document ITTC(XI)/7/Rev.3, 'ITTO Guidelines for the Conservation of Biological Diversity in Tropical Production Forests'.
57. ITTO document ITTC(IX)/D.1, 'Draft Report of the International Tropical Timber Council at its Ninth Session, Yokohama, Japan, 16–23 November 1990', 8 February 1991, para 34, p9.
58. ITTO document ITTC(X)/20, 'Draft Report of the International Tropical Timber Council at its Tenth Session, Quito, Ecuador, 29 May–6 June 1991', 15 August 1991, para 61, p12.
59. ITTO document ITTC(XI)/7 Rev.2, 'Third draft proposal for ITTO guidelines on the conservation of biological diversity in tropical production forests', para 1, pp3–4.
60. 'Presentation of the International Environmental NGOs to the 10th session of the ITTC'.
61. Survival International internal document, 'Report on the XIth meeting of the ITTC, Yokohama, 28 November–4 December 1991'. (A copy of the NGO draft resolution is appended to this document.)
62. ITTO (December 1990), op cit, principles 35 and 36, possible action 34, pp9–10.
63. Friends of the Earth and World Rainforest Movement, op cit, p37.
64. ITTO document PCF(VIII)/20, 'Draft Report to the International Tropical Timber Council, Eighth Session of the Permanent Committee on Reforestation and Forest Management', 6 June 1991, Appendix III, 'Memorandum of Understanding, Project: PD 34/88, Chimanes, Bolivia', pp37–8.
65. 'Position of the indigenous organizations represented by COICA at the tenth session of the International Tropical Timber Council', Quito, 4 June 1991, *ECO*, ITTO 10th Session, Quito, 29 May–6 June 1991, no 5, p1. (The COICA, based in Lima, Peru, unites 300 Amazonian peoples in 80 local federations centralized in five national bodies. *ECO* is a newspaper produced by NGO observers at ITTC sessions and other intergovernmental meetings.)
66. ITTO Mission, 'The Promotion of Sustainable Forest Management: A Case Study in Sarawak, Malaysia', April 1990, copy held by the Royal Geographical Society library, London. Also issued as ITTO document ITTC(VIII)/7, 7 May 1990.
67. A 60 per cent reduction in the cut would have been more appropriate according to Chris Elliott of WWF–International, cited in Fred Pearce, 'Sarawak Study Disappoints Rainforest Campaigners', *New Scientist*, 2 June 1990, p25.
68. Colchester, op cit, p171.
69. 'Statement by the Chief Minister of Sarawak, Malaysia, Rt Hon Datuk Patinggi Tan Sri Haji Abdul Mahmud at the Opening Ceremony of the 13th Session of the International Tropical Timber Council, Yokohama, 16–21 November 1992', p1.
70. ITTO document PCM, PCF, PCI(V)/1, 'Pre-Project Proposal, Labelling Systems for the Promotion of Sustainably-Produced Tropical Timber', 15 August 1989.
71. Chris C Park, *Tropical Rainforests* (London: Routledge, 1992), p147.
72. ITTO document PCM(V)/D.1, 'Report to the International Tropical Timber Council, Fifth Session of the Permanent Committee on Economic Information and Market Intelligence', 3 November 1989, p6.
73. ITTO document PCM,PCF,PCI(V)/1/Rev.2, 'Pre-Project Proposal, Incentives in Producer and Consumer Countries to Promote Sustainable Development of Tropical Forests', 6 November 1989.

74. Simon Counsell, Friends of the Earth, personal communication (interview), London, 16 February 1993.
75. ITTO document PCM,PCF,PCI(V)/1/Rev.3, 'Pre-Project Report on Incentives in Producer and Consumer Countries to Promote Sustainable Development of Tropical Forests', Oxford Forestry Institute and Timber Research and Development Association, Oxford, February 1991.
76. ITTO document ITTC(X)/12, 'Interim Report of the Expert Panel, Round Table on "The Agenda for Trade in Tropical Timber from Sustainably Managed Forests by the Year 2000", Report by the Chairman of the Expert Panel', 3 June 1991, para 8, p5.
77. ITTO document ITTC(X)/16, 'Decision 3(X), Sustainable Tropical Forest Management . . .' op cit, p1.
78. London Environmental Economics Centre, 'Draft Final Report: ITTO Activity PCM(IX)/4, The Economic Linkages between the International Trade in Tropical Timber and the Sustainable Management of Tropical Forests, Main Report, October 16, 1992'.
79. 'International Tropical Timber Council, XIII Session, 16–21 November 1992, Yokohama, Statement by Austria', p1.
80. WWF–International press release, 'Timber Labelling Scheme Seems Inevitable', 20 November 1992.
81. Anon, 'ASEAN criticizes Austrian labelling on tropical timber', *GATT Focus*, no 95, November–December 1992, p4. Other GATT contracting parties shared some of the ASEAN concerns, namely Argentina, Australia, Bolivia, Brazil, Canada, Chile, Colombia, Côte d'Ivoire, Hong Kong, India, Japan, South Korea, Mexico, Pakistan and Peru.
82. Ian Traynor, 'Another part of the forest', *Guardian*, 16 April 1993, p19; and Brian Chase, 'Tropical forests and trade policy: the legality of unilateral attempts to promote sustainable development under the GATT', *Third World Quarterly*, vol 14, no 4, 1993, pp760–3.
83. World Wide Fund for Nature, *ITTO: Tropical Forest Conservation and the International Tropical Timber Organization, Position Paper 1* (Gland, Switzerland: WWF–International, June 1988), p10.
84. Jean-Paul Jeanrenaud and Francis Sullivan, *Timber Certification and the Forest Stewardship Council: A WWF Perspective* (Godalming: WWF–UK, January 1994); and Peter Knight, 'Timber watchdog ready to bark', *Financial Times*, 6 October 1993, p18.
85. London Environmental Economics Centre, op cit, p v. See also Alistair Sarre, 'What is Timber Certification?', *Tropical Forest Update*, vol 3, no 4, August 1993, pp2–3. (*Tropical Forest Update*, formerly called *Tropical Forest Management Update* is an International Tropical Timber Organization newsletter.)
86. World Wide Fund for Nature, *Truth or Trickery?: Timber Labelling Past and Future* (Godalming: WWF–UK, March 1994), esp pp12–13; Debra Callister, 'Who Certifies the Certifiers?', *ECO*, ITTO 14th session, Kuala Lumpur, no 1, 11–19 May 1993, p1; Syed Abu Bakar and Pang Hin Yue, 'Scheme must cover all types of wood', *New Straits Times* (Malaysia), 14 May 1993, p3.
87. Anon, 'LEEC or just a load of old vegetables?', *ECO*, ITTO 13th session, Yokohama, Number 3, 20 November 1992, p2.
88. Bramble and Porter, in Hurrell and Kingsbury (eds), op cit, p345.
89. Article 1 of 'General Agreement on Tariffs and Trade, 1994', in GATT, 'The Uruguay Round, Trade Negotiations Committee: Final Act Embodying the Results of the Uruguay Round of Multilateral Trade Negotiations', Marrakesh, 15 April 1994, Annex 1A, p23.
90. GATT document DS21/R, 'United States — Restrictions on Imports of Tuna: Report of the Panel', 3 September 1991. Article XX(g) allows for exemptions from the GATT for measures 'relating to the conservation of exhaustible natural resources if such measures are made effective in conjunction with restrictions on domestic production or consumption'. See GATT, *The Text of the General Agreement on Tariffs and Trade* (Geneva: GATT, July 1986), p38.
91. Charles Arden-Clarke, *The General Agreement on Tariffs and Trade, Environmental Protection and Sustainable Development: A WWF International Discussion Paper* (Gland, Switzerland: WWF–International, November 1991), p17.
92. GATT (1986), op cit: Article II, pp3–5; Article III, pp6–7; Article XI, pp17–18.

93. David Baldock and Jonathan Hewett, *European Community Policy and Tropical Forests*, *WWF Discussion Paper* (Gland, Switzerland: WWF–International, October 1991), p8; Charles Arden-Clarke, *Conservation and Sustainable Management of Tropical Forests: The role of ITTO and GATT, WWF Discussion Paper* (Gland, Switzerland: WWF–International, November 1990), pp1, 8; and Chase, op cit, esp pp764–6.

94. Luis V Ople, information officer, GATT, personal communication (letter), 29 June 1992.

95. Chase, op cit, p764.

96. Arden-Clarke (November 1990), op cit, pp1, 11; Debra J Callister, *Illegal Tropical Timber Trade: Asia-Pacific, A TRAFFIC Network Report* (Cambridge: TRAFFIC International, 1992), p76; IUCN/UNEP/WWF, *Caring for the Earth: A Strategy for Sustainable Living* (Gland, Switzerland: IUCN/UNEP/WWF, 1991), p135.

97. For an evaluation of the Papua New Guinea NFAP, see James Mayers and Basil Peutalo, *NGOs in the Forest: Participation of NGOs in National Forestry Action Programmes: New Experiences in Papua New Guinea* (London: IIED, 1995).

98. ITTO document ITTC(X)/20, op cit, para 38, p8.

99. ITTO document PCI(VIII)/11 Rev.1, 'Report to the International Tropical Timber Council, Eighth Session of the Permanent Committee on Forest Industry', para 8, p3.

100. 'NGO Statement to the ITTO Council, November 16, 1988', section III, para F, p4.

101. 'Summary Report on the Fifteenth Meeting of the TFAP Forestry Advisers Group on Harmonizing International Forestry Development Cooperation, 30 November–4 December 1992, San José, Costa Rica', section 9, p8.

102. Caroline Sargent and Elaine Morrison, 'Towards Effective Cooperation between ITTO's Target 2000 Activities and TFAP', Executive Summary, Draft study', London, IIED, 1993.

103. Brian Johnson, *Expansion or Eclipse for ITTO?: A Look at the First Five Years of the International Tropical Timber Organization and its Potential, A WWF Discussion Paper* (Gland, Switzerland: WWF–International, May 1991), p6.

104. Elliott (November 1990), op cit, p4.

105. Anlage (Annex) 13, 'Conclusions of the German–Japanese Expert Meeting on Tropical Forests, September 16th and 17th, 1992', in Bundesministerium für Ernährung, Land-wirtschaft und Forsten, *Schutz und Bewirtschaftung der Tropenwälder: Tropenwäldbericht der Bundesregierung, 3 Bericht (Bericht–Nr B 245/93)* (Bonn: March 1993).

106. Government of the Netherlands, op cit, p62.

107. Dennis Thompson, 'Tropical Timber pact faces conservation call', *Financial Times*, 17 November 1987.

108. ITTO document PPR 23/91 (M), 'Pre-Project Study on the Conservation Status of Tropical Timbers in Trade', Report prepared by the World Conservation Monitoring Centre, February 1991, vol II, p303.

109. Conclusions of the workshop are cited in, Fred Pearce, 'Tropical Countries Veto Rainforest Protection Scheme', *New Scientist*, 3 April 1993, p11. Excerpts from this article are reproduced in *TRAFFIC Bulletin*, vol 14, no 1, 1993, p32.

110. At the meeting of the parties to CITES in Kyoto, March 1992, one species was placed on Appendix I listing, namely Brazilian Rosewood (*Dalbergia nigra*) while three were placed on Appendix II, namely Afrormosia (*Pericopsis elata*), Central American mahogany (*Swietenia mahogoni*) and Guaiacus officinale (*Lignum vitae*).

111. Robert Thomson, 'Politics holds sway over ecology at wildlife talks', *Financial Times*, 12 March 1992, p4.

112. ITTO document ITTC(XII)/17, 'Decision 6(XII), Actions to Improve Cooperation between the ITTO and CITES', 14 May 1992.

113. Ibid, and Debbie Callister of TRAFFIC Oceania, internal document, 'Report of the Twelfth Session of the International Tropical Timber Council and Tenth Sessions of the Permanent Committees, Yaounde, Cameroon, 6–14 May 1992', p3.

114. 'UK Tropical Forest Forum, Ninth Large Meeting, 25 July 1995, Royal Botanic Gardens, Kew: Record of Meeting', p26.

NOTES TO CHAPTER 4

1. This chapter is based on articles that were first published as 'The Forests Debate of the UNCED Process', *Paradigms: The Kent Journal of International Relations*, vol 7, no 1, Summer 1993, pp43–54; and 'The UNCED Process and Global Forest Conservation', in Frank R Pfetsch (ed) *International Relations and Pan-Europe: Theoretical Approaches and Empirical Findings* (Münster: LIT Verlag, 1993), pp589–608.
2. United Nations, *General Assembly, Official Records, Supplement No 49(A/44/49), Resolutions and Decisions Adopted by the General Assembly during its Forty-Fourth Session* (New York: United Nations, 1990), para 12 (d) of UNGA Resolution 44/228, p153.
3. The São Paulo Declaration of the IPCC–AFOS workshop, 9–11 January 1990, is reproduced in World Meteorological Organization/United Nations Environment Programme, *Climate Change: Intergovernmental Panel on Climate Change, The IPCC Response Strategies*, 1990, pp83–4.
4. The independent review noted that 'Article XIV of the FAO Constitution could, we understand, provide a legal basis for the proposed International Convention on Forests': Ola Ullsten et al, *Tropical Forestry Action Plan: Report of the Independent Review* (Kuala Lumpur: FAO, May 1990), p48.
5. Robert Winterbottom, *Taking Stock: The Tropical Forestry Action Plan After Five Years* (Washington DC: World Resources Institute, June 1990), p30.
6. EC document SN 60/90, 'Presidency Conclusions, European Council, Dublin, 25 and 26 June 1990', Annex II, 'The Environmental Imperative Declaration by the European Council', p8.
7. 'The Houston Declaration of the Group of Seven Industrialized Countries, July 1990', Article 67.
8. Barbara J Bramble and Gareth Porter, 'Non-Governmental Organizations and the Making of US International Environmental Policy', in Andrew Hurrell and Benedict Kingsbury (eds) *The International Politics of the Environment: Actors, Interests, and Institutions* (Oxford: Clarendon Press, 1992), p313.
9. Fact Sheet, The White House, Office of the Press Secretary (Houston, Texas), 'Proposed Global Forests Convention', July 11, 1990.
10. *Official Journal of the European Communities*, no C295, vol 33, 26 November 1990, 'Resolution on measures to protect the ecology of tropical forests', 25 October 1990, para 8, p196.
11. 'Conference Statement', in J Jäger and H L Ferguson (eds) *Climate Change: Science, Impacts and Policy, Proceedings of the Second World Climate Conference* (Cambridge: Cambridge University Press, 1991), p502.
12. 'Statement of Environmental Non-Governmental Organizations', in ibid, p550. Most of the NGOs were from the developed North, but notable exceptions include the Forum of Brazilian NGOs to the UNCED–92 and the Kenya Energy and Environmental Organization.
13. The ministerial declaration (ibid, p538) confined itself to noting that 'the protection and management of . . . forest eco-systems must be well-coordinated and preferably compatible with other possible types of action related to the reduction of emission of greenhouse gases'.
14. IUCN, *Resolutions and Recommendations: 18th Session of the General Assembly of IUCN — The World Conservation Union, Perth, Australia, 28 November–5 December 1990* (Gland, Switzerland: IUCN, 1991), Resolution 18.30, 'Legal Instrument for the Conservation of Forests', pp26–7.
15. Koy Thomson, *United Nations Conference on Environment and Development (UNCED): A User's Guide, No 1* (London: IIED, March 1991), p7.
16. Jeff Sayer, 'International initiatives for forest conservation', *IUCN Forest Conservation Newsletter*, no 10, June 1991, p1.

17. Chris Elliott, WWF–International, personal communication (telephone conversation), 20 July 1992.
18. FAO document (unnumbered), 'Report: Committee on Forestry, Tenth Session, Rome, 24–28 September 1990', pp3–4.
19. FAO, 'Possible Main Elements of an Instrument (Convention, Agreement, Protocol, Charter, etc) for the Conservation and Development of the World's Forests', draft, Rome, 18.10.90 (unpublished), Section V, 'Main Obligations of the Parties', Article V.8, p12.
20. Ibid, Article I, p3.
21. The three principles previously appeared in a document circulated at the tenth session of the COFO. See FAO document COFO–90/3(a), 'Item 4 of the Provisional Agenda, Proposal for an International Convention on Conservation and Development of Forests', September 1990, para 34, p7. However, the significance of the FAO draft is that it linked these concepts, making them the cornerstone of a proposed global bargain around which forest conservation could be based.
22. FAO draft, op cit, Article V.2 (Formulation of National Forest Policy), p7; and International Tropical Timber Organization, *ITTO Technical Series 5: ITTO Guidelines for the Sustainable Management of Natural Tropical Forests* (Yokohama: ITTO, December 1990), Possible Action 2 (Forest Policy), p2.
23. Article V.3 (National Forest Policy) of the FAO draft is a near-complete reproduction of Principle 10 (National Forest Service) of the ITTO Guidelines, and parts of the FAO's Article V.4 (Monitoring of Forest Resources: National Forest Inventory) repeat Principle 4 (National Forest Inventory) of the ITTO Guidelines.
24. 'Statement by Dr E Saouma, Director-General FAO' at the Second World Climate Conference, 6 November 1990, in Jäger and Ferguson (eds), op cit, p532. See also, Edouard Saouma, 'SWCC: An International Convention on Forest Conservation and Development', *GENEVA NEWS and International Report*, November 1990, pp23–4.
25. H W O Röbbel, assistant to the assistant director-general, Forestry Department of FAO, personal communication (letter), 9 July 1992.
26. The decision to establish the G15 was announced at the conclusion of the Ninth Non-Aligned Summit Meeting, Belgrade, September 1989. The original 15 members were Algeria, Argentina, Brazil, Egypt, India, Indonesia, Jamaica, Malaysia, Mexico, Nigeria, Peru, Senegal, Venezuela, Yugoslavia and Zimbabwe. At the Third Summit Meeting in Dakar, Senegal in November 1992 Chile was admitted as a G15 member, increasing the membership to 16.
27. Thomson, op cit.
28. 'Canadian Statement on Deforestation and a Proposal for a Global Forest Convention' to Working Group I, Nairobi, 16 August 1990 (corrected version).
29. UN document A/CONF.151/PC/27, 'Conservation and Development of Forests: Progress Report by the Secretary-General of the Conference', 5 February 1991.
30. European Community, 'Commission Staff Paper on the Drafting of an International Instrument for the Protection of Forests, For the second meeting of the UNCED PrepCom (Geneva 18 March–4 April 1991)', p4.
31. *Note on Nomenclature:* The literature and documentation on the UNCED forests debate is not consistent on the nomenclature of groups established by the UNCED secretariat and Working Group I. Note that there were two groups only. (a) The expert group referred to in this paper as the *UNCED Working Party on Forests* is referred to elsewhere as the Ad-Hoc Working Group on Forests, the UNCED Inter-Agency Working Group on Forests and the UNCED Working Panel on Forests. This group met three times: 17–18 December 1990 (Geneva); 16–17 April 1991 (Geneva); and just before the 10th World Forestry Congress, 13–14 September 1991 (Paris). (b) The group established by Working Group I during PrepCom 2 was originally referred to as the *Ad Hoc Subgroup on Forests*. Confusingly, when the group was re-established at PrepComs 3 and 4, and at the UNCED in Rio, it was called the *Contact Group on Forests*.
32. 'Terms of Reference, Ad Hoc Subgroup on Forests as approved by Working Group I on Friday, 22 March [1991], 5 p m', printed in: United Nations, *General Assembly Official Records: Forty-Sixth Session, Supplement No 48(A/46/48), Report of the Preparatory*

Committee for the United Nations Conference on Environment and Development, Volume 1 (New York: United Nations, 1991), p55.

33. The concept of 'opportunity cost foregone' appeared in FAO documents prior to PrepCom 2. See FAO document COFO–90/3(a), op cit, para 34, p7, and FAO draft, op cit, Article I.2.b, p3 and Article VI.2, p19. The concept had presumably been introduced to the FAO by governments of the South; in the UNCED forests debate the concept was frequently raised in discussions by the G77.

34. Maria Elena Hurtado, 'Malaysia turns the tables on the North', *Crosscurrents*, PrepCom 2, no 3, 28 March–3 April, 1991, p6. (*Crosscurrents* was an independent NGO newspaper produced during the UNCED PrepComs.)

35. 'US Statement on Forests' to PrepCom 2, Geneva, 25 March 1991.

36. 'E-Mail briefing note on PrepCom II of UNCED', igc:lgoree, en.unced.gener, 5.08 am, 26 March 1991, signed, 'Kimo'. ('Kimo' is the pseudonym of Langston James Goree VI, a journalist who works within the NGO community at major UN conferences. He is now the managing director of the *Earth Negotiations Bulletin*.)

37. UN document A/CONF.151/PC/WG.I/Misc.3, 'Ad Hoc Subgroup, Forests, Draft Synoptic List: Compiled by the Secretariat on the Basis of Informal Consultations', 26 March 1991, p4. These proposals included a freestanding instrument, existing mechanisms such as ITTO, an International Forests Charter, a GFC, and protocols under a Climate Change and/or a Biodiversity Convention.

38. 'E-Mail briefing note on PrepCom II of UNCED', op cit.

39. UN document A/CONF.151/PC/WG.I/L.18/Rev.1, 'Revised decision submitted by the Chairman on the basis of informal consultations', para 5, p3.

40. UN document A/CONF.151/PC/65, 'Guiding Principles for a Consensus on Forests', principle 3, p3. See also FAO draft, op cit, Article I.2.b, p3.

41. UN document A/CONF.151/PC/65, ibid, principle 4, p3. See also FAO draft, ibid, Article I.2.c, p3.

42. UN document A/CONF.151/PC/85, 'Beijing Ministerial Declaration on Environment and Development (adopted on 19 June 1991)', 13 August 1991, para 6, p3.

43. UN document A/CONF.151/PC/WG.I/L.20, 'Proposal on Forest Principles submitted by the United States of America', 14 August 1991, p2.

44. UN document A/CONF.151/PC/WG.I/L.24, 'Proposal submitted by Canada: Guiding principles towards a global consensus for the conservation and sustainable development of all types of forests worldwide', 21 August 1991, p3.

45. The Technical Workshop to Explore Options for Global Forestry Management was held in Bangkok from 24–30 April 1991 following an offer made by Thailand at the Second World Climate Conference. The workshop was a hybrid forum; it was not an inter-governmental forum, nor was it a strictly a technical one. Representatives varied from ambassadors, ministers, civil servants and forestry managers to NGOs and independent consultants.

46. 'Intervention Notes: H E Ting Wen Lian', in David Howlett and Caroline Sargent (eds) *Technical Workshop to Explore Options for Global Forestry Management, Bangkok, 1991, Proceedings* (London: IIED, 1991), p246.

47. PrepCom 3, Malaysia, 'Intervention Note on Agenda Item 3(a), Land Resources: Deforestation', p3.

48. Edward Kufuor, quoted by Pratap Chatterjee, 'G77 rejects "Stewardship"', *Crosscurrents*, PrepCom 3, no 8, 28–29 August 1991, p4.

49. Daniel Nelson, 'Malaysia "worried about development" in UNCED', *Crosscurrents*, PrepCom 3, no 5, 21–22 August 1991, p1.

50. UN document A/CONF.151/PC/85, op cit, para 7, p8.

51. PrepCom 3, Statement made by India, 'Forestry' (undated), p4.

52. Prepcom 3, Malaysia, 'Intervention Note. . .', op cit, Section (ii), 'Financial Cooperation', pp6–7.

53. UN document A/CONF.151/PC/WG.I/L.22, 'A non-legally binding authoritative statement of principles for a global consensus on the management, conservation and development of all types of forest', Proposal submitted by Ghana (on behalf of the Group of 77), 16 August 1991, principles 8 and 10.

54. 'Statement by the Representatives of Ghana on behalf of the Group of 77 in Working Group I on item 3(a): Land Resources: Deforestation, Geneva, 15 August, 1991', p2. The reference to 'common but differentiated responsibilities' in this statement is interesting; the principle was later written into the Convention on Climate Change. See Chapter 7.
55. UNCED secretary-general Maurice Strong notes the centrality of the financial and technology transfer issues, in 'ECO '92: Critical Challenges and Global Solutions', *Journal of International Affairs*, vol 44, no 2, Winter 1991, p291.
56. UN document A/CONF.151/PC/85, op cit, pp6–7. See also UN document A/CONF.151/PC/86, 'Proposal submitted by the Delegation of the People's Republic of China, The Green Fund', 15 August 1991, for full details of the proposal submitted at PrepCom 3.
57. As noted in UN document A/CONF.151/PC/86, ibid, para 10, p3.
58. The Global Environment Facility (GEF) was established in November 1990 and is jointly administered by the World Bank, UNDP and UNEP. The GEF was established to deal with four global environmental problems: global warming; pollution of international waterways; destruction of biological diversity; and ozone depletion. During the UNCED process, the North advocated the GEF as the mechanism for multilateral environmental funding.
59. PrepCom 3, Malaysia, 'Intervention Note. . .', op cit, p4; 'Statement by the Representative of Ghana. . .', op cit, p2; PrepCom 3, Statement made by India, op cit, p4.
60. UN document A/CONF.151/PC/WG.I/L.22, op cit, principles 12 and 13, p3.
61. PrepCom 3, Statement made by India, op cit, p3.
62. UN document A/CONF.151/PC/L.41, 'China and Ghana (on behalf of the Group of 77): draft decision, Financial resources', p2.
63. Jennifer White, 'US pushes green conditionality', *Crosscurrents*, PrepCom 3, no 9, 30–31 August 1991, p8.
64. UN document A/CONF.151/PC/WG.I/CRP.14.Rev.1, 'A non-legally binding authoritative statement of principles for a global consensus on the management, conservation and sustainable development of all types of forests', 3 September 1991.
65. 'Statement by the Representatives of the Netherlands on behalf of the European Community and its member states, Working Group I, Agenda Item 3a: Conservation and Development of Forests (Geneva, 14 August 1991)'. Among the EC governments, Germany was one of the strongest advocates for a GFC, and in a statement at PrepCom 3 argued that 'A convention should be ready for signature within a period of two years after UNCED '92': 'Statement by the German Delegation: 3rd Session of the UNCED PrepCom, Geneva, 12 August–4 September 1991, WG I, Agenda item 3 (a)', para 7.
66. UN document A/CONF.151/PC/120, Annex I, 'African Common Position on Environment and Development', adopted at the Second Regional African Ministerial Preparatory Conference for the UNCED, Abidjan, 11–14 November 1991, para 20, p10.
67. 'Address given by the President of Venezuela, Mr Carlos Andres Perez, at the opening ceremony of the Group of 15 summit meeting', Caracas, 27–29 November 1991, p4. (The author is grateful to Peter Willetts for drawing his attention to this document.)
68. 'Joint Communiqué, Second Meeting of the Summit Level Group for South–South Consultation and Cooperation (Group of 15)', Caracas, 27–29 November 1991, para 35 'Role of technology', p8.
69. *Earth Summit Bulletin*, 31 March 1992, vol 1, no 22, pp1–2. The *Earth Summit Bulletin* was an NGO newspaper produced at UNCED PrepCom 4 and at the UNCED in Rio by Langston James Goree VI. Since Rio, Goree has produced the *Earth Negotiations Bulletin* at selected UN conferences and other intergovernmental fora. On the subject of financial resources note also that the draft decision submitted by China and the G77 at PrepCom 3 on financial resources was reissued in a revised form: UN document A/CONF.151/PC/L.41/Rev.1, 'China and Pakistan (on behalf of the Group of 77): revised draft decision, Financial resources and mechanisms', 10 March 1992.
70. UN document A/CONF.151/6, 'Non-legally binding authoritative statement of principles for a global consensus on the management, conservation and sustainable development of all types of forests', 21 April 1991, principles 7(b), 9(a), 10, 11 and 12.
71. Press notice 1/92, High Commissioner for Malaysia, London, 'Speech by the Prime Minister of Malaysia, Dato' Seri Dr Mahathir Mohamad at the official opening of the

Second Ministerial Conference of Developing Countries on Environment and Development, Kuala Lumpur on Monday 27 April 1992', paras 7, 9 and 11.

72. Press notice 2/92, High Commissioner for Malaysia, London, 'The Kuala Lumpur Declaration on Environment and Development issued at the end of the Second Ministerial Conference of Developing Countries on Environment and Development on 29 April 1992', paras 4, 10, 11, 12, 14 and 15.

73. This account of the UNCED organization is drawn in part from the *Earth Summit Bulletin*, vol 2, no 13, 16 June 1992, and from Stanley Johnson, 'Introduction: Did We Really Save the Earth at Rio?', in Stanley Johnson (ed) *The Earth Summit: The United Nations Conference on Environment and Development* (London: Graham and Trotman, 1993), pp3–16.

74. US chief negotiator Curtis Bohlen, cited by John Vidal and Paul Brown, 'Deadlock in Talks about Aid Cash', *Guardian*, 10 June 1992, p8.

75. *Earth Summit Bulletin*, vol 2, no 13, 16 June 1992, pp7–9.

76. UN document A/CONF.151/6/Rev.1, 'A non-legally binding authoritative statement of principles for a global consensus on the management, conservation and sustainable development of all types of forests' (final version), Preamble (d), p1.

77. Ibid, principle 1(a), p2.

NOTES TO CHAPTER 5

1. The information in this section has been collated from the following sources: (a) *ITTO documents*: ITTO document PrepCom(I)/3, 'Informal Working Group on ITTA Renegotiation, Washington DC, 24–25 September 1992 (Report by the Chairman)', 14 October 1992 (this document became known within ITTO circles as the 'Issues and Options' paper); ITTO document PrepCom(I)/5, 'Report of the First Session of the Preparatory Committee for the Renegotiation of the ITTA, 1983', 5 January 1993: ITTO document PrepCom(II)/2, 'Chairman's Paper for Second Session of the Preparatory Committee', 5 January 1993. (b) *UN document*: TD/TIMBER.2/R.2, 'Note by the Secretariat', 26 February 1993. (c) *Unpublished statements or position papers by ITTO producers*: 'Draft of Producer Countries: Preamble', 13 November 1992; 'Statement by the producer spokesman on the key issues relating to the renegotiation of the International Tropical Timber Agreement (ITTA), 1983', Quito, Ecuador, 27 January 1993. (d) *Unpublished position paper by ITTO consumers*: 'Consumers: Preamble and Article 1', 13 November 1992. (e) *Unpublished statements or position papers by environmental NGOs*: 'First session of the Preparatory Committee for the renegotiation of the ITTA: Opening statement by WWF', Yokohama, 11 November 1992; 'Environmental NGOs' statement to the 13th Session of the ITTC, Yokohama, 16th November 1992'; 'NGO "Non-paper": Issues and Options for the renegotiation of ITTA relating to the incorporation of timber from all forest types', 22 November 1992; 'NGOs' statement to consumers on inclusion of temperate timber in the new Agreement' (undated); 'Renegotiation of the ITTA: A WWF–UK Briefing Paper', undated. (f) *Interviews with NGO campaigners*: Simon Counsell, Friends of the Earth, personal communication (telephone conversation), London, 20 October 1993; Francis Sullivan, World Wide Fund for Nature, personal communication (interview), Godalming, 15 March 1994. (g) *Published sources*: Debra Callister, *Renegotiation of the International Tropical Timber Agreement: Issues Paper, A Joint TRAFFIC/WWF Paper* (Cambridge: TRAFFIC/WWF–International, October 1992); Brian Johnson, 'Crisis in Yokohama', *Timber Trades Journal*, 28 November 1992, pp6–7.

2. Letter from George J Medley, Director, WWF–UK, Godalming to Baroness Chalker, Minister of State, Overseas Development Administration, London, 3 November 1992.

3. 'UK Tropical Forest Forum, ITTO Working Group, Minutes of Meeting Held on 27 October 1992 at the WWF Offices, Beauchamp Place, 2–5 pm', para 18, p6.

4. Zulkifli Talib, 'Indons to back timber proposal', *Star* (Malaysia), 4 December 1992.

5. WWF–International press release, 'Quito: The end of the road for ITTO?', 24 November 1992.

6. Arthur Morrell of the UK Timber Trade Federation speaking at the first PrepCom, ITTO document PrepCom(I)/5, op cit, para 42, p13.
7. 'Thirteenth Session of the ITTO Council and Eleventh Session of its Permanent Committees, Yokohama, Japan, 16–21 November 1992, Statement by the FAO Representative'.
8. In a comment that demonstrates the importance attached to the producers/NGO alliance, Simon Counsell of Friends of the Earth argued that by supporting mention of Target 2000 in the new agreement, NGOs seriously risked alienating the producers, and thus losing their support for their other lobbying objectives; Simon Counsell, Friends of the Earth, personal communication (telephone conversation), London, 20 October 1993. Despite the opposition of Friends of the Earth, most members of the NGO community favoured inclusion of Target 2000 in the new agreement.
9. Anon, 'ITTO must act on illegal timber trade', *ECO*, ITTO 13th session, Yokohama, 23–24 November 1992, no 4, pp1–2; 'Environmental NGO statement to the 13th Session of the ITTC...', op cit; WWF–International press release, 'ITTO ignoring illegal timber trade', 12 November 1992.
10. Letter from Takashi Kosugi (vice-president, GLOBE International), Wakako Hironako (vice-president, GLOBE–Japan), Akiko Domoto (vice-president, GLOBE–Japan) and Hemmo Muntingh (president, GLOBE–EC and vice-president, GLOBE International) to the executive director, ITTO, dated 2 December 1991, sent to the author by Lena Lindahl, GLOBE–Japan, personal communication (fax), 4 August 1992.
11. Nigel Dudley, *Forests in Trouble: A Review of the Status of Temperate Forests Worldwide* (Gland, Switzerland: WWF–International, October 1992).
12. Anon, 'Timber body must be expanded, Says Keng Yaik', *New Straits Times* (Malaysia), 11 November 1992, and Fadzil Ghazali, 'Lim on managing temperate forests: Broader Itto operations sought', *Business Times* (Malaysia), 11 November 1992.
13. Talib, op cit.
14. In reply to a written question (No 2615/92) from Hemmo Muntingh MEP the European Commission stated that Target 2000 'should be among the major goals of the next agreement'. See, *Official Journal of the European Communities*, no C 86, 26 March 1993, pp28–9. Muntingh's question was tabled on 27 October 1992; the reply of the commission was dated 10 December 1992.
15. UN document TD/TIMBER.2/1/Add.1, 'Provisional Agenda and Annotations', 24 February 1993, and ITTO document ITTC(XIV)/6, 'Report of the First Part of the United Nations Conference for the Negotiation of a Successor Agreement to the International Tropical Timber Agreement, 1983, 13–16 April 1993, Geneva, Switzerland', 22 April 1993, para 9, p4 and paras 13–14, p5.
16. See UN document TD(VIII)/Misc.4, 'A New Partnership for Development: The Cartagena Commitment', 27 February 1992. The Cartagena Commitment contained several references favourable to the producers' negotiating position including: 'improved market access for developing country exports in conjunction with sound environmental policies, would have a positive economic impact' (para 151, p46); 'Environmental policies should deal with the root causes of environmental degradation, thus preventing environmental measures from resulting in unnecessary restrictions to trade' (para 152, p46); and 'Donor countries are requested to continue contributing extrabudgetary resources to further strengthen UNCTAD's work on interlinkages between environment, trade and sustainable development' (para 155, p48).
17. Polly Ghazi, 'Global campaign to save rainforests in danger of collapse', *Observer*, 11 April 1993, p11.
18. ITTO document ITTC(XIV)/6, op cit, paras 17, 19, 20, 21, 23, 24, 26 and 27, pp7–8.
19. WWF–International press release, 'ITTA renegotiations broaden forest debate', 13 April 1993; and Anon (Reuters), 'Canada urged to join tropical-timber accord', *Toronto Globe and Mail*, 17 April 1993.
20. 'Chairman's Non-Paper, 'Informal consultations regarding a successor agreement to the International Tropical Timber Agreement, 1983, during XIV Session of ITTC', para 1. This document was subsequently circulated at the second part of the conference as UN

document TD/TIMBER.2/CRP.1, 'Note by the UNCTAD Secretariat: Chairman's Non-Paper', 2 June 1993.

21. For the composite text see, UN document TD/TIMBER.2/R.3, 'Note by the Secretariat', April 1993.

22. WWF–Malaysia press release, 'WWF–Malaysia defends stand on Sarawak: Reacts to personal attack on Executive Director', Kuala Lumpur, 13 May 1993.

23. This paper was attached as Annex I to UN document TD/TIMBER.2/CRP.1, op cit.

24. 'Opening Statement Presented by the Environmental NGOs at the Fourteenth Session of the International Tropical Timber Council, Kuala Lumpur, 15 May 1993'; and Anon, 'NGOs hit out at double standards of consumers', *New Sunday Times* (Malaysia), 16 May 1993.

25. Eldon Ross, head of the US delegation, cited by Debora MacKenzie, 'Timber: the beam in Europe's eye', *New Scientist*, 26 June 1993, p9.

26. See Al Gore, *Earth in the Balance: Ecology and the Human Spirit* (New York: Plume, 1993). Although this volume deals briefly with tropical rainforest destruction (pp116–20), it contains no mention of the ITTO.

27. 'Broadening the Scope of the International Tropical Timber Agreement to Encompass Timbers From All Types of Forests: A Rationale for US Support, A discussion paper prepared by the Global Forest Policy Project of the National Wildlife Federation, Sierra Club and Friends of the Earth, March 11, 1993', p9.

28. Letter from Kathryn S Fuller, president, WWF–US, to Ms Eileen Claussen, special assistant to the president of the United States of America and senior director for Global Environmental Affairs, dated April 12, 1992.

29. Letter from Al Gore, Office of the Vice President, Washington DC, to The Honourable Jimmy Carter et al, Fernback Museum of Natural History, Atlanta, Georgia, dated 7 May 1993.

30. Anon, 'US Rethinking on Forests: Will it Help a Global Timber Agreement?', *ECO*, ITTO 14th session, Kuala Lumpur, 11–19 May 1993, no 1, p2.

31. Francis Sullivan, World Wide Fund for Nature, personal communication (interview), Godalming, 15 March 1994, and 'NGO Statement, 24 June 1993'.

32. 'Negotiation of a Successor Agreement to the International Tropical Timber Agreement, 1983, President's Discussion Paper, 25 June 1993, 5 pm' (no document number).

33. UN document TD/TIMBER.2/CRP.2, 'Note by the UNCED Secretariat: President's Revised Discussion Paper', 24 September 1993, p1.

34. 'Comments on President's 25 June 1993 Discussion Paper regarding the Renegotiation of the ITTA', 4 August 1993, submitted by Christopher Elliott (WWF–International) and William E Mankin (coordinator of the Global Forest Policy Project of National Wildlife Federation, Sierra Club and Friends of the Earth–USA).

35. UN document TD/TIMBER.2/CRP.2, op cit, pp5, 9.

36. 'Speech by the Prime Minister Y A B Dato Seri Dr Mahathir Bin Mohamad at the Official Opening of the 14th Commonwealth Forestry Conference at the Shangri-La Hotel, Kuala Lumpur on Monday, 13th September 1993 at 10.00 am', pp4–5.

37. '14th Commonwealth Forestry Conference, Kuala Lumpur, 13–18 September 1993: Recommendations'. For an assessment of the conference see Frans Arentz, '14th Commonwealth Forestry Conference', *Tropical Forest Update*, vol 3, no 5, October 1993, p16.

38. Simon Counsell, Friends of the Earth, personal communication (telephone conversation), 20 October 1993.

39. UNCTAD press release, 'Australia announces commitment to year 2000 target for sustainable management of all its forests', 6 October 1993.

40. 'Canadian Statement, First Session of Executive Committee, ITTA INC 3, October 6, 1993'.

41. 'Draft Non-Paper (13.10.93): Statement by the Consumers'. There are certain advantages to be gained by actors who issue non-papers: they float ideas for discussion while their informal nature means that they do not express a commitment to a policy and are not formally attributable. Note that non-papers, once floated, may be issued at a later date as official UN documents; see endnote 20 above.

42. 'Third Session of Negotiations on a Successor Agreement to the International Tropical Timber Agreement, 1983, Statement on Behalf of the European Community and its Member States, Geneva, 15 October 1993'.

43. UN document TD/TIMBER.2/CRP.4, 'Proposal by the Consumers, Article 18, Objective Year 2000 Fund', 12 October 1993, para 5, p2.

44. UN document TD/TIMBER.2/CRP.3, 'Proposal by the Producers, Article 18, Financial Resources and Mechanism', 7 October 1993, para 1, p1.

45. 'Statement of the Global Forest Policy Project at the Closing of the Third Part of the Conference for the Negotiation of a Successor Agreement to the International Tropical Timber Agreement of 1983, 15 October 1993, Geneva'; 'WWF Closing Statement at the Closing of the Third Part of the Conference for the Negotiation of a Successor Agreement to the International Tropical Timber Agreement, 1983' (undated).

46. 'Statement of the Central Region of the Nuu-Chah-Nulth Tribal Council at the Opening of the Third Part of the Conference for the Negotiation of a Successor Agreement to the International Tropical Timber Agreement of 1983, October 6, 1993', p2.

47. 'Statement on behalf of the Netherlands Committee for IUCN, Geneva, 15 October 1993'.

48. UN document TD/TIMBER.2/CRP.5/Rev.1, 'Preparation of a Successor Agreement to the International Tropical Timber Agreement, 1983: Draft articles submitted by the President', 15 October 1993.

49. Following the final ratification by Germany, the Maastricht Treaty entered into legal effect on 1 November 1993. At the fourth part of the negotiating conference official UN documents referred not to the European Community, as was the case during the first to third parts, but to the European Union.

50. 'European Union, Statement on the Sustainable Management of Forests in the European Union, Geneva, 12 January 1994', para 3, p1. This was subsequently issued as UN document TD/TIMBER.2/L.7, 'Statement on the Sustainable Management of Forests in the European Union by the European Union', 24 January 1994, pp1–2.

51. Frances Williams, 'Divisions remain over tropical timber accord', *Financial Times*, 12 January 1994, p34.

52. UN document TD/TIMBER.2/L.6, 'Formal Statement by the Consumer Members', 21 January 1994, para 6, p2.

53. UN document TD/TIMBER.2/L.9, 'International Tropical Timber Agreement, 1994', 25 January 1994, Preamble, p5.

54. Ibid, Article 21, p20; and UN document TD/ TIMBER.2/CRP.4, 'Proposal by the Consumers. . .', op cit. Many important features of the consumers' proposal appear in Article 21 of the ITTA 1994 on the Bali Partnership Fund, most noticeably that the fund will be a third ITTO account solely for producers' pre-project and project work towards Objective 2000.

55. UN document TD/TIMBER.2/L.9, op cit, Article 35, p30.

56. Note that although it was due to expire on 31 March 1994, the ITTA 1983 may, by special vote of the ITTC, continue in effect until such time as the ITTA 1994 receives the requisite number of ratifications: UN document TD/TIMBER/11/Rev.1, 'International Tropical Timber Agreement, 1983', Article 42.3, p18.

57. Bernardo Zentilli, Senior Programme Advisor (Forestry), UNCED secretariat, personal communication (interview), Geneva, 24 September 1992.

58. GLOBE International, 'Model for a Convention for the Conservation and Wise Use of Forests', April 1992, 'Legal Text', Part II, and FAO, 'Possible Main Elements of an Instrument (Convention, Agreement, Protocol, Charter, etc) for the Conservation and Development of the World's Forests', draft, Rome, 18.10.90.

59. UN document TD/TIMBER.2/L.9, op cit, Article 1, para (l), p6.

60. This information has been obtained from: UN document TD/TIMBER.2/15, 'Summaries of statements made after the adoption of the Agreement at the closing plenary meeting of the Conference on 26 January 1994', 22 March 1994; Francis Sullivan, World Wide Fund for Nature, personal communication (interview), Godalming, 15 March 1994; Simon Counsell of Friends of the Earth, personal communication (telephone conversation), 15 April 1994; and a confidential briefing from an EU delegate, personal communication (telephone conversation), 15 April 1994.

61. At the second PrepCom in Quito the producers' spokesman stated that Target 2000 should be seen as 'a guiding target' and not 'a deadline or ultimatum which would be made legally-binding to members through its formal inscription in the new Agreement': 'Statement by the producer spokesman. . .', 27 January 1993, op cit, pp1–2.
62. Philip Hurst, *Rainforest Politics: Ecological Destruction in South-East Asia* (London: Zed Books, 1990), p210.
63. Paul Harrison, *The Third Revolution: Population, Environment and a Sustainable World* (London: Penguin, 1992), p97.

NOTES TO CHAPTER 6

1. 'Speech by the Prime Minister Y A B Dato Seri Dr Mahathir Bin Mohamad at the Official Opening of the 14th Commonwealth Forestry Conference at the Shangri-La Hotel, Kuala Lumpur on Monday, 13th September 1993 at 10.00 am', p3.
2. The group was originally called the Intergovernmental Working Group on Global Forests.
3. Langston James Goree VI, 'Canada and Malaysia Discuss Forests', *The Network*, May 1994, no 37, p4.
4. UN document E/CN.17/1995/26, 'Report of the Second Meeting of the Intergovernmental Working Group on Forests, held at Ottawa/Hull, Canada from 10 to 14 October 1994', 23 February 1995.
5. 'Workshop Towards Sustainable Forestry: Preparing for Commission on Sustainable Development 1995, New Delhi, July 25–27, 1994, Resolution', 1994, p1.
6. Nigel Dudley, Jean-Paul Jeanrenaud and Francis Sullivan, *Bad Harvest? The Timber Trade and the Degradation of the World's Forests* (London: Earthscan, 1995), p120.
7. UN document A/CONF.151/6/Rev.1, 'Non-legally binding authoritative statement of principles for a global consensus on the management, conservation and sustainable development of all types of forests', 13 June 1992, pp4–5.
8. United Nations, *Earth Summit Agenda 21: The United Nations' Programme of Action from Rio* (New York: United Nations, 1993), p94.
9. Not, of course, to be confused with the Helsinki process which followed the negotiation of the Helsinki Final Act at the Conference on Security and Cooperation in Europe in 1975.
10. Ministerial Conference on the Protection of Forests in Europe, *Ministerial Conference on the Protection of Forests in Europe: Documents* (Helsinki: Ministry of Agriculture and Forestry, 1993), p4.
11. Montreal Process, *Criteria and Indicators for the Conservation and Sustainable Management of Temperate and Boreal Forests* (Quebec: Canadian Forest Service, 1995), p5.
12. The four resolutions are: H1 General Guidelines for the Sustainable Management of Forests in Europe; H2 General Guidelines for the Conservation of the Biodiversity of European Forests; H3 Forestry Cooperation with Countries with Economies in Transition; and H4 Strategies for a Process of Long-term Adaptation of Forests in Europe to Climate Change; see Ministerial Conference on the Protection of Forests in Europe, op cit.
13. 'Statement of the State Secretary of Agriculture, Nature Management and Fisheries, J D Gabor, of the Netherlands, during the Second Ministerial Conference on the Protection of Forests in Europe, June 16–18, 1993 at Helsinki'.
14. Ministerial Conference on the Protection of Forests in Europe, *Interim Report on the Follow-Up of the Second Ministerial Conference on the Protection of Forests in Europe, 16–17 June 1993 in Helsinki* (Helsinki: Ministry of Agriculture and Forestry, 1995), p19.
15. Ministerial Conference on the Protection of Forests in Europe, *European List of Criteria and Most Suitable Quantitative Indicators* (Helsinki: Ministry of Agriculture and Forestry, 1994), p5.
16. *Earth Negotiations Bulletin*, 'A Review of Selected International Forest Policy Meetings', vol 13, no 1, 27 February 1995, p3.
17. J C Mercier, 'Closing Summary', in CSCE, 'Seminar of Experts on Sustainable Development of Boreal and Temperate Forests, Technical Report: Annex I', 1993, p2.

18. *Earth Negotiations Bulletin*, 'Commission on Sustainable Development: Year-End Update', vol 5, no 26, 16 December 1994, p5.
19. UN document E/CN.17/1995/34, 'Regional Workshop on the Definition of Criteria and Indicators for Sustainability of Amazonian Forests, Final Document, Tarapoto, Peru, February 25, 1995', 10 April 1995, p6. See also Montreal Process, op cit, p14.
20. UN document E/CN.17/1995/34, ibid, p8.
21. UN document E/CN.17/1995/10, 'Report of the Ad Hoc Inter-sessional Working Group on Sectoral Issues of the Commission on Sustainable Development (New York, 27 February–3 March 1995)', 31 March 1995, p7.
22. UN document E/CN.17/1995/26, op cit, p27.
23. Much of the information in this section was kindly supplied to the author by Kilaparti Ramakrishna of the Woods Hole Research Center; personal communication (e-mail), 1 September 1995.
24. WCFSD Organizing Committee, 'World Commission on Forests and Sustainable Development: Possible Mandate, Key Issues, Strategy and Work Plan, June 1993', p5.
25. Indeed at the Second Ministerial Conference on the Protection of Forests in Europe the delegation from the Netherlands explicitly stated its support for the WCFSD initiative noting that it hoped the commission 'will contribute towards achieving a Forest Convention, which we consider of the utmost importance'; 'Statement of the State Secretary of Agriculture, Nature Management and Fisheries, J D Gabor, of the Netherlands...', op cit, p3.
26. World Commission on Forests and Sustainable Development, 'Activities of the WCFSD in the Areas of Conflict Resolution, Policy Reform and Strengthening of Scientific Research: Decisions from the First Meeting of the WCFSD held in Geneva, Switzerland during 4–6 June 1995', p6.
27. Ibid, p4.
28. Chris Brown, 'Critical Theory and Postmodernism in International Relations', in A J R Groom and Margot Light (eds) *Contemporary International Relations: A Guide to Theory* (London: Pinter, 1994).
29. See, for example, Julian Saurin, 'Global Environmental Degradation, Modernity and Environmental Knowledge', *Environmental Politics*, vol 2, no 4, Winter 1993, pp46–64.
30. Tariq Banuri and Frédérique Apffel Marglin, 'A systems-of-knowledge analysis of deforestation, participation and management', in Tariq Banuri and Frédérique Apffel Marglin (eds) *Who will Save the Forests?: Knowledge, Power and Environmental Destruction* (London: Zed Books, 1993), p2.
31. *Earth Negotiations Bulletin*, 'A Review of Selected International Forest Policy Meetings...', op cit, p4.
32. *Earth Negotiations Bulletin*, 'Highlights of the 12th Session of the FAO Committee on Forestry, 13–15 March 1995', vol 13, no 2, 20 March 1995, pp1–2.
33. UN document E/CN.17/1995/26, op cit, p12.
34. Forest Stewardship Council, 'Forest Stewardship Principles and Criteria for Natural Forest Management, June 1994, Oaxaca, Mexico', p1.
35. Ibid, pp7–8.
36. Sustainable forest management 'means the stewardship and use of forests and forest lands in a way, and at a rate, that maintains their biodiversity, productivity, regenerative capacity, vitality and their potential to fulfil, now and in the future, relevant ecological, economic and social functions, at local, national and global levels, and that does not cause damage to other ecosystems'. See Ministerial Conference on the Protection of Forests in Europe (1994), op cit, p26.
37. Forest Stewardship Council, op cit, p1.
38. UN document E/CN.17/1995/28, 'Activities of the Food and Agriculture Organization of the United Nations and of the World Food Programme in sustainable development', 23 February 1995, p4.
39. Earth Negotiations Bulletin, 'Highlights of the 12th Session of the FAO Committee on Forestry...', op cit, p8.
40. FAO, 'Committee on Forestry, Twelfth Session, Rome 13–16 March 1995: Report', pp3–4.

41. FAO, 'Rome Statement on Forestry, Ministerial Meeting on Forestry, Rome, 16–17 March 1995', pp3–4.
42. Earth Negotiations Bulletin, 'Highlights of the 12th Session of the FAO Committee on Forestry. . .', op cit, p8.
43. Bundesministerium für Ernährung, Landwirtschaft und Forsten, *Schutz und Bewirtschaft der Tropenwälder: Tropenwäldbericht der Bundesregierung, 3 Bericht* [Third German Government Report on Tropical Forests] (Bonn: Bundesministerium für Ernährung, Landwirtschaft und Forsten, 1993), Anlage 13, 'Conclusions of the German–Japanese Expert-Meeting on Tropical Forests, Berlin, September 16 and 17, 1992'.
44. ITTO document PrepCom(I)/5, 'Report of the First Session of the Preparatory Committee for the Renegotiation of the ITTA, 1983 (Yokohama, Japan, November 1992)', 5 January 1993, p11; and Forestry Canada, 'Canada's Commitment to Sustainable Forestry is the Same at Home and Abroad', June 1993.
45. Ministerial Conference on the Protection of Forests in Europe (1993), op cit, pp9–10; see also Ministerial Conference on the Protection of Forests in Europe (1995), op cit, pp14–15.
46. GLOBE International, 'Action Agenda Towards a Global Regime for the Conservation and Wise Use of All Forests, Adopted unanimously by the 7th GLOBE International General Assembly, Tokyo, September 1, 1993'.
47. GLOBE International, 'Model for a Convention for the Conservation and Wise Use of Forests', April 1992, pii.
48. AIDEnvironment, 'Convention for the Conservation of Forests, Final Report, May 1992' (internal document), p5.
49. European Commission, 'Protocol on the Sustainable Management of Forest Resources' (Protocol to the Lomé IV Convention), 30 June 1995.
50. The quote is from FAO, 'Rome Statement on Forestry. . .', op cit, p3. For very similar formulations see: *Earth Negotiations Bulletin*, 'A Review of Selected International Forest Policy Meetings. . .', op cit, p4 (which provides the wording for the Antalya meeting); FAO, 'Committee on Forestry. . .', op cit, p2; and UN document E/CN.17/1995/36, 'Commission on Sustainable Development, Report on the Third Session (11–28 April 1995), Economic and Social Council Records, 1995, Supplement No 12', p52.
51. *Earth Negotiations Bulletin*, 'Highlights of the 12th Session of. . .', op cit, p2.

NOTES TO CHAPTER 7

1. Stephen D Krasner, 'Structural causes and regime consequences: regimes as intervening variables', in Stephen D Krasner (ed) *International Regimes* (Ithaca NY: Cornell University Press, 1983), p2.
2. Robert O Keohane, 'The Analysis of International Regimes: Towards a European–American Research Programme', in Volker Rittberger (ed) *Regime Theory and International Relations* (Oxford: Clarendon Press, 1993), p28.
3. Volker Rittberger, 'Research on Regimes in Germany: The Adaptive Internalization of an American Social Science Concept', in ibid, p9.
4. John Gerard Ruggie, 'International regimes, transactions and change: embedded liberalism in the postwar economic order', in Krasner (ed), op cit, pp195–231.
5. Karen Litfin, 'Eco-regimes: Playing Tug of War with the Nation-State', in Ronnie D Lipschutz and Ken Conca (eds) *The State and Social Power in Global Environmental Politics* (New York: Columbia University Press, 1993), p101.
6. Robert O Keohane, Peter M Haas and Marc A Levy, 'The Effectiveness of International Environmental Institutions', in Peter M Haas, Robert O Keohane and Marc A Levy (eds) *Institutions for the Earth: Sources of Effective International Environmental Protection* (Cambridge MA: MIT Press, 1993), p14.
7. David Humphreys, 'Hegemonic Ideology and the International Tropical Timber Organization', in John Vogler and Mark Imber (eds) *The Environment and International Relations* (London: Routledge, 1996).

8. Gareth Porter and Janet Welsh Brown, *Global Environmental Politics* (Boulder Colorado: Westview Press, 1991), Chapter 3, 'The Issues and Formation of International Regimes', pp69–105.

9. Stanley Johnson, 'Authoritative Statement of Forest Principles', in Stanley Johnson (ed) *The Earth Summit: The United Nations Conference on Environment and Development* (London: Graham and Trotman, 1993), p110.

10. Oran R Young, *International Cooperation: Building Regimes for Natural Resources and the Environment* (Ithaca NY: Cornell University Press, 1989), pp176–8.

11. Gail Osherenko and Oran R Young, 'The Formation of International Regimes: Hypotheses and Cases', in Oran R Young and Gail Osherenko (eds) *Polar Politics: Creating International Environmental Regimes* (Ithaca NY: Cornell University Press, 1993), p13.

12. Gail Osherenko and Oran R Young, *The Age of the Arctic: hot conflicts and cold realities* (Cambridge: Cambridge University Press, 1989), p264.

13. Oran R Young and Gail Osherenko, 'International Regime Formation: Findings, Research Priorities and Applications', in Young and Osherenko (eds), op cit, p238.

14. John P Renninger, 'The Failure to Launch Global Negotiations at the 11th Special Session of the General Assembly', in Johan Kaufmann (ed) *Effective Negotiation: Case Studies in Conference Diplomacy* (Dordrecht, Netherlands: Martinus Nijhoff, 1989), p249.

15. Osherenko and Young (1989), op cit, p265.

16. Andrew Hurrell, 'International Society and the Study of Regimes: A Reflective Approach', in Rittberger (ed), op cit, p58.

17. Robert O Keohane, *After Hegemony: Cooperation and Discord in the World Political Economy* (Princeton NJ: Princeton University Press, 1984), p104.

18. Mark W Zacher, 'Toward a Theory of International Regimes', *Journal of International Affairs*, vol 44, 1990, pp140–1.

19. Johnson in Johnson (ed), op cit, p110–11. Following the call for a convention on desertification at the UNCED in 1992, the International Negotiating Committee on Desertification was subsequently established by the UN General Assembly. Negotiation of the text of the Convention to Combat Desertification was finalized in Paris in June 1994.

20. Stanley Johnson, 'Rio's Forest Fiasco', *Geographical Magazine*, September 1992, p28.

21. UN document A/CONF.151/6/Rev.1, 'Non-legally binding authoritative statement of principles for a global consensus on the management, conservation and sustainable development of all types of forests', 13 June 1992, pp5–6, and UN document TD/TIMBER.1/L.9, 'International Tropical Timber Agreement, 1994', 25 January 1994, pp17–19.

22. Noted in 'New Delhi Workshop, 25–27 July 1994, Paper 1, Overview of UNCED Decisions and Commission for Sustainable Development (CSD) Requirements in Forestry, 1994', pp9–10. The Delhi meeting of September 1993 was the first ministerial-level meeting of the Forestry Forum for Developing Countries, which was originally formed to formulate and present the views of tropical country governments at FAO discussions on new institutional arrangements for the TFAP. See Chapter 2.

23. Christer Jönsson, 'Cognitive Factors in Explaining Regime Dynamics', in Rittberger (ed), op cit, p210.

24. Osherenko and Young, 'The Formation of International Regimes: Hypotheses and Cases', in Young and Osherenko (eds), op cit, p14.

25. UN document TD/TIMBER.2/R.2, 'Note by the UNCTAD Secretariat', 26 February 1993, p18.

26. 'Statement by Mr Wisber Loeis, President of the UN Conference for the Negotiation of a Successor Agreement to the ITTA, 1983, before the Fourteenth Session of the International Tropical Timber Council, Kuala Lumpur, Monday, 17 May 1993'.

27. ITTO document ITTC(XIII)/5 Add.1, 'Assessments by Producing Members of Resources Needed to Attain Sustainable Management of their Tropical Forests by the Year 2000, Indonesia', 11 November 1992, p3.

28. ITTO document ITTC(XII)/7 Rev.1, 'Report from the Expert Panel on Estimation of Resources Needed by Producer Countries to Attain Sustainability by the Year 2000', 13 April 1992, p30.

29. ITTO document ITTC(XII)/7, 'Working document for ITTO Expert Panel on "Resources needed by producer countries to achieve sustainable management by the Year 2000"', by I S Ferguson and J Muñoz-Reyes Navarro, 9 March 1992. p22.

30. Delphine Borione and Jean Ripert, 'Exercising Common but Differentiated Responsibility', in Irving M Mintzer and J A Leonard (eds) *Negotiating Climate Change: The Inside Story of the Rio Convention* (Cambridge: Cambridge University Press, 1994) pp77–96; and Sten Nilsson and David Pitt, *Protecting the Atmosphere: The Climate Change Convention and its Context* (London: Earthscan, 1994), p23.

31. As noted in 'Chairman's Summary of the High Level Segment of the third session of the Commission on Sustainable Development', 28 April 1995, p4, and *Earth Negotiations Bulletin*, 'Summary of the Third Session of the UN Commission on Sustainable Development, 11–28 April 1995', vol 5, no 42, 1 May 1995, p3.

32. Gail Osherenko and Oran R Young, 'The Formation of International Regimes: Hypotheses and Cases', in Young and Osherenko (eds), op cit, pp5–6.

33. Bernardo Zentilli, UNCED secretariat, personal communication (interview), Geneva, 24 September 1992.

34. Anon, 'UNCED ignores ten critical issues', *Third World Resurgence*, nos 24/25, August/September 1992, p12. The contributors to *Third World Resurgence* are among the leading NGO campaigners for full TNC accountability. See the articles in two special editions of *Third World Resurgence*: 'Transnational Corporations Hijack Earth Summit', no 20, April 1992; and 'TNCs Rule OK', no 40, December 1993.

35. World Wide Fund for Nature, *UNCED: The Way Forward* (Gland, Switzerland: WWF–International, September 1992), p10.

36. ITTO Mission, 'The Promotion of Sustainable Forest Management, A Case Study in Sarawak, Malaysia', April 1990 (unpublished), copy held by the Royal Geographical Society library, London, p3.

37. Marcus Colchester, 'The International Tropical Timber Organization: Kill or Cure for the Rainforests?', *The Ecologist*, vol 20, no 5, September/October 1990, p171.

38. Earl of Cranbrook, personal communication (interview), University of Keele, 31 August 1993. The mission's terms of reference are contained in ITTO document ITTC(VI)/14, 'Resolution 1(VI), The Promotion of Sustainable Forest Management: A Case Study in Sarawak, Malaysia', 24 May 1989.

39. FAO document CL 99/22, 'Outcome of Meeting of Ad Hoc Group of Experts on TFAP', Appendix B, paras 8 and 9, p B2.

40. David Potter, 'Environmental Problems in their Political Context', in Pieter Glasbergen and Andrew Blowers (eds) *Environmental Policy in an International Context: Perspectives on Environmental Problems* (London: Arnold, 1995), pp105–6. See also J Camilleri and J Falk, *The End of Sovereignty: The Politics of a Shrinking and Fragmenting World* (Aldershot: Edward Elgar, 1992).

Annex A:
Possible Main Elements of an Instrument (Convention, Agreement, Protocol, Charter) for the Conservation and Development of the World's Forests

FOOD AND AGRICULTURE ORGANIZATION OF THE UNITED NATIONS, ROME — DRAFT, 18 OCTOBER 1990

I. Preamble

In accordance with general usage, the preamble could set forth the *motives* and the *basic principles* of the Instrument.

1. *Motives* should include:

a) *Resources.* Forests are a precious resource for mankind for the production of food, fibre, fuel and shelter. The use of forests plays a major role in the economies of many countries. There are close links between forest conservation and forest development in contributing to sustainable socioeconomic development and to the satisfaction of basic human needs.

b) *Biological Diversity.* Forests are a major source of biological diversity. They contain not only woody species and wild animals but — especially in the moist and seasonal tropics and in some parts of the sub-tropics — a wealth of other plant species of actual or potential socioeconomic value, including gene pools of wild relatives and primitive cultivars of our main food crops.[1]

c) *Social and Cultural Diversity.* Forests are important for the maintenance of social and cultural diversity, particularly for local peoples, including indigenous and forest peoples and other communities that depend on forests. They are also of educational, scientific, cultural and spiritual importance for many peoples — including urban peoples who do not directly depend upon forests for their livelihood.

---

1. It is estimated that the forest ecosystems of seven countries — Australia, Brazil, Colombia, Indonesia, Madagascar, Mexico and Zaire — contain at least 40 per cent of all the mammalian species and 79 per cent of the primate species of the planet, 60 per cent of the bird species and 50 per cent of all vegetal species.

d) *Protecting Watersheds.* Forests are important for maintaining and protecting watersheds and for the protection of other biosystems, including riverine and coastal areas, as well as protecting urban areas from flooding.

e) *Maintaining Soil Cover.* Natural and replanted forests are important for maintaining soil cover, arresting desertification and for the long-term sustainability of agricultural land; they provide other essential environmental services on a local or regional level.

f) *Climatic Stability.* The conservation of the world's forests is essential for global climatic stability, particularly having regard to the important contribution of forest destruction to global warming through the emission of carbon dioxide, methane and other trace gases and to the role of forests as reservoirs and sinks of greenhouses gases.[2]

2. *Principles*

At least three basic principles could be recognized.

a) *Sovereignty.* The sovereignty of states over the forest resources and potential resources under their jurisdiction.

b) *Stewardship.* The stewardship of those resources in such a manner as to ensure the attainment and continued satisfaction of human needs for present and future generations.

c) *Burden-sharing.* There should be an equitable sharing by the international community of the burden of forest conservation and development, including the application of financial compensation mechanisms to offset opportunity costs of forest conservation undertaken for the purpose of contributing to international environmental goals (for example preserving biological diversity and climatic stability).

II. Definitions

1. 'Forests' — to include all forests: tropical, temperate and boreal.

2. 'Sustainable use/development' — Sustainable use/development means the management and conservation of the national resource base, and the orientation of technological and institutional change in such a manner as to ensure the attainment and continued satisfaction of human needs for present and future generations. Such sustainable use/development in the forestry sector conserves land, water, plant and animal genetic resources, is environmentally non-degrading, technically appropriate, economically viable and socially acceptable.

3. Interpretation of 'Party' — 'Party' means a State or any regional economic integration organization constituted by sovereign States which has competence in respect of the negotiation, conclusion and application of international agreements in matters covered by this Convention for which this Convention is in force.

   In matters within their competence, the regional economic integration organizations which are Parties to this Convention shall in their own name exercise the

---

2. Forests contribute approximately 20 per cent of the global anthropogenic carbon dioxide release into the atmosphere. Conversion of forests to agricultural and urban lands accompanied by burning of the forest materials and oxidation of soils during land conversion constitute the bulk of the contribution.

rights and fulfil the responsibilities which this Convention attributes to their member States. In such cases the member States of these organizations shall not be entitled to exercise such rights individually.

III. General Goals of the Instrument

1. These should apply to all types of forest — tropical, temperate and boreal — and include:

a) *Conservation.* The conservation and protection of natural forest ecosystems and the maintenance of biodiversity.

b) *Sustainable management.* The sustainable management of forests for wood or non-wood products or services.

c) *Rehabilitation.* The rehabilitation of degraded forests to improve their ecological functions and increase their productivity.

d) *Afforestation.* Afforestation of low productivity areas and waste or barren lands.

e) *Watershed management.* Management and expansion of catchment forests and other tree vegetation types that contribute directly to agricultural production, erosion control or water regime stabilization.

f) *Reduction of damage by pollution.* The protection of forests from damage from other national and international causes — particularly air pollution and acid deposition.

g) *Global environmental services.* The reduction of net biotic emissions of carbon dioxide and other greenhouse gases from the forest sector.

IV. Specific Goals of the Instrument

1. The Parties should agree on the following specific goals of the Instrument.

a) *No loss of global forested area.* In view of their vital contribution to national economic and social development, to the maintenance of biodiversity and to regional and global climatic stability, there should be no loss of the global forested area and that area should be increased whenever and wherever possible.

Parties could endorse the target (included in the Noordwijk Declaration) of a net global forest growth of 12 million hectares per year, through conservation of existing forests and through aggressive programmes of reforestation and afforestation.

b) *Reduce global biotic emissions of CO_2.* In view of the critical role of forests as an absorber and contributor of greenhouse gases, a global goal could be established of reducing net biotic emissions from the forest sector to zero by the year 2000.

V. Main Obligations of the Parties

The forest instrument should reflect the consensus of the Parties on general rules of forest conservation and development. These general rules would aim at gradually optimizing the various types of forest land use, as well as the forestry components of other land uses, so as to reconcile ecological, economic and social requirements.

Explicit commitment of the Parties should be sought for the following specific measures and tasks:

1. Integration of Forest Considerations into General Development Policy

The importance of forest conservation, the sustainable use of forests, as well as reforestation and afforestation, should be fully recognized in national development planning and forestry considerations should be integrated in all sectors of policy.

Parties should recognize that national policies must be developed to achieve spatial patterns of settlement, economic activity and administrative services that will sustain investment in, and the productivity of, forest resources and provide the maintenance or establishment of a permanent forest base.

In particular, the Parties could agree to use the following types of policies to integrate responses for conserving forests with other development factors.

a) *Investment and fiscal policies* so as to provide incentives and disincentives conducive to the sustainable management of forest resources.[3]
b) *Integrate forest policies with other policies* aimed at reducing poverty and increasing the living standards of the population, so as to decrease population pressure on forests.
c) *Promote sustainable agriculture/agroforestry* outside currently forested areas.
d) *Orient development policies* to support small, rural producers.
e) *Use transport systems* that have less impact on forests.

2. Formulation of National Forest Policy

A national forest policy, forming an integral part of a national land use policy aiming at the sustainable use of all natural resources, should be formulated by means of a process seeking the consensus of all the actors involved: government, local populations and the private sector.

In this context, Parties would undertake to adopt clear targets or objectives for the conservation, reforestation, afforestation and/or sustainable development of forests, together with measures to achieve this at the national level.

Parties would also agree to formulate or revise national forest conservation and development plans covering at least the following main elements:

a) the preservation of primary forests by means of establishing designated protection areas, including natural reserve areas, parks, landscape protection areas, ethnological protection areas and resource protection areas;
b) the protection of other ecologically sensitive forest ecosystems;
c) the sustainable utilization of forests in the interests of socioeconomic development and environmental protection;
d) the scope and methods for reforestation and afforestation and for the rehabilitation of degraded areas; and
e) the monitoring of forest resources.

---

3. Public policies affect where populations move and what they do where they settle. Public investments displace (for example dams), channel (for example roads) and attract (for example industrial and administrative centres) people. Taxes, wages, prices, subsidies, and regulatory policies affect the relative advantages for people of different activities, intensities of investment and places for settlement and livelihood.

3. National Forest Service

Parties should commit themselves to establishing a national agency capable of managing the government forest estate, and assisting in the management of private and customarily held forests, according to the objectives laid down in the national forest policy.

4. Monitoring of Forest Resources: National Forest Inventory

a) Regular monitoring of forest resources should cover important parameters such as area, health and major uses and functions.
b) A national forest inventory should establish the importance of all forests independently of their ownership status. Such inventories should be responsive to a great variety of user needs and applications, addressing global climate change, biodiversity, biomass, ecological zone influences, wildlife habitat, desertification in dry areas and carbon uptake, as well as the suitability of forested land for the production of timber and non-timber products. The most appropriate use should be made of satellite and computer technologies.
c) National forestry action plans should give priority to the strengthening of the institutional infrastructure and technical capabilities to support the implementation of national continuous forest resources monitoring, as part of a coherent system of global monitoring of forest resources.

5. Sustainable Use/Development of Forests

Parties should agree to ensure the sustainable use/development of their forests.

In order to achieve sustainable use/development of forests, Parties should agree to adopt or promote improved management practices, such as:

a) providing for an overall land use scheme that balances the needs for all kinds of land uses;
b) encouraging development projects that improve the sustainability and economic return of forest extraction (for example extensive systems based on harvesting non-timber forest products) and promoting policies that increase and diversify the use of forest products (fruits, timber, game, fibre);
c) following forestry policies to strengthen the small forest landowner, including incentives and funds to reinvest in forest management;
d) promotion of programmes and policies of replanting and forest regeneration after logging that compensate the rate of use;
e) using more species and more size classes for timber, so that intensive harvest and management are more feasible, and less extensive areas will be logged, as well as developing technologies and marketing methods to promote greater use of the lesser-known species that have good timber properties and expanding marketing potential of secondary timber species through government purchasing programmes;
f) promotion of multipurpose forest management compatible with genetic resource conservation;
g) definition of areas for wood production (or permanent forest estate with guaranteed long term forestry use);

h) setting of national management objectives for each forestry unit;
i) choosing silvicultural practices aimed at sustained yield at minimum cost, while respecting the need for ecological sustainability;
j) timber harvesting based on a detailed forest inventory and exactly specified plans, so as to minimize damage and maximize natural regeneration;
k) adopting reliable methods for controlling timber yield in order to ensure a sustained production of timber from each forest management unit;
l) preparing working plans which guarantee the respect of environmental standards in field operations;
m) specifying the conditions under which an Environmental Impact Assessment (EIA) should be required and providing for qualified staff to carry out EIAs.

6. Protected Areas

Parties should pledge to implement their commitment to, *inter alia*, the designation of protected areas (parks, reserves) and establishment of measures for their protection and control.

Certain forests could be nominated by Governments for inclusion on a list of Forests of International Importance and, if the nomination is accepted (mechanism to be specified — it could be analogous to those provided for under the Ramsar and World Heritage Conventions), the Government could be eligible for 'additional' financial support (See Section VI.2 on Financial Mechanisms).

7. Protection of Indigenous Populations

Parties should pledge:

a) Respect for the basic rights and needs of local populations, including indigenous and forest peoples and other communities that depend on the forests;
b) the promotion of the rights and sustainable livelihood and cultural integrity of indigenous people through policies and laws that recognize and protect their land, economic, intellectual and cultural rights;
c) direct and effective participation of local peoples in planning and decision-making related to forest areas.

8. Protection of Biodiversity

Parties should make specific commitments regarding the role of forests in the conservation of biodiversity. Such commitments could include:

a) initial and periodic surveys and inventories of biological diversity in forest areas within their jurisdiction;
b) creating specially protected areas in those forest areas under their jurisdiction which are important for the conservation of biological diversity and which may be at risk due to human activities;
c) undertaking to apply as far as forest areas falling under their jurisdiction are concerned the provisions of the Convention on Biological Diversity.[4]

---

4. This provision would be otiose in a situation where a Contracting Party to the Forest Convention has already ratified and applied the Biodiversity Convention. However, where

9. Controls on Activities likely to Jeopardize Sustainable Use of Forests

Parties should agree to control activities that degrade the potential use of a forest as an economic and ecological resource, particularly:

a) agricultural practices that result in permanent impoverishment of the soil and vegetation;
b) the clearing of forests for grazing purposes only;
c) extractive industries (such as mining and logging); and
d) major public or private development schemes (such as roads and dams).

In the case of the last two categories, Parties should agree to insist that a full Environment Impact Assessment (EIA) is carried out before the operation starts and that there is regular monitoring of effects once the operation is under way and following its completion.

10. Mitigation of Climate Change Effects through Maintenance of Existing Forest Areas

The Parties could recognize that forestry contributes approximately 20 per cent of the global anthropenic carbon dioxide release into the atmosphere. Conversion of forests to agricultural and urban lands accompanied by burning of the forest materials and oxidation of soils during land conversion constitute the bulk of the contributions.

Each Party could pledge that it will permit no net reduction of forest areas under its jurisdiction, as well as no loss of sustainability of any existing forest areas.

Countries unable or unwilling to pledge themselves to the above targets ('no net reduction' and 'no loss of sustainability') should nevertheless commit themselves to alternative targets and notify them to the Instrument's Secretariat, for publication and dissemination.

Achievement of the targets ('no net reduction' and 'no loss of sustainability') would be taken into account for the purpose of assessing any compensation to be made to Parties for global environmental services rendered, in accordance with agreed principles. (See Section VI.2 on Financial Mechanisms.)

Where alternative targets have been adopted by a Contracting Party and where such targets, though falling short of the 'no net reduction' objective, are recognized nevertheless as contributing significantly to global environmental objectives, compensation from international sources should still be available.[5]

11. Commitment to Increase Forest Areas

The Parties may further agree to increase their existing forest areas through vigorous programmes of reforestation and afforestation, through urban tree planting and the creation of urban forest reserves, and other similar methods.

Parties who decide to increase their existing forest areas may notify the Instru-

---

this is not the case, the articles of the Forestry Convention relating to biodiversity could provide useful added value.

5. If a Climate Convention has been ratified and has entered into force, it may be appropriate for the Instrument's Secretariat to seek the comments of the Climate Convention Secretariat on any climate-related forestry targets which Parties may adopt.

ment's Secretariat of their targets and objectives, as well as any other details concerning the plan or programme. This information would be published and disseminated. Progress made in achieving stated targets would be taken into account for 'compensation' purposes (as described in the previous paragraph).

12. Strengthening of Education, Public Information and Institution Building Efforts in Forestry and Related Fields

Parties should pledge themselves to action aimed at removing the institutional constraints impeding the conservation and development of forest resources by strengthening public forest administrations and related government agencies, integrating forestry concerns into development planning, providing institutional support for private and local organizations, developing professional technical and vocational training, and improving extension and research.

13. Fuelwood and Energy

Parties should pledge themselves to action aimed at restoring fuelwood supplies in the countries affected by shortages through support for national fuelwood and wood energy programmes, development of wood-based energy systems for rural and industrial development, regional training and demonstration, and intensification of research and development.

14. Forest-based Industrial Development

Countries should pledge themselves to actions aimed at promoting appropriate forest-based industries by intensifying resource management and development, promoting appropriate raw material harvesting, establishing and managing appropriate forest industries, reducing waste, and developing the marketing of forest industry products.

15. Prevention of Pollution

Parties should pledge to increase their efforts to combat damage to forests and woodlands from pollution, notably air pollution and acid deposition, having regard to the 'critical load' of pollutants not to be exceeded in such ecosystems. They should take positive urgent steps to curb pollutants emitted from industrial installations and motor vehicles.

They should engage themselves to ratify as early as possible relevant regional conventions and agreements relation to air pollution liable to cause forest damage (for example sulphur dioxide, No_x and ozone) and, where such conventions do not exist, to use their best efforts to bring them into being as soon as possible.

They should also pledge themselves to conclude, where appropriate, bilateral agreements with neighbouring countries designed to limit the transboundary effects of air pollution on forest ecosystems.[6]

---

6. It is now clear that forests in industrialized countries, especially but not exclusively those in Eastern Europe, have been severely damaged by emissions of sulphur dioxide, ozone and other industrial wastes. The annual loss of production from forest depletion is estimated at US$ 30,000 million in Europe alone.

16. Research Relating to Forest Conservation and Development

The Parties should recognize the need for the expansion of research systems and programmes on forest conservation and management, reforestation, afforestation, agroforestry, forest products utilization and related fields such as the forest sector's contribution and response to expected climate change.

Critical areas of research include:

a) basic research on the national function of the forests, including cycles of carbon and nutrients, as well as biotic interactions;
b) species life histories, distribution and community compositions;
c) a more complete inventory of forests species and ecosystems;
d) aquatic ecosystems and wetlands;
e) water yield;
f) forest fire frequency and severity;
g) insect and disease outbreak;
h) wildlife and domestic livestock — habitat relations;
i) the contribution of the forest sector to climate change;
j) the impact of climate change on forests — tropical, temperate, boreal — and on woodlands and savannahs; and
k) the impact of pollution on forests and trees.

Preferably, such research should be done by national and local research institutions. As relevant, research should be carried out at the regional level and at the international level through such organizations as the International Union of Forestry Research Organizations (IUFRO) and the Consultative Group on International Agricultural Research (CGIAR).

VI. International Cooperation

1. Commitment by Parties to Support Bilateral and Multilateral Development Aid and Investment that Promote the Conservation and Sustainable Development of Forests

The Parties should pledge to support bilateral and multilateral development aid, and

a) to encourage investments that provide activities having a positive impact upon forest conservation and sustainable development (including activities outside the forestry sector); and
b) refrain from activities that have a negative impact upon forest conservation and sustainable development or which have not been subject to an assessment of that impact.

The Parties should encourage the international organizations and financial institutions active in forestry and other related fields and of which they are members, notably the World Bank, the International Monetary Fund (IMF), the Regional Development Banks, the European Economic Community (EEC), the United Nations Development Programme (UNDP), the Food and Agriculture Organization of the United Nations (FAO) and the International Tropical Timber Organization (ITTO), to review their policies and programmes of technical assistance, capital investment, or debt management, as appropriate, with a view to eliminating existing impediments for countries to make progress in achieving the objectives of the Instrument.

More specifically, Parties should encourage multilateral and bilateral international cooperation agencies in particular to:

a) strengthen national institutional, scientific, technical and economic capabilities with regard to forest conservation and sustainable development, in particular through existing arrangements such as the Tropical Forestry Action Programme (TFAP). These arrangements would be reformed or adapted as required;

b) encourage participation in a worldwide network to develop, sustain and utilize suitable technical tools, such as the proposed Global Forest Resources Monitoring Programme, to obtain regular and accurate information on the extent, health, rate of conversion to other uses and economic potentials of the forests on a worldwide basis;

c) support activities directly related to forest conservation and sustainable development conducted by other international governmental and non-governmental institutions, such as the International Union for the Conservation of Nature and Natural Resources (IUCN) and the World Wide Fund for Nature (WWF).

2. Commitment by Parties from the Developed World to Increase International Resource Flows to Parties from the Developing World in Order to Help Them Meet their Obligations and Engagements under the Instrument

Parties from the developed world should pledge to help *increase* international resource flows to Parties from the developing world ('additionality') in order to compensate them for the global environmental services rendered and the opportunity costs foregone to other users of the forest.

Such global environmental services would include, *inter alia*, those provided by forest conservation and development programmes intended to mitigate climate change effects through the maintenance of existing forest areas (see Section V.10).

Actions to be taken to meet this pledge might include:

a) increasing aid and investment flows through existing mechanisms (for example multilateral and bilateral funding), special attention being given to the role of the TFAP in providing a framework for the coordination of these 'additional' flows;

b) new sources of funds, for example the World Bank's Global Environment Facility, which might be available for the purpose;

c) general debt relief operations;

d) structural adjustment packages; and

e) debt for nature swaps.

(The principles for allocating such 'additional' resources as compensation to developing countries for environmental services rendered would need to include, *inter alia*, the extent to which the country has met or is meeting its stated forestry targets, and possible other factors such as level of economic development or area available for forestry. Account would need to be taken of parallel initiatives to provide 'compensation for environmental services', for example under the Vienna Ozone Convention and its CFC's protocol or in the context of the reduction of energy-related CO_2 emissions under the future Climate Convention. Innovative proposals on funding are also expected from the Secretariat of the United Nations Conference on Environment and Development.)

3. Trade Policy that Encourages the Conservation and Sustainable Development of Forests

Parties should pledge to carry out trade policies that encourage the conservation and sustainable development of forests, and to improve the terms of trade for forest productions coming from sustainably managed forest resources.

In particular, the Parties agree to the following:

a) timber price should be used as a policy mechanism to promote sustainable resource management;

b) to work for the reduction or elimination of tariffs against imports of tropical crop products and to negotiate improvements in the terms of trade for such products. This action should include not only tropical forest products but also agricultural products, with the objective, *inter alia*, of encouraging greater agricultural production intensity and thus reducing the pressure to develop new farmlands which may lead to increased deforestation; and

c) to work for the reduction or removal of non-tariff barriers which are used specifically to restrict trade, such as quantitative restrictions, as well as other non-tariff barriers which may have other primary aims (such as certain health, safety and technical standards) but which either intentionally or unintentionally also restrict trade.

Restrictions on trade of certain forest products are, however, provided for under the Convention on International Trade in Endangered Species of Wild Flora and Fauna (CITES) and Parties should confirm their commitment to that Convention.

Parties could also, individually or collectively, undertake to apply restrictive measures against imports from a country which does not ratify the Instrument and which permits the exploitation of its forests in a manner that is clearly inconsistent with the basic principles of the Instrument.

In the case of flagrant abuses, such restrictions could by a decision of the Parties to the Convention be imposed on a general and mandatory basis.

VII. Institutional Arrangements

This aspect needs to be fully developed only when a sufficient level of consensus and maturity has been reached internationally on the substantive content of the forest Instrument. At this stage, it seems premature to advance any suggestion other than some points for consideration based on the nature of the Instrument and on general usage for legal instruments of this type:

■ a Conference of the Parties would be necessary as a decision-making organ and supreme authority. For greater efficiency, decisions could be taken by majority vote, with votes being weighted according to agreed principles (for example forest area, GNP and participation in international financial mechanisms);

■ a Secretariat of the Instrument would be needed to prepare and service the meetings of the Conference, to receive reports and other information, and to inform governments and institutions concerned, as well as the general public, on matters related to the Instrument;

■ the Secretariat of the Instrument could be entrusted to an existing inter-governmental organization, or to a new body.

Specific funding arrangements would be required to service the forest instrument, to allow for the functioning of its Statutory Bodies and Secretariat, and for certain central reporting and information tasks. The funds required could be raised through mandatory assessments of the Parties to the Instrument based on an equitable assessment method. Entry into force of the Instrument should depend on ratifications by an appropriate number of countries from the major regions of the world covering all types of forest — tropical, temperate and boreal — accounting for, say, two-thirds of the (weighted) votes.

Annex B:
Non-Legally Binding Authoritative Statement of Principles for a Global Consensus on the Management, Conservation and Sustainable Development of All Types of Forests

UNITED NATIONS CONFERENCE ON ENVIRONMENT AND
DEVELOPMENT, RIO DE JANEIRO, JUNE 1992

Preamble

a) The subject of forests is related to the entire range of environmental and development issues and opportunities, including the right to socio-economic development on a sustainable basis.

b) The guiding objective of these principles is to contribute to the management, conservation and sustainable development of forests and to provide for their multiple and complementary functions and uses.

c) Forestry issues and opportunities should be examined in a holistic and balanced manner within the overall context of environment and development, taking into consideration the multiple functions and uses of forests, including traditional uses, and the likely economic and social stress when these uses are constrained or restricted, as well as the potential for development that sustainable forest management can offer.

d) These principles reflect a first global consensus on forests. In committing themselves to the prompt implementation of these principles, countries also decide to keep them under assessment for their adequacy with regard to further international cooperation on forest issues.

e) These principles should apply to all types of forests, both natural and planted, in all geographic regions and climatic zones, including austral, boreal, sub-temperate, temperate, subtropical and tropical.

f) All types of forests embody complex and unique ecological processes which are the basis for their present and potential capacity to provide resources to satisfy human needs as well as environmental values, and as such their sound management and conservation is of concern to the Governments of the countries to which they belong and are of value to local communities and to the environment as a whole.

g) Forests are essential to economic development and the maintenance of all forms of life.

h) Recognizing that the responsibility for forest management, conservation and sustainable development is in many States allocated among federal/national, state/provincial and local levels of government, each State, in accordance with its constitution and/or national legislation, should pursue these principles at the appropriate level of government.

Principles/Elements

1. (a) 'States have, in accordance with the Charter of the United Nations and the principles of international law, the sovereign right to exploit their own resources pursuant to their own environmental policies and have the responsibility to ensure that activities within their jurisdiction or control do not cause damage to the environment of other States or of areas beyond the limits of national jurisdiction.'

   (b) The agreed full incremental cost of achieving benefits associated with forest conservation and sustainable development requires increased international cooperation and should be equitably shared by the international community.

2. (a) States have the sovereign and inalienable right to utilize, manage and develop their forests in accordance with their development needs and level of socio-economic development and on the basis of national policies consistent with sustainable development and legislation, including the conversion of such areas for other uses within the overall socio-economic development plan and based on rational land-use policies.

   (b) Forest resources and forest lands should be sustainably managed to meet the social, economic, ecological, cultural and spiritual human needs of present and future generations. These needs are for forest products and services, such as wood and wood products, water, food, fodder, medicine, fuel, shelter, employment, recreation, habitats for wildlife, landscape diversity, carbon sinks and reservoirs, and for other forest products. Appropriate measures should be taken to protect forests against harmful effects of pollution, including airborne pollution, fires, pests, and diseases in order to maintain their full multiple value.

   (c) The provision of timely, reliable and accurate information on forests and forest ecosystems is essential for public understanding and informed decision-making and should be ensured.

   (d) Governments should promote and provide opportunities for the participation of interested parties, including local communities and indigenous people, industries, labour, non-governmental organizations and individuals, forest dwellers and women, in the development, implementation and planning of national forest policies.

3. (a) National policies and strategies should provide a framework for increased efforts, including the development and strengthening of institutions and pro-

grammes for the management, conservation and sustainable development of forests and forest lands.

(b) International institutional arrangements, building on those organizations and mechanisms already in existence, as appropriate, should facilitate international cooperation in the field of forests.

(c) All aspects of environmental protection and social and economic development as they relate to forests and forest lands should be integrated and comprehensive.

4. The vital role of all types of forests in maintaining the ecological processes and balance at the local, national, regional and global levels through, *inter alia*, their role in protecting fragile ecosystems, watersheds and freshwater resources and as rich storehouses of biodiversity and biological resources and sources of genetic material for biotechnology products, as well as photosynthesis, should be recognized.

5. (a) National forests policies should recognize and duly support the identity, culture and the rights of indigenous people, their communities and other communities and forest dwellers. Appropriate conditions should be promoted for these groups to enable them to have an economic stake in forest use, perform economic activities, and achieve and maintain cultural identity and social organization, as well as adequate levels of livelihood and well-being, through, *inter alia*, those land tenure arrangements which serve as incentives for the sustainable management of forests.

(b) The full participation of women in all aspects of the management, conservation and sustainable development of forests should be actively promoted.

6. (a) All types of forests play an important role in meeting energy requirements through the provision of a renewable source of bio-energy, particularly in developing countries, and the demands for fuelwood for household and industrial needs should be met through sustainable forest management, afforestation and reforestation. To this end, the potential contribution of plantations of both indigenous and introduced species for the provision of both fuel and industrial wood should be recognized.

(b) National policies and programmes should take into account the relationship, where it exists, between the conservation, management and sustainable development of forests and all aspects related to the production, consumption, recycling and/or final disposal of forest products.

(c) Decisions taken on the management, conservation and sustainable development of forest resources should benefit, to the extent practicable, from a comprehensive assessment of economic and non-economic values of forest goods and services and of the environmental costs and benefits. The development and improvement of methodologies for such evaluations should be promoted.

(d) The role of planted forests and permanent agricultural crops as sustainable and environmentally sound sources of renewable energy and industrial raw material should be recognized, enhanced and promoted. Their contribution to the maintenance of ecological processes, to offsetting pressure on primary/

old-growth forest and to providing regional employment and development with the adequate involvement of local inhabitants should be recognized and enhanced.

(e) Natural forests also constitute a source of goods and services, and their conservation, sustainable management and use should be promoted.

7. (a) Efforts should be made to promote a supportive international economic climate conducive to sustained and environmentally sound development of forests in all countries, which include, *inter alia*, the promotion of sustainable patterns of production and consumption, the eradication of poverty and the promotion of food security.

(b) Specific financial resources should be provided to developing countries with significant forest areas which establish programmes for the conservation of forests including protected natural forest areas. These resources should be directed notably to economic sectors which would stimulate economic and social substitution activities.

8. (a) Efforts should be undertaken towards the greening of the world. All countries, notably developed countries, should take positive and transparent action towards reforestation, afforestation and forest conservation, as appropriate.

(b) Efforts to maintain and increase forest cover and forest productivity should be undertaken in ecologically, economically and socially sound ways through the rehabilitation, reforestation and re-establishment of trees and forests on unproductive, degraded and deforested lands, as well as through the management of existing forest resources.

(c) The implementation of national policies and programmes aimed at forest management, conservation and sustainable development, particularly in developing countries, should be supported by international financial and technical cooperation, including through the private sector, where appropriate.

(d) Sustainable forest management and use should be carried out in accordance with national development policies and priorities and on the basis of environmentally sound national guidelines. In the formulation of such guidelines, account should be taken, as appropriate and if applicable, of relevant internationally agreed methodologies and criteria.

(e) Forest management should be integrated with management of adjacent areas so as to maintain ecological balance and sustainable productivity.

(f) National policies and/or legislation aimed at management, conservation and sustainable development of forests should include the protection of ecologically viable representative or unique examples of forests, including primary/old-growth forests, cultural, spiritual, historical, religious and other unique and valued forests of national importance.

(g) Access to biological resources, including genetic material, shall be with due regard to the sovereign rights of the countries where the forests are located and to the sharing on mutually agreed terms of technology and profits from biotechnology products that are derived from these resources.

(h) National policies should ensure that environmental impact assessments should be carried out where actions are likely to have significant adverse impacts on important forest resources, and where such actions are subject to a decision of a competent national authority.

9. (a) The efforts of developing countries to strengthen the management, conservation and sustainable development of their forest resources should be supported by the international community, taking into account the importance of redressing external indebtedness, particularly where aggravated by the net transfer of resources to developed countries, as well as the problem of achieving at least the replacement value of forests through improved market access for forest products, especially processed products. In this respect, special attention should also be given to the countries undergoing the process of transition to market economies.

(b) The problems that hinder efforts to attain the conservation and sustainable use of forest resources and that stem from the lack of alternative options available to local communities, in particular the urban poor and poor rural populations who are economically and socially dependent on forests and forest resources, should be addressed by Governments and the international community.

(c) National policy formulation with respect to all types of forests should take account of the pressures and demands imposed on forest ecosystems and resources from influencing factors outside the forest sector, and intersectoral means of dealing with these pressures and demands should be sought.

10. New and additional financial resources should be provided to developing countries to enable them to sustainably manage, conserve and develop their forest resources, including through afforestation, reforestation and combating deforestation and forest and land degradation.

11. In order to enable, in particular, developing countries to enhance their endogenous capacity and to better manage, conserve and develop their forest resources, the access to and transfer of environmentally sound technologies and corresponding know-how on favourable terms, including on concessional and preferential terms, as mutually agreed, in accordance with the relevant provisions of Agenda 21, should be promoted, facilitated and financed, as appropriate.

12. (a) Scientific research, forest inventories and assessments carried out by national institutions which take into account, where relevant, biological, physical, social and economic variables, as well as technological development and its application in the field of sustainable forest management, conservation and development, should be strengthened through effective modalities, including international cooperation. In this context, attention should also be given to research and development of sustainably harvested non-wood products.

(b) National and, where appropriate, regional and international institutional capabilities in education, training, science, technology, economics, anthropology and social aspects of forests and forest management are essential to the conservation and sustainable development of forests and should be strengthened.

(c) International exchange of information on the results of forest and forest management research and development should be enhanced and broadened, as appropriate, making full use of education and training institutions, including those in the private sector.

(d) Appropriate indigenous capacity and local knowledge regarding the conservation and sustainable development of forests should, through institutional and financial support, and in collaboration with the people in the local communities concerned, be recognized, respected, recorded, developed and, as appropriate, introduced in the implementation of programmes. Benefits arising from the utilization of indigenous knowledge should therefore be equitably shared with such people.

13. (a) Trade in forest products should be based on non-discriminatory and multilaterally agreed rules and procedures consistent with international trade law and practices. In this context, open and free international trade in forest products should be facilitated.

(b) Reduction or removal of tariff barriers and impediments to the provision of better market access and better prices for higher value-added forest products and their local processing should be encouraged to enable producer countries to better conserve and manage their renewable forest resources.

(c) Incorporation of environmental costs and benefits into market forces and mechanisms, in order to achieve forest conservation and sustainable development, should be encouraged both domestically and internationally.

(d) Forest conservation and sustainable development policies should be integrated with economic, trade and other relevant policies.

(e) Fiscal, trade, industrial, transportation and other policies and practices that may lead to forest degradation should be avoided. Adequate policies, aimed at management, conservation and sustainable development of forests, including where appropriate, incentives, should be encouraged.

14. Unilateral measures, incompatible with international obligations or agreements, to restrict and/or ban international trade in timber or other forest products should be removed or avoided, in order to attain long-term sustainable forest management.

15. Pollutants, particularly airborne pollutants, including those responsible for acidic deposition, that are harmful to the health of forest ecosystems at the local, national, regional and global levels should be controlled.

Annex C:
International Tropical Timber
Agreement, 1994[1]

GENEVA, JANUARY 1994

PREAMBLE

The Parties to this agreement,

Recalling the Declaration and the Programme of Action on the Establishment of a New International Economic Order; the Integrated Programme for Commodities; the New Partnership for Development: the Cartagena Commitment and the relevant objectives contained in the Spirit of Cartagena,

Recalling the International Tropical Timber Agreement, 1983, and *recognizing* the work of the International Tropical Timber Organization and its achievements since its inception, including a strategy for achieving international trade in tropical timber from sustainably managed sources,

Recalling further the Rio Declaration on Environment and Development, the Non-Legally Binding Authoritative Statement of Principles for a Global Consensus on the Management, Conservation and Sustainable Development of all Types of Forests, and the relevant Chapters of Agenda 21 as adopted by the United Nations Conference on Environment and Development in June 1992, in Rio de Janeiro; the United Nations Framework Convention on Climate Change; and the Convention on Biological Diversity,

Recognizing the importance of timber to the economies of countries with timber producing forests,

Further recognizing the need to promote and apply comparable and appropriate guidelines and criteria for the management, conservation and sustainable development of all types of timber producing forests,

Taking into account the linkages of tropical timber trade and the international timber market and the need for taking a global perspective in order to improve transparency in the international timber market,

Noting the commitment of all members, made in Bali, Indonesia, in May 1990, to

---

1. The International Tropical Timber Agreement, 1994 was negotiated under the auspices of the United Nations Conference on Trade and Development.

achieve exports of tropical timber products from sustainably managed sources by the year 2000 and *recognizing* Principle 10 of the Non-Legally Binding Authoritative Statement of Principles for a Global Consensus on the Management, Conservation and Sustainable Development of all Types of Forests which states that new and additional financial resources should be provided to developing countries to enable them to sustainably manage, conserve and develop their forests, including through afforestation, reforestation and combating deforestation and forest and land degradation,

Noting also the statement of commitment to maintain, or achieve by the year 2000, the sustainable management of their respective forests made by consuming members who are Parties to the International Tropical Timber Agreement, 1983 at the fourth session of the United Nations Conference for the Negotiation of a Successor Agreement to the International Tropical Timber Agreement, 1983 in Geneva on 21 January 1994,

Desiring to strengthen the framework of international cooperation and policy development between members in finding solutions to the problems facing the tropical timber economy,

Have agreed as follows:

CHAPTER I. OBJECTIVES

Article 1
Objectives

Recognizing the sovereignty of members over their natural resources, as defined in Principle 1 (a) of the Non-Legally Binding Authoritative Statement of Principles for a Global Consensus on the Management, Conservation and Sustainable Development of all Types of Forests, the objectives of the International Tropical Timber Agreement, 1994 (hereinafter referred to as 'this Agreement') are:

a) To provide an effective framework for consultation, international cooperation and policy development among all members with regard to all relevant aspects of the world timber economy;

b) To provide a forum for consultation to promote non-discriminatory timber trade practices;

c) To contribute to the process of sustainable development;

d) To enhance the capacity of members to implement a strategy for achieving exports of tropical timber and timber products from sustainably managed sources by the year 2000;

e) To promote the expansion and diversification of international trade in tropical timber from sustainable sources by improving the structural conditions in international markets, by taking into account, on the one hand, a long-term increase in consumption and continuity of supplies and, on the other, prices which reflect the costs of sustainable forest management and which are remunerative and equitable for members, and the improvement of market access;

f) To promote and support research and development with a view to improving forest management and efficiency of wood utilization as well as increasing the capacity to conserve and enhance other forest values in timber producing tropical forests;

g) To develop and contribute towards mechanisms for the provision of new and additional financial resources and expertise needed to enhance the capacity of producing members to attain the objectives of this Agreement;

h) To improve market intelligence with a view to ensuring greater transparency in the international timber market, including the gathering, compilation and dissemination of trade related data, including data related to species being traded;

i) To promote increased and further processing of tropical timber from sustainable sources in producing member countries with a view to promoting their industrialization and thereby increasing their employment opportunities and export earnings;

j) To encourage members to support and develop industrial tropical timber reforestation and forest management activities as well as rehabilitation of degraded forest land, with due regard for the interests of local communities dependent on forest resources;

k) To improve marketing and distribution of tropical timber exports from sustainably managed sources;

l) To encourage members to develop national policies aimed at sustainable utilization and conservation of timber producing forests and their genetic resources and at maintaining the ecological balance in the regions concerned, in the context of tropical timber trade;

m) To promote the access to, and transfer of, technologies and technical cooperation to implement the objectives of this Agreement, including on concessional and preferential terms and conditions, as mutually agreed; and

n) To encourage information-sharing on the international timber market.

CHAPTER II. DEFINITIONS

Article 2
Definitions

For the purpose of this Agreement:

1. 'Tropical timber' means non-coniferous tropical wood for industrial uses, which grows or is produced in the countries situated between the Tropic of Cancer and the Tropic of Capricorn. The term covers logs, sawnwood, veneer sheets and plywood. Plywood which includes in some measure conifers of tropical origin shall also be covered by this definition;

2. 'Further processing' means the transformation of logs into primary wood products, semi-finished and finished products made wholly or almost wholly of tropical timber;

3. 'Member' means a Government or an intergovernmental organization referred to in article 5 which has consented to be bound by this Agreement whether it is in force provisionally or definitively;

4. 'Producing member' means any country with tropical forest resources and/or a net exporter of tropical timber in volume terms which is listed in annex A and which becomes a party to this Agreement, or any country with tropical forest resources and/or a net exporter of tropical timber in volume terms which is not

so listed and which becomes a party to this Agreement and which the Council, with the consent of that country, declares to be a producing member;

5. 'Consuming member' means any country listed in annex B which becomes a party to this Agreement, or any country not so listed which becomes a party to this Agreement and which the Council, with the consent of that country, declares to be a consuming member;

6. 'Organization' means the International Tropical Timber Organization established in accordance with article 3;

7. 'Council' means the International Tropical Timber Council established in accordance with article 6;

8. 'Special vote' means a vote requiring at least two-thirds of the votes cast by producing members present and voting and at least 60 per cent of the votes cast by consuming members present and voting, counted separately, on condition that these votes are cast by at least half of the producing members present and voting and at least half of the consuming members present and voting;

9. 'Simple distributed majority vote' means a vote requiring more than half of the votes cast by producing members present and voting and more than half of the votes cast by consuming members present and voting, counted separately;

10. 'Financial year' means the period from 1 January to 31 December inclusive;

11. 'Freely usable currencies' means the Deutsche mark, the French franc, the Japanese yen, the pound sterling, the United States dollar and any other currency that has been designated from time to time by a competent international monetary organization as being in fact widely used to make payments for international transactions and widely traded in the principal exchange markets.

CHAPTER III. ORGANIZATION AND ADMINISTRATION

Article 3
Headquarters and structure of the International Tropical Timber Organization

1. The International Tropical Timber Organization established by the International Tropical Timber Agreement, 1983 shall continue in being for the purposes of administering the provisions and supervising the operation of this Agreement.

2. The Organization shall function through the Council established under article 6, the committees and other subsidiary bodies referred to in article 26 and the Executive Director and staff.

3. The headquarters of the Organization shall be in Yokohama, unless the Council, by special vote, decides otherwise.

4. The headquarters of the Organization shall at all times be located in the territory of a member.

Article 4
Membership in the Organization

There shall be two categories of membership in the Organization, namely:
a) Producing; and
b) Consuming.

Article 5
Membership by intergovernmental organizations

1. Any reference in this Agreement to 'Governments' shall be construed as including the European Community and any other intergovernmental organization having responsibilities in respect of the negotiation, conclusion and application of international agreements, in particular commodity agreements. Accordingly, any reference in this Agreement to signature, ratification, acceptance or approval, or to notification of provisional application, or to accession shall, in the case of such intergovernmental organizations, be construed as including a reference to signature, ratification, acceptance or approval, or to notification of provisional application, or to accession, by such intergovernmental organizations.

2. In the case of voting on matters within their competence, such intergovernmental organizations shall vote with a number of votes equal to the total number of votes attributable to their member States in accordance with article 10. In such cases, the member States of such intergovernmental organizations shall not be entitled to exercise their individual voting rights.

CHAPTER IV. INTERNATIONAL TROPICAL TIMBER COUNCIL

Article 6
Composition of the International Tropical Timber Council

1. The highest authority of the Organization shall be the International Tropical Timber Council, which shall consist of all the members of the Organization.
2. Each member shall be represented in the Council by one representative and may designate alternates and advisers to attend sessions of the Council.
3. An alternate representative shall be empowered to act and vote on behalf of the representative during the latter's absence or in special circumstances.

Article 7
Powers and functions of the Council

1. The Council shall exercise all such powers and perform or arrange for the performance of all such functions as are necessary to carry out the provisions of this Agreement.
2. The Council shall, by special vote, adopt such rules and regulations as are necessary to carry out the provisions of this Agreement and as are consistent therewith, including its own rules of procedure and the financial rules and staff regulations of the Organization. Such financial rules and regulations shall, *inter alia*, govern the receipt and expenditure of funds under the Administrative Account, the Special Account and the Bali Partnership Fund. The Council may, in its rules of procedure, provide for a procedure whereby it may, without meeting, decide specific questions.
3. The Council shall keep such records as are required for the performance of its functions under this Agreement.

Article 8
Chairman and Vice-Chairman of the Council

1. The Council shall elect for each calendar year a Chairman and a Vice-Chairman, whose salaries shall not be paid by the Organization.
2. The Chairman and the Vice-Chairman shall be elected, one from among the representatives of producing members and the other from among the representatives of consuming members. These offices shall alternate each year between the two categories of members, provided, however, that this shall not prohibit the re-election of either or both, under exceptional circumstances, by special vote of the Council.
3. In the temporary absence of the Chairman, the Vice-Chairman shall act in his place. In the temporary absence of both the Chairman and the Vice-Chairman, or in the absence of one or both of them for the rest of the term for which they were elected, the Council may elect new officers from amount the representatives of the producing members and/or from among the representatives of the consuming members, as the case may be, on a temporary basis or for the rest of the term for which the predecessor or predecessors were elected.

Article 9
Sessions of the Council

1. As a general rule, the Council shall hold at least one regular session a year.
2. The Council shall meet in special session whenever it so decides or at the request of:
   a) The Executive Director, in agreement with the Chairman of the Council; or
   b) A majority of producing members or a majority of consuming members; or
   c) Members holding at least 500 votes.
3. Sessions of the Council shall be held at the headquarters of the Organization unless the Council, by special vote, decides otherwise. If on the invitation of any member the Council meets elsewhere than at the headquarters of the Organization, that member shall pay the additional cost of holding the meeting away from headquarters.
4. Notice of any sessions and the agenda for such sessions shall be communicated to members by the Executive Director at least six weeks in advance, except in cases of emergency, when notice shall be communicated at least seven days in advance.

Article 10
Distribution of votes

1. The producing members shall together hold 1000 votes and the consuming members shall together hold 1000 votes.
2. The votes of the producing members shall be distributed as follows:
   a) Four hundred votes shall be distributed equally among the three producing regions of Africa, Asia-Pacific and Latin America. The votes thus allocated to each of these regions shall then be distributed equally among the producing members of that region;
   b) Three hundred votes shall be distributed among the producing members in accordance with their respective shares of the total tropical forest resources of all producing members; and
   c) Three hundred votes shall be distributed amount the producing members in pro-

portion to the average of the values of their respective net exports of tropical timber during the most recent three-year period for which definitive figures are available.

3. Notwithstanding the provisions of paragraph 2 of this article, the total votes allocated to the producing members from the African region, calculated in accordance with paragraph 2 of this article, shall be distributed equally among all producing members from the African region. If there are any remaining votes, each of these votes shall be allocated to a producing member from the African region: the first to the producing member which is allocated the highest number of votes calculated in accordance with paragraph 2 of this article, the second to the producing member which is allocated the second highest number of votes, and so on until all the remaining votes have been distributed.

4. For purposes of the calculation of the distribution of votes under paragraph 2 (b) of this article, 'tropical forest resources' means productive closed broadleaved forests as defined by the Food and Agriculture Organization (FAO).

5. The votes of the consuming members shall be distributed as follows: each consuming member shall have 10 initial votes: the remaining votes shall be distributed amount the consuming members in proportion to the average volume of their respective net imports of tropical timber during the three-year period commencing four calendar years prior to the distribution of votes.

6. The Council shall distribute the votes for each financial year at the beginning of its first session of that year in accordance with the provisions of this article. Such distribution shall remain in effect for the rest of that year, except as provided for in paragraph 7 of this article.

7. Whenever the membership of the Organization changes or when any member has its voting rights suspended or restored under any provision of this Agreement, the Council shall redistribute the votes within the affected category or categories of members in accordance with the provisions of this article. The Council shall, in that event, decide when such redistribution shall become effective.

8. There shall be no fractional votes.

Article 11
Voting procedure of the Council

1. Each member shall be entitled to cast the number of votes it holds and no member shall be entitled to divide its votes. A member may, however, cast differently from such votes any votes that it is authorized to cast under paragraph 2 of this article.

2. By written notification to the Chairman of the Council, any producing member may authorize, under its own responsibility, any other producing member, and any consuming member may authorize, under its own responsibility, any other consuming member, to represent its interests and to cast its votes at any meeting of the Council.

3. When abstaining, a member shall be deemed not to have cast its votes.

Article 12
Decisions and recommendations of the Council

1. The Council shall endeavour to take all decisions and to make all recommend-

ations by consensus. If consensus cannot be reached, the Council shall take all decisions and make all recommendations by a simple distributed majority vote, unless this Agreement provides for a special vote.

2. Where a member avails itself of the provisions of article 11, paragraph 2, and its votes are cast at a meeting of the Council, such member shall, for the purposes of paragraph 1 of this article, be considered as present and voting.

Article 13
Quorum for the Council

1. The quorum for any meeting of the Council shall be the presence of a majority of members of each category referred to in article 4, provided that such members hold at least two-thirds of the total votes in their respective categories.
2. If there is no quorum in accordance with paragraph 1 of this article on the day fixed for the meeting and on the following day, the quorum on the subsequent days of the session shall be the presence of a majority of members of each category referred to in article 4, provided that such members hold a majority of the total votes in their respective categories.
3. Representation in accordance with article 11, paragraph 2, shall be considered as presence.

Article 14
Cooperation and co-ordination with Other Organizations

1. The Council shall make arrangements as appropriate for consultations and cooperation with the United Nations and its organs, including the United Nations Conference on Trade and Development (UNCTAD) and the Commission on Sustainable Development (CSD), intergovernmental organizations, including the General Agreement on Tariffs and Trade (GATT) and the Convention on International Trade in Endangered Species of Wild Fauna and Flora (CITES), and non-governmental organizations.
2. The Organization shall, to the maximum extent possible, utilize the facilities, services and expertise of existing intergovernmental, governmental or non-governmental organizations, in order to avoid duplication of efforts in achieving the objectives of this Agreement and to enhance the complementarity and the efficiency of their activities.

Article 15
Admission of observers

The Council may invite any non-member Government or any of the organizations referred to in article 14, article 20 and article 29, interested in the activities of the Organization to attend as observers any of the meetings of the Council.

Article 16
Executive Director and staff

1. The Council shall, by special vote, appoint the Executive Director.
2. The terms and conditions of appointment of the Executive Director shall be determined by the Council.

b) in the event of the headquarters of the Organization being moved from the country of the host Government; or

c) in the event of the Organization ceasing to exist.

CHAPTER VI. FINANCE

Article 18
Financial accounts

1. There shall be established:
a) the Administrative Account;
   b) The Special Account;
   c) The Bali Partnership Fund; and
   d) Such other accounts as the Council shall deem appropriate and necessary.
2. The Executive Director shall be responsible for the administration of these accounts and the Council shall make provision therefore in the financial rules of the Organization.

Article 19
Administrative Account

1. The expenses necessary for the administration of this Agreement shall be brought into the Administrative Account and shall be met by annual contributions paid by members in accordance with their respective constitutional or institutional procedures and assessed in accordance with paragraphs 3, 4 and 5 of this article.
2. The expenses of delegations to the Council, the committees and any other subsidiary bodies of the Council referred to in article 26 shall be met by the members concerned. In cases where a member requests special services from the Organization, the Council shall require that member to pay the costs of such services.
3. Before the end of each financial year, the Council shall approve the administrative budget of the Organization for the following financial year and shall assess the contribution of each member to that budget.
4. The contribution of each member to the administrative budget for each financial year shall be in the proportion which the number of its votes at the time the administrative budget for that financial year is approved bears to the total votes of all the members. In assessing contributions, the votes of each member shall be calculated without regard to the suspension of any member's voting rights or any redistribution of votes resulting therefrom.
5. The initial contribution of any member joining the Organization after the entry into force of this Agreement shall be assessed by the Council on the basis of the number of votes to be held by that member and the period remaining in the current financial year, but the assessment made upon other members from the current financial year shall not thereby be altered.
6. Contributions to administrative budgets shall become due on the first day of each financial year. Contributions of members in respect of the financial year in which they join the Organization shall be due on the date on which they become members.

7. If a member has not paid its full contribution to the administrative budget within four months after such contribution becomes due in accordance with paragraph 6 of this article, the Executive Director shall request that member to make payment as quickly as possible. If that member has still not paid its contribution within two months after such request, that member shall be requested to state the reasons for its inability to make payment. If at the expiry of seven months from the due date of contribution, that member has still not paid its contribution, its voting rights shall be suspended until such time as it has paid in full its contribution, unless the Council, by special vote, decides otherwise. If, on the contrary, a member has paid its full contribution to the administrative budget within four months after such contribution becomes due in accordance with paragraph 6 of this article, the member's contribution shall receive a discount as may be established by the Council in the financial rules of the Organization.
8. A member whose rights have been suspended under paragraph 7 of this article shall remain liable to pay its contribution.

Article 20
Special Account

1. There shall be established two sub-accounts under the Special Account:
   a) The Pro-Project Sub-Account; and
   b) The Project Sub-Account.
2. The possible sources of finance for the Special Account may be:
   a) The Common Fund for Commodities;
   b) Regional and international financial institutions; and
   c) Voluntary contributions.
3. The resources of the Special Account shall be used only for approved pre-projects or projects.
4. All expenditures under the Pre-Project Sub-Account shall be reimbursed from the Project Sub-Account if projects are subsequently approved and funded. If within six months of the entry into force of this Agreement the Council does not receive any funds for the Pre-Project Sub-Account, it shall review the situation and take appropriate action.
5. All receipts pertaining to specific identifiable pre-projects or projects under the Special Account shall be brought into that Account. All expenditures incurred on such pre-projects or projects, including remuneration and travel expenses of consultants and experts, shall be charged to the same Account.
6. The Council shall, by special vote, establish terms and conditions on which it would, when and where appropriate, sponsor projects for loan financing, where a member or members have voluntarily assumed full obligations and responsibilities for such loans. The Organization shall have no obligations for such loans.
7. The Council may nominate and sponsor any entity with the consent of that entity, including a member or members, to receive loans for the financing of approved projects and to undertake all the obligations involved, except that the Organization shall reserve to itself the right to monitor the use of resources and to follow up on the implementation of projects so financed. However, the Organization shall not be responsible for guarantees voluntarily provided by individual members or other entities.

8. No member shall be responsible by reason of its membership in the Organization for any liability arising from borrowing or lending by any other member or entity in connection with projects.
9. In the event that voluntary unearmarked funds are offered to the Organization, the Council may accept such funds. Such funds may be utilized for approved pre-projects and projects.
10. The Executive Director shall endeavour to seek, on such terms and conditions as the Council may decide, adequate and assured finance for pre-projects and projects approved by the Council.
11. Contributions for specified approved projects shall be used only for the projects for which they were originally intended, unless otherwise decided by the Council in agreement with the contributor. After the completion of a project, the Organization shall return to each contributor for specific projects the balance of any funds remaining *pro rata* to each contributor's share in the total of the contributions originally made available for financing that project, unless otherwise agreed to by the contributor.

Article 21
The Bali Partnership Fund

1. A Fund for sustainable management of tropical timber producing forests is hereby established to assist producing members to make the investments necessary to achieve the objective of article 1 (d) of this Agreement.
2. The Fund shall be constituted by:
   a) Contributions from donor members;
   b) Fifty per cent of income earned as a result of activities related to the Special Account; and
   c) Resources from other private and public sources which the Organization may accept consistent with its financial rules.
3. Resources of the Fund shall be allocated by the Council only for pre-projects and projects for the purpose set out in paragraph 1 of this article and approved in accordance with article 25.
4. In allocating resources of the Fund, the Council shall take into account:
   a) The special needs of members whose forestry sectors' contribution to their economies is adversely affected by the implementation of the strategy for achieving the exports of tropical timber and timber products from sustainably managed sources by the year 2000; and
   b) The needs of members with significant forest areas who establish conservation programmes in timber producing forests.
5. The Council shall examine annually the adequacy of the resources available to the Fund and endeavour to obtain additional resources needed by producing members to achieve the purpose of the Fund. The ability of members to implement the strategy referred to in paragraph 4 (a) of this article will be influenced by the availability of resources.
6. The Council shall establish policies and financial rules for the operation of the Fund, including rules covering the settlement of accounts on termination or expiry of this Agreement.

Article 22
Forms of payment

1. Contributions to the Administrative Account shall be payable in freely usable currencies and shall be exempt from foreign-exchange restrictions.
2. Financial contributions to the Special Account and the Bali Partnership Fund shall be payable in freely usable currencies and shall be exempt from foreign-exchange restrictions.
3. The Council may also decide to accept other forms of contributions to the Special Account or the Bali Partnership Fund, including scientific and technical equipment or personnel, to meet the requirements of approved projects.

Article 23
Audit and publication of accounts

1. The Council shall appoint independent auditors for the purpose of auditing the accounts of the Organization.
2. Independently audited statements of the Administrative Account, of the Special Account, and of the Bali Partnership Fund shall be made available to members as soon as possible after the close of each financial year, but not later than six months after that date, and be considered for approval by the Council at its next session, as appropriate. A summary of the audited accounts and balance sheet shall thereafter be published.

CHAPTER VII. OPERATIONAL ACTIVITIES

Article 24
Policy work of the Organization

In order to achieve the objectives set out in article 1, the Organization shall undertake policy work and project activities in the areas of Economic Information and Market Intelligence, Reforestation and Forest Management and Forest Industry, in a balanced manner, to the extent possible integrating policy work and project activities.

Article 25
Project activities of the Organization

1. Bearing in mind the needs of developing countries, members may submit pre-project and project proposals to the Council in the fields of research and development, market intelligence, further and increased wood processing in producing member countries, and reforestation and forest management. Pre-projects and projects should contribute to the achievement of one or more of the objectives of this Agreement.
2. The Council, in approving pre-projects and projects, shall take into account:
   a) their relevance to the objectives of this Agreement;
   b) their environmental and social effects;
   c) the desirability of maintaining an appropriate geographical balance;
   d) the interests and characteristics of each of the developing producing regions;

e) the desirability of equitable distribution of resources among the fields referred to in paragraph 1 of this article;

f) their cost-effectiveness; and

g) the need to avoid duplication of efforts.

3. The Council shall establish a schedule and procedure for submitting, appraising, and prioritizing pre-projects and projects seeking funding from the Organization, as well as for their implementation, monitoring and evaluation. The Council shall decide on the approval of pre-projects and projects for financing or sponsorship in accordance with article 20 or article 21.

4. The Executive Director may suspend disbursement of the Organization's funds to a pre-project or project if they are being used contrary to the project document or in cases of fraud, waste, neglect or mismanagement. The Executive Director will provide to the Council at its next session a report for its consideration. The Council shall take appropriate action.

5. The Council may, by special vote, terminate its sponsorship of any pre-project or project.

Article 26
Establishment of Committees

1. The following are hereby established as Committees of the Organization:
   a) Committee on Economic Information and Market Intelligence;
   b) Committee on Reforestation and Forest Management;
   c) Committee on Forest Industry; and
   d) Committee on Finance and Administration.

2. The Council may, by special vote, establish such other committees and subsidiary bodies as it deems appropriate and necessary.

3. Participation in each of the committees shall be open to all members. The rules of procedure of the committees shall be decided by the Council.

4. The committees and subsidiary bodies referred to in paragraphs 1 and 2 of this article shall be responsible to, and work under the general direction of, the Council. Meetings of the committees and subsidiary bodies shall be convened by the Council.

Article 27
Functions of the Committees

1. The Committee on Economic Information and Market Intelligence shall:
   a) Keep under review the availability and quality of statistics and other information required by the Organization;
   b) Analyse the statistical data and specific indicators as decided by the Council for the monitoring of international timber trade;
   c) Keep under continuous review the international timber market, its current situation and short-term prospects on the basis of the data mentioned in subparagraph (b) above and other relevant information, including information related to undocumented trade;
   d) Make recommendations to the Council on the need for, and nature of, appropriate studies on tropical timber, including prices, market elasticity, market substitutability, marketing of new products, and long-term prospects of the

international tropical timber market, and monitor and review any studies commissioned by the Council;

e) Carry out any other tasks related to the economic, technical and statistical aspects of timber assigned to it by the Council; and

f) Assist in the provision of technical cooperation to developing member countries to improve their relevant statistical services.

2. The Committee on Reforestation and Forest Management shall:

a) Promote cooperation between members as partners in development of forest activities in member countries, *inter alia*, in the following areas:
(i) Reforestation;
(ii) Rehabilitation; and
(iii) Forest management.

b) Encourage the increase of technical assistance and transfer of technology in the fields of reforestation and forest management to developing countries;

c) Follow up ongoing activities in this field, and identify and consider problems and possible solutions to them in cooperation with the competent organizations;

d) Review regularly the future needs of international trade in industrial tropical timber and, on this basis, identify and consider appropriate possible schemes and measures in the field of reforestation, rehabilitation and forest management;

e) Facilitate the transfer of knowledge in the field of reforestation and forest management with the assistance of competent organizations; and

f) Co-ordinate and harmonize these activities for cooperation in the field of reforestation and forest management with relevant activities pursued elsewhere, such as those under the auspices of the Food and Agriculture Organization (FAO), the United Nations Environment Programme (UNEP), the World Bank, the United Nations Development Programme (UNDP), regional development banks and other competent organizations.

3. The Committee on Forest Industry shall:

a) Promote cooperation between member countries as partners in the development of processing activities in producing member countries, *inter alia,* in the following areas:
(i) Product development through transfer of technology;
(ii) Human resources development and training;
(iii) Standardization of nomenclature of tropical timber;
(iv) Harmonization of specifications of processed products;
(v) Encouragement of investment and joint ventures; and
(vi) Marketing including the promotion of lesser known and lesser used species.

b) Promote the exchange of information in order to facilitate structural changes involved in increased and further processing in the interests of all member countries, in particular developing member countries;

c) Follow up on-going activities in this field, and identify and consider problems and possible solutions to them in cooperation with the competent organizations; and

d) Encourage the increase of technical cooperation for the processing of tropical timber for the benefit of producing member countries.

4. In order to promote the policy and project work of the Organization in a balanced manner the Committee on Economic Information and Market Intelligence, the Committee on Reforestation and Forest Management and the Committee on Forest Industry shall each:

   a) Be responsible for ensuring the effective appraisal, monitoring and evaluation of pre-projects and projects;

   b) Make recommendations to the Council relating to pre-projects and projects;

   c) Follow up the implementation of pre-projects and projects and provide for the collection and dissemination of their results as widely as possible for the benefit of all members;

   d) Develop and advance policy ideas to the Council;

   e) Review regularly the results of project and policy work and make recommendations to the Council on the future of the Organization's programme;

   f) Review regularly the strategies, criteria and priority areas for programme development and project work contained in the Organization's Action Plan and recommend revisions to the Council;

   g) Take account of the need to strengthen capacity building and human resource development in member countries; and

   h) Carry out any other task related to the objectives of this Agreement assigned to them by the Council.

5. Research and development shall be a common function of the Committees referred to in paragraphs 1, 2, and 3 of this article.

6. The Committee on Finance and Administration shall:

   a) Examine and make recommendations to the Council regarding the approval of the Organization's administrative budget proposals and the management operations of the Organization;

   b) Review the assets of the Organization to ensure prudent asset management and that the Organization has sufficient reserves to carry on its work;

   c) Examine and make recommendations to the Council on the budgetary implications of the Organization's annual work programme, and the actions that might be taken to secure the resources needed to implement it;

   d) Recommend to the Council the choice of independent auditors and review the independent audited statements;

   e) Recommend to the Council any modifications it may judge necessary to the Rules of Procedure or the Financial Rules;

   f) Review the Organization's revenues and the extent to which they constrain the work of the Secretariat.

CHAPTER VIII. RELATIONSHIP WITH THE COMMON FUND FOR COMMODITIES

Article 28
Relationship with the Common Fund for Commodities

The Organization shall take full advantage of the facilities of the Common Fund for Commodities.

CHAPTER IX. STATISTICS, STUDIES AND INFORMATION

Article 29
Statistics, studies and information

1. The Council shall establish close relationships with relevant intergovernmental, governmental and non-governmental organizations, in order to help ensure the availability of recent reliable data and information on the trade in tropical timber, as well as relevant information on non-tropical timber and on the management of timber producing forests. As deemed necessary for the operation of this Agreement, the Organization, in cooperation with such organizations, shall compile, collate and, where relevant, publish statistical information on production, supply, trade, stocks, consumption and market prices of timber, the extent of timber resources and the management of timber producing forests.
2. Members shall, to the fullest extent possible not inconsistent with their national legislation, furnish, within a reasonable time, statistics and information on timber, its trade and the activities aimed at achieving sustainable management of timber producing forests as well as other relevant information as requested by the Council. The Council shall decide on the type of information to be provided under this paragraph and on the format in which it is to be presented.
3. The Council shall arrange to have any relevant studies undertaken of the trends and of short and long-term problems of the international timber markets and of the progress towards the achievement of sustainable management of timber producing forests.

Article 30
Annual report and review

1. The Council shall, within six months after the close of each calendar year, publish an annual report on its activities and such other information as it considers appropriate.
2. The Council shall annually review and assess:
   a) The international timber situation; and
   b) Other factors, issues and developments considered relevant to achieve the objectives of this Agreement.
3. The review shall be carried out in the light of:
   a) Information supplied by members in relation to national production, trade, supply, stocks, consumption and prices of timber;

b) Other statistical data and specific indicators provided by members as requested by the Council;

c) Information supplied by members on their progress towards the sustainable management of their timber producing forests; and

d) Such other relevant information as may be available to the Council either directly or through the organizations in the United Nations system and inter-governmental, governmental or non-governmental organizations.

4. The Council shall promote the exchange of views among member countries regarding:

a) The status of sustainable management of timber producing forests and related matters in member countries; and

b) Resources flows and requirements in relation to objectives, criteria and guidelines set by the Organization.

5. Upon request, the Council shall endeavour to enhance the technical capacity of member countries, in particular developing member countries, to obtain the data necessary for adequate information-sharing, including the provision of resources for training and facilities to members.

6. The results of the review shall be included in the reports of the Council's deliberations.

CHAPTER X. MISCELLANEOUS

Article 31
Complaints and disputes

Any complaint that a member has failed to fulfil its obligations under this Agreement and any dispute concerning the interpretation or application of this Agreement shall be referred to the Council for decision. Decisions of the Council on these matters shall be final and binding.

Article 32
General obligations of members

1. Members shall, for the duration of this Agreement, use their best endeavours and cooperate to promote the attainment of its objectives and to avoid any action contrary thereto.

2. Members undertake to accept and carry out the decisions of the Council under the provisions of this Agreement and shall refrain from implementing measures which would have the effect of limiting or running counter to them

Article 33
Relief from obligations

1. Where it is necessary on account of exceptional circumstances or emergency or *force majeure* not expressly provided for in this Agreement, the Council may, by special vote, relieve a member of an obligation under this Agreement if it is satisfied by an explanation from that member regarding the reasons why the obligation cannot be met.

2. The Council, in granting relief to a member under paragraph 1 of this article, shall state explicitly the terms and conditions on which, and the period for which, the member is relieved of such obligation, and the reasons for which the relief is granted.

Article 34
Differential and remedial measures and special measures

1. Developing importing members whose interests are adversely affected by measures taken under this Agreement may apply to the Council for appropriate differential and remedial measures. The Council shall consider taking appropriate measures in accordance with section III, paragraphs 3 and 4, of resolution 93 (IV) of the United Nations Conference on Trade and Development.
2. Members in the category of least developed countries as defined by the United Nations may apply to the Council for special measures in accordance with section III, paragraph 4, of resolution 93 (IV) and with paragraphs 56 and 57 of the Paris Declaration and Programme of Action for the Least Developed Countries for the 1990s.

Article 35
Review

The Council shall review the scope of this Agreement four years after its entry into force.

Article 36
Non-discrimination

Nothing in this Agreement authorizes the use of measures to restrict or ban international trade in, and in particular as they concern imports of and utilization of, timber and timber products.

CHAPTER XI. FINAL PROVISIONS

Article 37
Depositary

The Secretary-General of the United Nations is hereby designated as the depositary of this Agreement.

Article 38
Signature, ratification, acceptance and approval

1. This Agreement shall be open for signature, at United Nations Headquarters from 1 April 1994 until one month after the date of its entry into force, by Governments invited to the United Nations Conference for the Negotiation of a Successor Agreement to the International Tropical Timber Agreement, 1983.
2. Any Government referred to in paragraph 1 of this article may:
   a) At the time of signing this Agreement, declare that by such signature it expresses its consent to be bound by this Agreement (definitive signature); or

b) After signing this Agreement, ratify, accept or approve it by the deposit of an instrument to that effect with the depositary.

Article 39
Accession

1. This agreement shall be open for accession by the Governments of all states upon conditions established by the Council, which shall include a time-limit for the deposit of instruments of accession. The Council may, however, grant extensions of time to Governments which are unable to accede by the time-limit set in the conditions of accession.
2. Accession shall be effected by the deposit of an instrument of accession with the depositary.

Article 40
Notification of provisional application

A signatory Government which intends to ratify, accept or approve this Agreement, or a Government for which the Council has established conditions for accession but which has not yet been able to deposit its instrument, may, at any time, notify the depositary that it will apply this Agreement provisionally either when it enters into force in accordance with article 41, or, if it is already in force, at a specified date.

Article 41
Entry into force

1. This Agreement shall enter into force definitively on 1 February 1995 or on any date thereafter, if 12 governments of producing countries holding at least 55 per cent of the total votes as set out in annex A to this Agreement, and 16 Governments of consuming countries holding at least 70 per cent of the total votes as set out in annex B to this Agreement have signed this Agreement definitively or have ratified, accepted or approved it or acceded thereto pursuant to article 38, paragraph 2, or article 39.
2. If this Agreement has not entered into force definitively on 1 February 1995, it shall enter into force provisionally on that date or on any date within six months thereafter, if, 10 Governments of producing countries holding at least 50 per cent of the total votes as set out in annex A to this Agreement, and 14 governments of consuming countries holding at least 65 per cent of the total votes as set out in annex B to this Agreement, have signed this Agreement definitively or have ratified, accepted or approved it pursuant to article 38, paragraph 2, or have notified the depositary under article 40 that they will apply this Agreement provisionally.
3. It the requirements for entry into force under paragraph 1 or paragraph 2 of this article have not been met on 1 September 1995, the Secretary-General of the United Nations shall invite those Governments which have signed this Agreement definitively or have ratified, accepted or approved it pursuant to article 38, paragraph 2, or have notified the depositary that they will apply this Agreement provisionally, to meet at the earliest time practicable to decide whether to put this Agreement into force provisionally or definitively among themselves in whole or in part. Governments which decide to put this Agreement into force provision-

ally among themselves may meet from time to time to review the situation and decide whether this Agreement shall enter into force definitively among themselves.

4. For any Government which has not notified the depositary under article 40 that it will apply this Agreement provisionally and which deposits its instrument of ratification, acceptance, approval or accession after the entry into force of this Agreement, this Agreement shall enter into force on the date of such deposit.

5. The Executive Director of the Organization shall convene the Council as soon as possible after the entry into force of this Agreement.

Article 42
Amendments

1. The Council may, by special vote, recommend an amendment of this Agreement to members.

2. The Council shall fix a date by which members shall notify the depositary of their acceptance of the amendment.

3. An amendment shall enter into force 90 days after the depositary has received notifications of acceptance from members constituting at least two-thirds of the producing members and accounting for at least 75 per cent of the votes of the producing members, and from members constituting at least two-thirds of the consuming members and accounting for at least 75 per cent of the votes of the consuming members.

4. After the depositary informs the Council that the requirements for entry into force of the amendment have been met, and notwithstanding the provisions of paragraph 2 of this article relating to the date fixed by the Council, a member may still notify the depositary of its acceptance of the amendment, provided that such notification is made before the entry into force of the amendment.

5. Any member which has not notified its acceptance of an amendment by the date on which such amendment enters into force shall cease to be a party to this Agreement as from that date, unless such member has satisfied the Council that its acceptance could not be obtained in time owing to difficulties in completing its constitutional or institutional procedures, and the Council decides to extend for that member the period for acceptance of the amendment. Such member shall not be bound by the amendment before it has notified its acceptance thereof.

6. If the requirements for the entry into force of the amendment have not been met by the date fixed by the Council in accordance with paragraph 2 of this article, the amendment shall be considered withdrawn.

Article 43
Withdrawal

1. A member may withdraw from this Agreement at any time after the entry into force of this Agreement by giving written notice of withdrawal to the depositary. That member shall simultaneously inform the Council of the action it has taken.

2. Withdrawal shall become effective 90 days after the notice is received by the depositary.

3. Financial obligations to the Organization incurred by a member under this Agreement shall not be terminated by its withdrawal.

Article 44
Exclusion

If the Council decides that any member is in breach of its obligations under this Agreement and decides further that such breach significantly impairs the operation of this Agreement , it may, by special vote, exclude that member from this Agreement. The Council shall immediately so notify the depositary. Six months after the date of the Council's decision, that member shall cease to be a party to this Agreement.

Article 45
Settlement of accounts with withdrawing or excluded members or members unable to accept an amendment

1. The Council shall determine any settlement of accounts with a member which ceases to be a party to this Agreement owing to:
   a) Non-acceptance of an amendment to this Agreement under article 42;
   b) Withdrawal from this Agreement under article 43; or
   c) Exclusion from this Agreement under article 44.
2. The Council shall retain any contribution paid to the Administrative Account, to the Special Account, or to the Bali Partnership Fund by a member which ceases to be a party to this Agreement.
3. A member which has ceased to be a party to the Agreement shall not be entitled to any share of the proceeds of liquidation or the other assets of the Organization. Nor shall such member be liable for payment of any part of the deficit, if any, of the Organization upon termination of this Agreement.

Article 46
Duration, extension and termination

1. This Agreement shall remain in force for a period of four years after its entry into force unless the Council, by special vote, decides to extend, renegotiage or terminate it in accordance with the provisions of this article.
2. The Council may, by special vote, decide to extend this Agreement for two periods of three years each.
3. If, before the expiry of the four-year period referred to in paragraph 1 of this article, or before the expiry of an extension period referred to in paragraph 2 of this article, as the case may be, a new agreement to replace this Agreement has been negotiated but has not yet entered into force either definitely or provisionally, the Council may, by special vote, extend this Agreement until the provisional or definitive entry into force of the new agreement.
4. If a new agreement is negotiated and enters into force during any period of extension of this Agreement under paragraph 2 or paragraph 3 of this article, this Agreement, as extended, shall terminate upon the entry into force of the new agreement.
5. The Council may at any time, by special vote, decide to terminate this Agreement with effect from such date as it may determine.
6. Notwithstanding the termination of this Agreement, the Council shall continue in being for a period not exceeding 18 months to carry out the liquidation of the Organization, including the settlement of accounts, and, subject to relevant deci-

sions to be taken by special vote, shall have during that period such powers and functions as may be necessary for these purposes.

7. The Council shall notify the depositary of any decision taken under this article.

Article 47
Reservations

Reservations may not be made with respect to any of the provisions of this Agreement.

Article 48
Supplementary and transitional provisions

1. This Agreement shall be the successor to the International Tropical Timber Agreement, 1983.
2. All acts by or on behalf of the Organization or any of its organs under the International Tropical Timber Agreement, 1983, which are in effect on the date of entry into force of this Agreement and the terms of which do not provide for expiry on that date shall remain in effect unless changed under the provisions of this Agreement.

IN WITNESS WHEREOF the undersigned, being duly authorized thereto, have affixed their signatures under this Agreement on the dates indicated.

DONE at Geneva, on the twenty-six day of January, one thousand nine hundred and ninety-four, the text of this Agreement in the Arabic, Chinese, English, French, Russian and Spanish languages being equally authentic.

ANNEX A OF THE INTERNATIONAL TROPICAL TIMBER AGREEMENT, 1994

List of producing countries with tropical forest resources and/or net exporters of tropical timber in volume terms, and allocation of votes for the purposes of article 41

Bolivia	21
Brazil	133
Cameroon	23
Colombia	24
Congo	23
Costa Rica	9
Côte d'Ivoire	23
Dominican Republic	9
Ecuador	14
El Salvador	9
Equatorial Guinea	23
Gabon	23
Ghana	23
Guyana	14
Honduras	9
India	34
Indonesia	170
Liberia	23
Malaysia	139
Mexico	14
Myanmar	33
Panama	10
Papua New Guinea	28
Paraguay	11
Peru	25
Philippines	25
Tanzania, United Republic of	23
Thailand	20
Togo	23
Trinidad & Tobago	9
Venezuela	10
Zaire	23

Total   1000

ANNEX B OF THE INTERNATIONAL TROPICAL TIMBER AGREEMENT, 1994

List of consuming countries and allocation of votes for the purposes of article 41

Afghanistan	10
Algeria	13
Australia	18
Austria	11
Bahrain	11
Bulgaria	10
Canada	12
Chile	10
China	36
Egypt	14
European Union	(302)
Belgium/Luxembourg	26
Denmark	11
France	44
Germany	35
Greece	13
Ireland	13
Italy	35
Netherlands	40
Portugal	18
Spain	25
United Kingdom	42
Finland	10
Japan	320
Nepal	10
New Zealand	10
Norway	10
Republic of Korea	97
Russian Federation	13
Slovakia	11
Sweden	10
Switzerland	11
United States of America	51

Total   1000

Annex D:
Draft Text for a Convention for the Conservation and Wise Use of Forests

GLOBAL LEGISLATORS ORGANISATION FOR A BALANCED ENVIRONMENT,[1] APRIL 1992

PART I. INTRODUCTION AND GENERAL MATTERS

A. Preamble

The Parties to this Convention,

Prompted by the desire to address the special significance of forests in the environment and as prime natural resources and developmental assets in urgent need of protection and at the same time to enable the most profound understanding of the conservation and wise use of forests to be represented in a global agreement;

Conscious that forests are some of the richest sources of biological diversity on Earth and that while nature warrants respect regardless of its worth to man, forests are also of worth for the survival and well-being of all humanity.

Recognizing the vital contribution of forests and forest protection in maintaining local and global climatic stability;

Convinced of the fundamental connection between the conservation and wise use of forests and the individual and collective human rights of forest peoples;

Recognizing that factors outside the forestry sector play a large part in determining the state of forests;

Whereas international action is necessary to promote conservation of nature, natural resources, and the environment to stimulate and support national action;

Bearing in mind General Assembly Resolutions 42/184, 42/186 and 42/187 of 11 December 1987, General Assembly Resolution 43/53 of 27 January 1989, the Declaration of the United Nations Conference on the Human Environment and in particular Principle 21;

---

1. Prepared for the Global Legislators Organisation for a Balanced Environment (GLOBE) by the Centre for International Environmental Law (now the Foundation for International Environmental Law and Development) with advice and input from AIDEnvironment, the International Institute for Environment and Development and the Gaia Foundation.

Noting parallel international legal initiatives to negotiate a Climate Change Convention and a Convention on Biological Diversity;

Believing that the creation and progressive development of law relating to the conservation and wise use of forests will contribute to the strengthening of cooperation and friendly relations between all members of international society in conformity with the principles of justice and equity;

Convinced that the economic and social advancement of all peoples of the world can only occur in harmony if the need to protect and preserve the environmental health of the planet is fully addressed;

Affirming that matters not regulated by this Convention continue to be governed by the rules and principles of general international law,

Have agreed as follows:

Article 1
Definitions

1. For the purposes of this Convention:

a) 'Conservation' means the directing of human behaviour towards the preservation, restoration and enhancement of the biosphere in order that the needs and aspirations of present and future generations may be satisfied and the inherent value and integrity of all life which is part of the biosphere be respected.

b) 'Sustainable development' means the progressive economic and social development of human society through maintaining the security of livelihood for all peoples and by enabling them to meet their present needs, together with a quality of life in accordance with their dignity and well-being, without compromising the ability of future generations to do likewise.

c) 'Wise use' means the use of nature, natural resources and the environment in a way which, based on the imperatives of conservation is capable of achieving sustainable development.

(d) 'Precautionary principle' means the principle establishing a duty to take such measures that anticipate, prevent and attack the causes of environmental degradation where there is sufficient evidence to identify a threat of serious or irreversible harm to the environment even if there is not yet scientific proof that the environment is being harmed.

e) 'Natural resources' refers to natural assets which are either renewable (such as forests, water, flora and fauna and soils) or non-renewable (such as oil, coal and minerals).

f) 'Forests' refers to biotic communities characterized by the predominance of woody vegetation in all climatic zones (boreal, temperate and tropical) including both closed canopy forests and open canopy woodlands; in the context of this Convention the term 'forests' includes *forest lands* and therefore includes their natural resources; the term covers 'natural forest', 'modified forest', and 'planted forest' as hereinafter defined.

g) 'Natural forest' refers to forests which have not been subjected to significant anthropogenic disturbances resulting in severe alteration of the forest structure, composition and physiognomy for the past 250 years.

h) 'Modified forest' refers to forests which have been subjected to significant anthropogenic disturbances resulting in alteration of the forest structure by, for example, logging or shifting cultivation but retains a tree or shrub cover of indigenous species; modified forests, also known as secondary forests, include a wide range of conditions from those that have been selectively logged to those that have been heavily modified.

i) 'Planted forest' refers to forests in which all or most trees (51 per cent or more) have been planted or sown by people.

j) 'Forest peoples' refers to tribal and indigenous peoples who live in forests and/or make claims to forest and all those who live in or near forests whose long established ways of life and livelihoods are closely and directly dependent on forests.

k) 'Deforestation' means the permanent removal of forest vegetation and/or the conversion of any forest (natural, modified or planted) to non-forest uses such as cropland, pasture or urban areas.

l) 'Forest degradation' means the reduction of the biomass, productivity and/or biological diversity of a forest to such an extent that its capacity to renew itself is significantly reduced and its contribution to hydrological cycles and local and global climate stability significantly altered and 'degraded forests' shall be interpreted accordingly.

m) 'Protected forest areas' means areas of forest that are subject to legal and/or administrative measures that limit or regulate human use of the natural resources within that area.

n) 'Protected forest area' means an area established by a Party in accordance with paragraph (1) of Article 8.

o) 'Buffer zones' means areas peripheral to protected forest areas which have use restrictions to give added protection but where activities that are compatible with the objectives of the protected forest area are permitted and provide compensation to local people for the loss of access to the natural resources of the protected forests areas.

p) 'National forest estate' means all the forest to be found in the area under the national jurisdiction of a Party whether or not currently classified as such.

q) 'Permanent forest estate' refers to forests designated for permanent conservation or production purposes pursuant to paragraph (1) of Article 4.

r) 'Transboundary forest resource' refers to a forest that physically crosses the boundary between an area under the national jurisdiction of one Party and an area under the national jurisdiction another Party (or State) or an area beyond the limits of national jurisdiction to the extent that its conservation or use in an area under the jurisdiction of one Party may affect its conservation and use in an area under the jurisdiction of another Party (or State) or in an area beyond the limits of national jurisdiction or visa versa.

s) 'Participation' means the democratic right, both individual and collective, to contribute to and be consulted by the political and legal decision-making processes which exercise control over the use of natural resources and the protection of nature and the environment including therefore, access to the public administration, the regulatory authorities, and the legal system.

t) 'Water pollution' means the introduction by peoples, directly or indirectly, of substances or energy into either underground, surface or coastal waters resulting in deleterious effects of such a nature as either to endanger human health, harm living natural resources or their ecosystems or material property and impair or interfere with amenities and other legitimate uses of nature, natural resources and the environment and 'water pollutants' shall be interpreted accordingly.

u) 'Air pollution' means the introduction by humans, directly or indirectly, of substances or energy into the air resulting in deleterious effects of such a nature as either to endanger human health, harm living natural resources or their ecosystems or material property and impair or interfere with amenities and other legitimate uses of the environment and 'air pollutants' shall be interpreted accordingly.

v) 'Long-range transboundary air pollution' means air pollution whose physical origin is situated wholly or in part within the area under the national jurisdiction of one Party and which has adverse effects in the area under the jurisdiction of another Party at such a distance that it is not generally possible to distinguish the contributions of individual emission sources or groups of sources.

w) 'Biological diversity' means the diversity of species living on the Earth, or in respect of any Party the diversity of indigenous species living in the area under the national jurisdiction of that Party; biological diversity includes genetic diversity, which is the diversity of genes and genotypes within each species; taxonomic diversity, which is the diversity between species; and ecological diversity, which is the diversity of the different types of community formed by living organisms and the relations between them.

x) 'Adverse effects' means changes in the physical environment or biota, including changes in climate, which have significant deleterious effects on human health or on the composition, resilience or productivity of natural and managed ecosystems, or on materials useful to mankind.

y) 'Parties' means the State Parties to this Convention including any regional economic integration organization constituted by sovereign States of a given region which has competence in respect of matters governed by this Convention or its protocols and has been duly authorized, in accordance with its internal procedures, to sign, ratify, accept, approve or accede to the instruments concerned.

z) 'the Authority' refers to the Multilateral Forest Authority established in accordance with Article 26.

aa) 'the Fund' refers to the Multilateral Forest Fund established in accordance with Article 35.

bb) 'the Conference' means the Conference of the Parties to this Convention.

Article 2
General Principles and Common Objectives

1. To strengthen common understanding Parties hereby agree that they shall be guided in the application, interpretation and implementation of the obligations contained in this Convention by the following general principles for natural resources management:

a) Guardianship and sovereignty

   (i) All members of international society bear the responsibility to protect nature, natural resources and the environment for the benefit of present and future generations.

   (ii) As bearers of the rights and obligations of sovereignty States have a special responsibility to ensure that the natural resources of the earth and especially representative and/or significant samples of natural ecosystems are safe-guarded for the benefit of present and future generations.

(iii) States have, in accordance with the Charter of the United Nations, principles of international law and subject to this Convention, the sovereign right to exploit their own resources pursuant to their own environmental policies, together with the responsibility to ensure that activities within their own jurisdiction and control do not cause avoidable damage to the environment and in particular to the environment of other States or areas beyond the limits of national jurisdiction.

b) Conservation and wise use of natural resources

(i) Parties shall maintain essential ecological processes and life support systems and preserve biological diversity through conservation and the wise use of nature, the environment and all natural and human resources.

(ii) Parties shall, in the formulation of all development plans, programmes and policies, give full consideration to ecological factors as well as economic and social ones and ensure that conservation plays an integral part in the implementation of these.

(iii) Parties shall ensure that the securing of individual, group and collective rights of persons to their livelihoods and to an environment adequate for their health and well-being is an integral part of the conservation and wise use of natural resources.

c) Precaution and protection.

(i) Parties shall recognize and undertake to apply the precautionary principle in every aspect of governance where natural resources or the environment are affected and shall accept that the burden of proving that any present or planned human economic activity does not threaten to cause avoidable serious or irreversible harm to the environment is fixed on the Party within whose territory that activity is continuing or is planned.

d) Partnership and cooperation.

(i) Parties shall collaborate and work together in partnership to encourage cooperation between and among States and to promote the work of inter-governmental and non-governmental organizations on relevant matters concerning the protection of nature, natural resources and the environment.

(ii) Accepting the need to take into account the differing needs and ability of countries to respond to environmental and ecological concerns, Parties recognize the need to provide additional and adequate resources to assist the fulfilment of the Common Objectives of this Convention.

e) Responsibility and liability.

(i) Parties shall be liable to make reparation for the consequences of the breach of their international legal obligations in accordance with international law.

2. Parties undertake to apply the General Principles set out in paragraph 1 in order to achieve the Common Objectives of this Convention:

a) Common Objectives

(i) Conservation and wise use of forests and dependent and associated biological diversity, soils and watersheds to ensure continued performance of essential.

(ii) Reduction and prevention of air and water pollution and other negative impacts threatening the existence of forests, reducing biodiversity or productive capacity and leading to long-term forest degradation.

(iii) Adoption of measures to arrest deforestation, particularly of natural forest, and immediate serious or irreversible forest degradation.

(iv) Respect of the collective and individual rights of forest peoples and local communities to use customary lands according to social and cultural traditions which promote wise use and to secure their livelihoods and way of life.

(v) Participation at local, national, and international levels in the preparation, implementation, and evaluation of measures affecting forest peoples, local communities, and all other forest users including the provision of information on measures affecting forests to all groups in an appropriate form.

(vi) Effective coordination of policy, planning and implementation at all levels and among all sectors related to primary land use.

(vii) Reforestation and regeneration of the productivity and biological diversity of suitable degraded forests (reason: reforestation stands on itself, productivity or biodiversity can not be reforested.)

(viii) Afforestation of appropriate lands to ease the burden on and secure the stability of existing forests.

(ix) Establishment at the national, regional and international levels of facilities to further scientific research and assessment, education and training, monitoring and provision of information and the implementation of any other infrastructure support necessary to implement the obligations of this Convention and its protocols.

PART II. OBLIGATIONS

Article 3
National Development and Land Use Policy

1. Parties shall take all necessary measures, within the framework of their respective national laws, to ensure that forests are regarded as a national asset by all sectors and that the conservation and wise use of forests is treated as an integral part of development policy planning at all stages and at all levels.

2. Parties shall identify or create and maintain the appropriate coordinating administrative machinery necessary to review and formulate national policies to ensure that representatives from the whole range of sectors which use, convert or otherwise affect the state of forests take into account:

a) an evaluation of all the benefits which forests can provide;

b) the need to treat forests as capital to ensure that everything is done to avoid irreversible reduction of the potential of forests;

c) the need to ensure that harvests do not exceed sustainable levels;

d) the need to maintain biological diversity; and

e) the needs of forests people living in and around forests.

3. Parties shall review energy policies in particular to ensure where there is considerable dependence on wood fuel sources supplies of these are restored and

maintained and complemented by policies promoting alternative energy sources.

4. Parties shall encourage the optimum use of land resources and in accordance with the precautionary principle take special care to minimize irreversible and unsustainable land use changes particulary in respect of natural forests.

5. Parties shall review incentives for any activity that affects forests, such as tax concessions, credits, grants or indirect incentives such as provision of infrastructure, and shall ensure that future incentives are carefully designed to ensure optimum, sustained production of a range of products and services and their equitable distribution.

Article 4
National Forest Policy for the Wise Use of Forests

1. Parties shall review and formulate a National Forest Policy covering the whole range of forest values and not merely timber, the principal aim of which shall be the maintenance of their forest capital by the establishment of a Permanent Forest Estate which shall be dedicated to conservation or production.

2. The Permanent Forest Estate shall be designated in the light of information collected through the assessments and surveys undertaken by Parties pursuant to paragraphs (1) and (3) of Article 6 after Protected Forest Areas necessary for safeguarding the areas described in paragraph 1 of Article 8 have been appropriately designated and zoned within the Permanent Forest Estate.

3. Parties shall ensure that the forest capital in the Permanent Forest Estate is used wisely and managed to provide multiple benefits where possible and appropriate, including the sustainable production of wood, a variety of forest products, ecological benefits such as watershed protection and biological diversity conservation, and leisure and recreational facilities.

4. Parties shall, after taking due consideration of any international guidelines, standards, and codes of conduct developed by the Authority for the wise use and management of forest resources pursuant to paragraph 3 of Article 21, take all necessary legal, administrative and regulatory measures within the framework of their national laws, to formulate, implement and enforce national guidelines, standards and codes of conduct for the wise use and management of their forest resources.

Article 5
Regeneration, Reforestation and Afforestation

1. Parties shall take urgent measures for the regeneration, reforestation and afforestation of suitable lands, paying particular attention to degraded forests or, where available, agricultural land surplus to food security requirements, and shall ensure that local peoples are consulted and involved in the development and management of such measures.

Article 6
Forest Assessments and Surveys

1. Where they are not already available, Parties shall undertake to prepare as soon as possible national resource surveys of all their forests which shall provide facts

on population, land use, climate, topography and land form, soil, flora and fauna, mineral resources and hydrology with special attention paid to degraded forest, fragile or sensitive areas and those where there is intense pressure of people on natural resources.

2. Until detailed surveys are available, Parties shall make the greatest use of other techniques to provide such information including the use of remote sensing.

3. Parties shall undertake to assess the capability of forests lands for possible uses, benefits and values, including leisure and tourism and the establishment of Protected Forest Areas pursuant to paragraph 1 of Article 8.

4. Parties shall review, revise and update their surveys and assessments every five years paying particular attention to new scientific knowledge, improved technology or changing social priorities.

5. Parties shall undertake to consult with forest peoples and local peoples when carrying out such assessments and surveys.

6. Information collected through forest surveys and assessments shall be publicly available in a timely manner and in an appropriate form in accordance with the provisions of Article 23.

7. Principles, criteria, standardized scientific and technical techniques and procedures are to be used in such surveys and assessments to ensure that these are comparable to the greatest extent possible and these shall be drawn up by the Authority and presented to the Conference at its first ordinary meeting or as early as possible thereafter for adoption.

Article 7
National Forest Authority

1. Where they have not already done so, Parties shall commit themselves to establishing a National Forest Authority capable of implementing the provisions of this Convention and where several governmental institutions are involved, create the necessary coordinating mechanism for the authorities dealing with different aspects of forest conservation and management.

2. Parties shall allocate sufficient funds to the Authority to enable it to carry out its tasks as well as sufficient qualified personnel with adequate enforcement powers.

Article 8
Establishment of Protected Forest Areas

1. Each Party shall, after consideration of information obtained as a result of forest assessments and surveys pursuant to paragraphs (1) and (3) of Article 6, establish appropriate protected forest areas within the area under their jurisdiction for the purpose of safeguarding:

a) the ecological, physical and biological processes essential to national, transboundary and regional ecosystems and watersheds, and local and global climatic stability;

b) representative samples of all types of forest ecosystems found in areas under their jurisdiction or control especially those of an exceptional or special significance;

c) areas of particular importance because of their scientific, educational, aesthetic, or cultural interest;

d) areas exceptionally rich in biological diversity or constituting the critical habitats of endangered or rare species, or species endemic to small areas, or species that migrate;

e) satisfactory population levels for the largest possible number of species belonging to any of these ecosystems.

2. Parties shall ensure that all customary rights are respected and shall take all measures necessary to ensure that the process of designation and establishment of protected forest areas is based on the effective participation of forest peoples and all local people who may be affected.

3. Specific objectives, criteria and guidelines to assist Parties in the selection, establishment and management of protected areas which may form the basis of a coordinated network of Protected Forest Areas throughout the world shall be drawn up by the Authority and presented to the Conference at its first ordinary meeting or as early as possible thereafter for adoption.

Article 9
Management of Protected Forest Areas

1. In respect of any Protected Forest Area established pursuant to this Convention the designating Party shall prepare a management plan and manage the Area in accordance with that plan reviewing and revising such plan from time to time as necessary.

2. Parties shall prohibit so far as possible:

a) those activities within the Protected Forest Area which are likely to cause disturbance or damage to its ecosystems or may undermine the effectiveness of the implementation of the management plan of that area;

b) the use or release of toxic substances or pollutants which could cause disturbances or damage to the Protected Forests Area or the species it contains including the introduction of exotic species of flora and fauna; and

c) to the maximum extent possible, prohibit or control any activity exercised outside the Protected Forest Area where such an activity is likely to cause disturbance to the ecosystems or species that such an area purport to protect.

3. Parties shall, in respect of Protected Forest Areas, establish where appropriate a buffer zone in which all activities that may have harmful consequences on the ecosystems that such Areas purport to protect shall be prohibited or regulated and activities consistent with the purpose of such Area promoted.

4. Where, with the consent of forest peoples, Protected Forest Areas are established on land subject to customary rights, the formulation and implementation of the management plan shall so far as possible be in the hands of or involve the effective participation of forest peoples and Parties shall ensure that assistance, financial or otherwise, is provided to such groups to enable the management plan to be effective.

5. Parties shall ensure that the National Forest Authority, established or designated pursuant to paragraph 1 of Article 7, is specifically obliged to safeguard protected areas and has powers and sufficient financial resources to do so.

6. Parties shall promote, through the adoption of appropriate measures including the use of management plans, the conservation and wise use of forest areas by private owners, community or local authorities.

Article 10
Forest Peoples' Rights

1. Parties shall recognize and respect customary systems of land ownership and use and shall take all steps necessary to demarcate and grant title to such lands (including the right to hold land communally) on the basis of traditional claims and taking into account the needs of such peoples to satisfy adequately their current and future needs on a sustainable basis and shall take all necessary steps to protect the boundaries of such lands from encroachment.

2. Parties shall take all necessary steps to ensure the effective participation of forest peoples at the local, national and international level in the formulation, implementation and evaluation of plans and programmes directly affecting them and, in particular, the designation of Protected Forest Areas pursuant to paragraph 1 of Article 8 and the Permanent Forest Estate pursuant to paragraph 1 of Article 4.

3. Parties shall ensure that their national development and forest policies respect and actively secure the cultural, social and economic, and political rights of forest peoples as recognized in international declarations and covenants to allow such peoples the maximum possible autonomy in the use of their customary lands.

4. Where specifically requested by forest peoples Parties shall take measures to ensure the provision of bilingual education, health facilities, and technical and scientific training (which is available to other citizens), to forest peoples including the provision of support, financial or otherwise, to local and forest peoples' organizations.

Article 11
Air Pollution

1. Parties shall, in view of the possible adverse effects on forests in the short and long term of air pollution, including long-range transboundary air pollution, take all appropriate measures to protect forests against air pollution and shall prevent and gradually reduce air pollution which is contributing to forest degradation.

2. Parties shall develop without delay in a separate Protocol to this Convention national, regional and international policies and strategies to combat the discharge of air pollutants taking into account the need to establish means of exchanges of information, consultation, research, monitoring and regular assessments bearing in mind efforts already made at national and international level.

Article 12
Water Resources and Water Pollution

1. Parties shall take into account the role of forests in the maintenance of hydrological cycles in their National Forest Policy and shall take all appropriate measures, for the conservation of their underground, surface and coastal waters which are an integral part of or dependent on or associated with forest ecosystems. Such measures could include regulation and control of water utilization with a view to achieving sufficient and continuous supplies of water for, inter alia, the conservation of such forest ecosystems.

2. Parties shall, in view of the possible adverse effects on forests in the short and

long term of water pollution, including transboundary effects, take all appropriate measures to prevent and reduce water pollution.

3. Parties shall cooperate in concluding bilateral or multilateral agreements between or among themselves to develop national, regional and international policies and strategies to combat water pollution affecting forests taking into account the need to establish means of exchanges of information, consultation, research, monitoring and regular assessments bearing in mind efforts already made at national and international level.

Article 13
Transboundary Forest Resources

1. Parties sharing transboundary forest resources shall cooperate in respect of their conservation and wise use, taking into account the rights and interests of each concerned Party.
2. Parties sharing transboundary forest resources shall use these resources in a reasonable and equitable manner with a view to avoiding to the maximum extent possible and reducing to the minimum extent possible adverse environmental effects in areas beyond the limits of their national jurisdiction.
3. Parties shall ensure that plans, programmes, policies and projects with respect to transboundary forest resources which may create a risk of significantly affecting the environment of another Party shall be subject to the provisions of Article 22.
4. Parties sharing transboundary forest resources shall:
a) notify in advance the other Party or Parties of the pertinent details of such plans, programmes, policies or projects to initiate, or make a change in, the conservation or utilization of the forests in question which can reasonably be expected to adversely affect nature, the natural resources or the environment of the other Party or Parties; and
b) upon request of the other Party or Parties, to enter into consultations concerning the above-mentioned plan, programme, policy or project; and
c) provide upon request to the other Party or Parties, specific additional pertinent information concerning the same.
5. Parties shall, in emergency situations arising from the utilization of transboundary forest resources or grave natural events such as fire or infestation, which may have sudden harmful repercussions on forest resources or the environment of another Party or Parties, urgently notify such Party or Parties.

Article 14
Protection of Biological Diversity

1. Without prejudice to the provisions of, or obligations deriving from, any treaty, convention or international agreement which may be concluded between Parties in respect of biological diversity, Parties shall, recognizing the important role of forests as sources of biological diversity, commit themselves within the context of the provisions and obligations of this Convention, and their obligations under existing international law, to taking appropriate measures for the conservation of biological diversity contained in or dependent on or associated with forests.

Article 15
Climate Change

1. Without prejudice to the provisions of, or obligations deriving from, any treaty, convention or international agreement which may be concluded between Parties in the future in respect of climate change, Parties shall, recognizing the contribution of forests to local and global climatic stability, commit themselves within the context of the provisions and obligations of this Convention, and their obligations under existing international law, to taking appropriate measures to prevent climate change.

Article 16
Multilateral Development Banks and Aid Organizations

1. Parties shall take appropriate measures to ensure that multilateral development banks and aid organizations to which they are members incorporate, in accordance with their competence and jurisdiction, the Convention's principles and obligations in their policies, projects and field operations and shall use their voting rights accordingly.

Article 17
International Cooperation

1. Parties shall cooperate on a regional and global basis directly or indirectly or through competent international organizations with a view to coordinating their activities in the field of conservation and wise use of forests and assisting each other in fulfilling their obligations under this Convention particularly in respect of the following:
   a) research, observation and monitoring, and with a view to using comparable or standardized research techniques and procedures and obtaining comparable data on matters relevant for the implementation of the obligations of this Convention;
   b) exchange of appropriate scientific, technical and other information and experience on a regular basis;
   c) formulation of measures including adoption of annexes and protocols in addition to the annexes and protocols opened for signature at the same time as this Convention and, where appropriate, national legislation and administrative measures which implement, harmonize or promote policies in accordance with the provisions of this Convention or its protocols;
   d) promotion of investment in activities which have a positive impact on forest conservation and sustainable development including non-forestry sectors;
   e) strengthening of regional and international organizations which promote or facilitate the achievement of Convention obligations; and
   f) any other matter to which the Conference may give special priority.

Article 18
Scientific Research

1. Parties shall, individually and in cooperation with other Parties and through appropriate international organizations, initiate and support research and scien-

tific assessments on matters relating to the conservation and wise use of forests including relevant social and economic research on inter alia:

a) matters relating to preparation of forest assessments and surveys under paragraphs (1) and (3) of Article 6;

b) the role of forests in relation to climate stability;

c) hydrological research with a view to ascertaining the characteristics of watersheds; and

d) matters relating to regeneration, reforestation and afforestation with a specific focus on native species.

2. To facilitate international cooperation and coordination of research efforts and the collection, validation, and transmission of the results of this research Parties shall, taking into account the need for improvement of the capability of developing countries to research, collect and assess scientific research and information relevant to the conservation and wise use of forests, take appropriate measures to establish and support the Authority's International Forests Information Network pursuant to paragraph 1 of Article 28.

Article 19
Observation and Monitoring

1. Parties shall collect and analyse information from all their respective national authorities concerning forests, in particular their composition, deforestation, degradation, effects of pollution, and any other relevant developments significantly affecting forests, whether in protected forest areas or otherwise.

2. Parties shall monitor and keep under surveillance the effects of any activities which they engage in or permit, whether in forests or non-forest areas, in order to determine whether these activities are causing or contributing or likely to cause and contribute to forest degradation or deforestation over and above those effects which the Parties took into account when deciding to engage in or permit such activities and reviewing, where appropriate, the continuance of such activities.

3. Parties shall ensure that information collected pursuant to this Article is utilized to further the compliance with this Convention and its protocols in accordance with the provisions of Article 33.

Article 20
Education and Public Awareness

1. Parties shall, individually and in cooperation with other Parties and through appropriate national, regional or international bodies take such steps as are necessary to promote through education programmes in educational establishments at all levels, and by other means, public awareness of the significance of the measures adopted for the purposes of the conservation and wise use of forests focusing inter alia on:

a) consumer behaviour promoting the conservation and wise use of forests including authentication and certification of timber and other forest products from forests managed in accordance with national guidelines, standards and codes of conduct formulated pursuant to paragraph 4 of Article 4;

b) energy efficiency and, where appropriate, efficient use and recycling of relevant forest products;

c) sound land use and agricultural practices; and
d) cultural values which respect to nature and forests.

Article 21
Trade and International Relations

1. Parties shall take appropriate steps to review and formulate their trade and foreign relations policies to ensure these are in harmony with and support measures taken by Parties to fulfil their obligations under this Convention.
2. Parties shall, in cooperation with other Parties and appropriate international organizations, introduce economic incentives in favour of forest products from forests managed in accordance with national guidelines, standards and codes of conduct formulated pursuant to paragraph 4 of Article 4.
3. The Authority shall formulate international guidelines, standards and codes of conduct for the wise use and management of forest resources which shall be presented to the Conference at its first ordinary meeting or as soon as possible thereafter, for adoption by the Conference.

Article 22
Environmental Impact Assessment

1. Parties shall require that proposals for any activity subject to a decision of a competent authority in accordance with a national procedure, which may significantly affect forests, forest peoples, local populations or forests peoples' customary land use rights shall be subject to an assessment of the likely or potential environmental and social impacts including the direct, indirect, cumulative short-term and long-term effects.
2. For the purposes of this Article the term 'activity' shall be used, to the extent appropriate, to cover policies, plans and programmes.
3. Parties shall take into consideration the results of such assessments in reaching a decision to authorize or not authorize the activity in question and shall ensure that where such any such activity is authorized it shall be undertaken on the basis that it is planned and carried out so as to overcome or minimize any assessed adverse effects and that such effects are regularly monitored with a view to taking remedial action as necessary.
4. The provisions of this Article shall apply to any proposed activity within the jurisdiction and control of the Party where it has been proposed, or in areas under the jurisdiction and control of another Party or in areas beyond the limits of national jurisdiction and whether the proponent is a State, any other public body, an international organization or a private person.
5. Information provided by the proponent of an activity for the purposes of an assessment shall be publicly available. Such information shall be disclosed in a timely and appropriate manner to persons and organizations who may be affected by the proposal.
6. Detailed guidelines, procedures and criteria for the application of environmental impact assessments covering, inter alia, criteria for assessing the size, location and nature of activities to be subject to such assessments and procedures to facilitate putting into effect the provision of this Article shall be drawn up by the

Authority for presentation to the first ordinary meeting of the Conference, or as soon as possible thereafter, for adoption.

Article 23
Access to Environmental Information

1. Parties shall ensure that all information relating to the forest environment, including in particular, information collected pursuant to paragraphs 1 and 3 of Article 6, paragraph 1 of Article 19, paragraph 4 of Article 22 and paragraph 2 of Article 33 is made available by the relevant public authorities to any natural or legal person resident or located in the area under their jurisdiction at his or her request and without he or she having to prove an interest.
2. Parties shall define the practical arrangements under which such information is effectively made available.
3. Parties shall ensure that information provided by public authorities shall be in a form which is accessible and intelligible to the public and in the case of forest peoples that translation facilities are available.
4. A person who considers that his or her request for information has been unreasonably refused or ignored, or has been inadequately answered by a public authority, may seek judicial review of the decision in accordance with the relevant national legal system.
5. Parties may provide for a request to be refused, subject to the provision of written reasons for refusal, where it affects:
a) matters which are sub judice;
b) confidentiality of personal data;
c) legal professional privilege; and
d) national security.
6. The foregoing obligations shall apply equally to appropriate information collected by the Authority which shall formulate and present to the first ordinary meeting of the Conference or as soon as possible thereafter, rules, regulations and practical procedures determining the availability of such information to the public for adoption.

PART III. IMPLEMENTATION

Article 24
Conference of the Parties

1. The Conference of the Parties (hereinafter referred to as 'the Conference') shall be the principle decision-making organ of this Convention.
2. Each Party shall have one vote and decisions by the Conference shall be by simple majority vote of the Parties present and voting except as otherwise provided in this Convention.
3. By written notification to the Chairman of the Conference, a Party may authorize, under its own responsibility, any other Party to represent its interest or to cast its vote at any meeting of the Conference.
4. Subject to the foregoing for the purposes of this paragraph, the phrase 'Parties

present' means the Parties present at the meeting at the time of the vote and the phrase 'Parties present and voting' means Parties present and casting an affirmative or negative vote. Parties who abstain from voting shall not be considered as voting.

5. The Depository shall call a meeting of the Conference of the Parties not later than one year after the entry into force of this Convention. Thereafter ordinary meetings of the Conference shall be held at intervals of not more than two years unless the Conference decides. Extraordinary meetings may be called at any time on the written request of at least one third of the Parties.

6. The Conference may create such subsidiary bodies as it deems necessary for the performance of its functions.

7. The Conference shall adopt its own rules of procedure.

8. The United Nations, its Specialized Agencies, the International Atomic Energy Agency, any other intergovernmental agencies and bodies, multilateral development banks, any State not a Party to the present Convention, as well as any other bodies designated by the Conference of the Parties may be represented at meetings of the Conference by observers.

9. Any body or agency technically qualified in matters dealt with in this Convention, in the following categories, which has informed the Authority of its desire to be represented at the meetings of the Conference by observers, shall be admitted unless at least one-third of the Parties present object:

a) international non-governmental agencies and bodies, and national governmental agencies and bodies; and

b) national non-governmental agencies and bodies which have been approved for this purpose by the Party in whose jurisdiction and control they are located.

10. Observers shall have the right to participate according to the rules of procedure adopted by the Conference, but not to vote.

11. Expenses incurred in the operation of the Conference shall be borne by the Fund.

Article 25
Functions of the Conference

1. At each of its meetings the Conference shall be responsible for reviewing the implementation of this Convention and shall:

a) keep under review the provisions of this Convention, including its Annexes and Protocols;

b) provide overall guidance, supervision and approval of the Authority's work and make such provision as may be necessary to enable the Authority to carry out its duties including the adoption of financial provisions;

c) receive and consider any reports presented by the Authority or by any Party or specially requested by the Conference concerning the workings of the Convention and its protocols and, where appropriate, examine any modifications necessary and consider and recommend adoption of amendments to this Convention, its protocols and Annexes;

d) make recommendations to a Party or Parties or international organizations for improving the effectiveness of the present Convention including proposals for the conclusion, with the States that are not Contracting Parties to the Convention, of agreements that would enhance the effective implementation of this Convention; and

e) consider and adopt at its first ordinary meeting or as soon as possible thereafter and, where subsequently necessary revise, upon recommendation of the Authority:

  (i) principles, objectives, criteria and guidelines for the selection, establishment and management of protected forest areas pursuant to paragraph (1) of Article 8;

  (ii) principles, criteria, standardized scientific techniques and procedures for the preparation of surveys and assessments pursuant to paragraph (1) of Article 6;

  (iii) guidelines, standards and codes of conduct for the wise use and management of forest resources pursuant to paragraph (1) of Article 21;

  (iv) rules, regulations, procedures and criteria for the application and preparation of environmental impact assessments pursuant to paragraph (1) of Article 22;

  (v) rules, regulations and practical arrangements relating to public access to information in respect of information collected by the Authority;

  (vi) criteria for eligibility and guidelines for determining priorities in respect of requests for assistance from the Fund pursuant to Article 40.

f) recommend appropriate measures to keep the public informed about the activities undertaken within the framework of this Convention;

g) upon recommendation of the Authority adopt and, where necessary, revise a Long-term Action Plan to assist in determining the short- and long-term priorities for the conservation and wise use of forests;

h) establish working groups as required to consider any matters related to this Convention; and

i) keep in review the financial regulation of the Convention, including budgets for the Authority and the Fund, at each ordinary Conference, and adopt budgets for the next financial period. Adoption shall be by a two-thirds majority of the Parties present and voting.

2. The Conference shall perform any other task which has been entrusted to it by this Convention.

Article 26
Establishment of the Authority

1. For the purposes of assisting the Parties to fulfil their obligations of this Convention and its protocols there is hereby established the Authority whose headquarters shall be at [...].

2. The Authority shall have such legal capacity as is provided in the Statute set forth in Annex I. The Authority constituted and operating in accordance with the administrative rules, regulations and procedures set forth in Annex I shall be subject to the control, supervision and financial regulation of the Conference.

3. The Authority shall be provided with such funds from the Fund as it may require to carry out its functions.

Article 27
Functions of the Authority

1. The functions of the Authority shall be to:

a) draw up for consideration and adoption by the first ordinary meeting of the Conference or as soon as possible thereafter:

  (i) principles, objectives, criteria and guidelines for the selection, establishment and management of protected forests areas pursuant to paragraph (1) of Article 8;

(ii) principles, criteria, standardized scientific techniques and procedures for the preparation of surveys and assessments pursuant to paragraph (1) of Article 6;

(iii) guidelines, standards and codes of conduct for the wise use and management of forest resources pursuant to paragraph (1) of Article 1;

(iv) rules, regulations, procedures and criteria for the application and preparation of environmental impact assessments pursuant to paragraph (1) of Article 22;

(v) rules, regulations and practical arrangements relating to public access to information in respect of information collected by the Authority;

(vi) criteria for eligibility and guidelines for determining priorities in respect of requests for assistance from the Fund pursuant to Article 40.

b) provide objective, reliable and comparable scientific, technical and other inform-ation thereby enabling measures for the conservation and wise use of forests to be assessed and to keep Parties and the public informed about the state of the world's forests;

c) monitor compliance with this Convention and its protocols by the Parties;

d) provide the necessary secretarial, administrative, scientific and technical support to the Conference;

e) assist the Conference in performing its tasks by providing any further advice or information requested by the Conference;

f) prepare a Long-term Action Plan to assist the Conference in determining the short- and long-term priorities for the conservation and wise use of forests;

g) seek under the Article 96 of the Charter of the United Nations through the General Assembly, Security Council or, if so authorized by the General Assembly in its own right, advisory opinions from the International Court of Justice on matters within the Authority's competence; and

h) assist the Conference with the administration of the Fund.

2. The Authority shall perform any other function specifically requested by the Conference.

Article 28
The Organs of the Authority

1. The Authority shall perform its functions through the following organs:

a) the International Compliance and Monitoring Body;

b) the International Forest Information Network;

c) the International Forests Products Council; and

d) the Secretariat.

2. In order to establish these organs the Authority shall, to the maximum extent possible, utilize the facilities, services and expertise of the United Nations and its organs and existing intergovernmental, governmental or non-governmental organizations, in order to avoid duplication of efforts in fulfilling its functions and to enhance the complementarity and efficiency of their activities.

3. Each organ of the Authority so established shall be responsible for carrying out the tasks specifically set out for it in this Convention and its protocols and shall function within the powers and in accordance with administrative rules, regu-

lations, and procedures devised for it by the Authority and approved by the Conference.

Article 29
International Compliance and Monitoring Body

1. The responsibility of the International Compliance and Monitoring Body shall be to investigate and report on the implementation of the Convention and its protocols by all Parties in order to report to the Conference any discrepancy between their obligations under the Convention and any protocol to which they are party and their laws and practices.

2. In carrying out this function the International Compliance and Monitoring Body shall:

a) publish guidelines setting out the form and regularity of reports to be submitted by the Parties' National Compliance and Monitoring Bodies;

b) examine and assess such reports as are submitted by the National Compliance and Monitoring Bodies established by Parties pursuant to paragraph 2 of Article 33;

c) receive and consider information relating to implementation by a Party received from any other Party, any body which has observer status at the Conference, or any national non-governmental organization or forest peoples group or local community group;

d) investigate complaints relating to the implementation by a Party received from another Party, any body which has observer status at the Conference, or any national non-governmental organization or forest peoples group or local community group that is directly interested in or has a bona fide interest in the subject matter of the complaint;

e) request the Party which is the object of the complaint to respond to the complaint and where necessary request further clarification on responses received in the light of the information or complaint submitted under paragraphs (c) and (d) above;

f) with the consent of the Party concerned, send a visiting commission should further enquiries in the area under the jurisdiction of the party concerned be necessary; and

g) in any event report the information, complaints, responses and its assessments and findings to the all Parties through the publication of a report whose form and regularity shall, subject to the approval of the Conference, be determined by the Authority.

3. The foregoing shall be without prejudice to arrangements which may exist between Parties concerning the settlement of disputes arising from this Convention and its protocols.

Article 30
International Forests Information Network

1. The responsibility of the International Forests Information Network (hereafter referred to as 'the Network') shall be to:

a) provide objective, reliable and comparable data at the international level to

enable the Parties to take the requisite measures for the conservation and wise use of forests;
b) record, collate and assess data on the results of such measures to ensure that the Parties and the public are properly informed about the effectiveness of such measures;
c) ensure broad dissemination of information collected pursuant paragraphs (a) and (b) above;
d) develop methods of assessing the costs of damage to forests and the environment including the costs of precautionary, preventative and protection policies including costs of reforestation, afforestation and restoration of degraded forests; and
e) stimulate exchange of scientific, technical and other relevant information and technologies ensure exchange of information and coordination between Parties on research and training programmes;
2. The Network shall comprise:
a) the main component elements of a national information network;
b) national focal points; and
c) topic centres
3. To enable the Network to be rapidly set up Parties shall, within 18 months from the entry into force of this Convention, designate:
a) one or more national institutions which in their judgement, by acting as national information networks could contribute to the work of the Network taking into account the need to ensure the fullest possible geographical coverage of the area under their national jurisdiction;
b) a national focal point for coordinating and/or transmitting the information to be provided at a national level to the Network and to other institutions forming part of the Network; and
c) institutions which could cooperate with the Network directly as regards certain specific areas or 'topics' of interest.
4. On completion of the designation process the Network shall come into operation in accordance with the provisions prescribed by the Authority and shall present to the ordinary meetings of the Conference a programme of action to fulfil the responsibilities set out in paragraph 1 above.

Article 31
International Forests Products Council

1. The International Forests Products Council shall be responsible for the formulation of international guidelines, standards and codes of conduct for the production, distribution and trade in forests products pursuant to paragraph 3 of Article 21.

Article 32
Secretariat

1. The Secretariat shall be responsible for the arrangement and servicing of the meetings of the Conference, and any other subsidiary bodies established or authorized by the Conference and shall provide the necessary administrative, secretarial and coordination support and skills to enable each institution of the Authority to carry out its tasks. The Secretariat shall inter alia:

a) maintain liaison with the bodies designated by the Parties under other international conventions and agreements which deal with matters concerning the conservation and wise use of forest;
b) maintain an updated register of all the protected areas designated by Parties pursuant to Article 8;
c) maintain and service the Fund;
d) prepare a budget for the Authority for approval by the ordinary meetings of the Conference; and
e) perform any other function entrusted to it under this Convention and its protocols or by the Conference.
2. The Secretariat functions shall be carried out on an interim basis by [...] until completion of the first ordinary meeting of the Conference when Parties shall make permanent arrangements.

Article 33
Compliance

1. Parties shall take all necessary legislative, regulatory and administrative measures to implement and enforce the provisions of this Convention. Parties shall abstain from any measure which could jeopardize the achievement of the Common Objectives.
2. Each Party shall nominate a body hereinafter referred to as 'the National Compliance and Monitoring Body' which shall be responsible for the monitoring and compliance with this Convention and its protocols. The National Compliance and Monitoring Body shall:
a) monitor compliance with this Convention and any Protocols to which it is also a Party within the area under the national jurisdiction of the Party; and
b) report to the International Compliance and Monitoring Body established pursuant to paragraph 1 of Article 28 in such form and at such intervals as the International Compliance and Monitoring Body shall determine on, inter alia, the following:
   (i) legal, regulatory and administrative measures taken by the Party to implement this Convention and its protocols;
   (ii) the adequacy and success of such measures; and
   (iii) problems encountered in the implementation of the Convention and its Protocols;
c) assist the International Compliance and Monitoring Body in the better performance of its duties.
3. Each Party may, whenever it deems necessary, draw the attention of the Conference to any activity, which in its opinion, affects the achievement of its obligations under this Convention and its protocols or the compliance by another Party with its obligations under this Convention and its protocols.
4. The Conference shall draw the attention of any State or regional economic integration organization which are not Parties to this Convention to any activity undertaken by that State or regional economic integration organization, their agencies or instrumentalities, natural or juridical persons, ships, or other craft which, in the opinion of the Conference, affects the achievement of the obligations under this Convention and its protocols.

5. Parties shall support and encourage the establishment of non-governmental organizations active in the field of forest conservation and the promotion of the welfare of forest peoples and local communities, and shall provide for their participation in the work of the National Compliance and Monitoring Body and other decision-making processes relating to the conservation and wise use of forests.

6. Where there is a risk of serious, irreversible damage to their forests arising from an action undertaken or proposed to be undertake in violation of their domestic legislation relating to the conservation and wise use of forests, Parties shall take immediate action to prevent such action from being taken, or bring about its cessation and shall introduce administrative and judicial procedures which may be necessary to give effect to this provision.

7. The provisions of this Convention and its protocols shall not prevent any Party from maintaining or adopting more stringent measures concerning the conservation and wise use of forests compatible with this Convention.

Article 34
Transfer of Technology and Technical Skills

1. Parties shall, consistent with their national laws, regulations and practices and taking into account the needs of developing countries, cooperate in promoting through the Authority or other competent international bodies the development and transfer of technology, knowledge, technical assistance and skills to facilitate participation in and implementation of this Convention.

2. Such cooperation shall be carried out by, inter alia:

a) gathering and exchange of information and know-how of local and forests peoples;

b) facilitating the provision and acquisition of such equipment, supply of special manuals and guides concerning silviculture practices and forest management systems which endorse the principles of conservation and wise use of forests;

c) training of appropriately qualified staff; and

d) provision of equipment needed for research and systematic observations projects, plans and programmes.

3. Parties may submit requests for assistance in respect of any of the above to the Fund.

Article 35
Establishment of the Fund

1. There shall be established by the Parties the Fund. The seat of the Fund shall be at the same place as the headquarters of the Authority.

2. The Fund shall be recognized by each Party as having the necessary legal personality under the laws of that Party to enable it to discharge its functions under the Convention and shall enjoy in all Parties' jurisdiction exemption from all taxes, levies and duties whatsoever.

3. The Fund shall be administered by the Conference.

4. The Conference may appoint a Board for the purposes of administering the Fund, and may delegate to the Board any powers it considers necessary for its operation. The Conference may appoint a Director of the Board.

5. The Board shall be composed of no more than 15 persons who shall be appointed by the Conference for a period of four years and may at the expiry of this period be eligible for one further consecutive period of four years. In appointing the members of the Board, the Parties shall pay due regard to the need to ensure an adequate representation of the different geographical regions and forest types of the world.

6. The Board shall discharge the tasks and responsibilities specified in its terms of reference, as determined by the Conference, and cooperate where required or as appropriate with the IBRD, UNEP, UNDP, FAO, inter-governmental developmental aid agencies and bodies, multilateral development banks and other appropriate agencies and bodies.

7. The Conference shall have the power to:

a) determine the financial conditions attached to loans made by the Fund;

b) specify the terms of reference and regulations necessary for the proper functioning of the Fund;

c) appoint auditors and approve the accounts of the Fund;

d) give directions to the Director concerning the administration of the Fund; and

e) perform such other functions as may from time to time be necessary for the proper operation of the Fund.

8. Parties to this Convention shall give their assistance to international fund-raising campaigns organized for the Fund.

9. Parties to this Convention shall consider and encourage the establishment of national, public and private foundations or associations whose purpose is to invite donations for the conservation and wise use of forest.

10. The Fund shall bear its own administrative costs.

Article 36
Purpose of the Fund

1. The purpose of the Fund shall be to advance the Common Objectives of this Convention and its protocols by assisting Parties to fulfil their obligations under this Convention or other related uses as determined by the Conference of the Parties.

Article 37
Resources of the Fund

1. The resources of the Fund shall consist of:

a) contributions collected pursuant to paragraph 1 of Article 38;

b) any interest due to the resources of the Fund;

c) funds borrowed by the Fund;

d) voluntary contributions of any kind made by Parties or any other legal or natural person;

e) funds raised by collections and receipts from events organized for the benefit for the Fund; and

f) all other resources authorized by the Conference.

2. Contributions of the kind referred to in paragraphs (d), (e) and (f) above may be subject to conditions that they be used only for a certain project, or in respect of a specific area, and such conditions shall be respected, provided that such project or assistance in a specific area is consistent with the purpose of the Fund.

Article 38
Contributions

1. Parties shall ensure that the contributions to the Fund referred to in paragraph (1)(a) of Article 37, calculated and collected in accordance with the provisions set forth in Annex II to this Convention are promptly paid to the Fund.

2. A Party that fails to pay its contributions shall have no vote at the Conference if the amount of its arrears equals or exceeds the amount of the contributions due from it in the preceding two full years. The Conference may, nevertheless, permit such a Party to vote if it is satisfied that the failure to pay is due to conditions beyond the control of the member.

3. A Party in arrears for two full years and if it has not satisfied the Conference that such failure is due to conditions beyond its control shall not be eligible to make applications for assistance under paragraph 1 of Article 40 until such arrears have been paid to the Fund.

Article 39
Payments from the Fund

1. The resources of the Fund shall be used to finance in the following order, the costs of:
a) the Conference;
b) the Authority;
c) the Board, if the Conference has approved the establishment of a Board to administer the Fund;
d) projects eligible for assistance pursuant to Article 40; and
e) any other costs specifically authorized by the Conference or the Board within its terms of reference.

2. Assistance from the Fund may take any form approved by the Conference or the Board as the case may, be including inter alia:
a) granting of cash payments;
b) low interest or interest free loans;
c) provision of experts, trained staff and specialists; and
d) supply of equipment.

3. Assistance given by the Fund shall be with the agreement of the beneficiary Party or Parties or other organizations approved under paragraph 5 of Article 40.

Article 40
Eligibility for Assistance

1. All Parties shall be entitled to submit a request for assistance to the Conference or the Board, as the case may be, for activities in accordance with the provisions of this Article the purpose of which is to further the Common Objectives of this Convention.

2. For the purposes of this Article the term 'activity' shall be used to cover proposed projects and to the extent appropriate, policies, plans and programmes.

3. The Conference or the Board, as the case may be, shall approve in full or in part or may not approve the assistance so requested, or may approve in full or in part subject to certain conditions the assistance so requested.

3. The Executive Director shall be the chief administrative officer of the Organization and shall be responsible to the Council for the administration and operation of this Agreement in accordance with decisions of the Council.

4. The Executive Director shall appoint the staff in accordance with regulations to be established by the Council. The Council shall, by special vote, decide the number of executive and professional staff the Executive Director may appoint. Any changes in the number of executive and professional staff shall be decided by the Council by special vote. The staff shall be responsible to the Executive Director.

5. Neither the Executive Director nor any member of the staff shall have any financial interest in the timber industry or trade, or associated commercial activities.

6. In the performance of their duties, the Executive Director and staff shall not seek or receive instructions from any member or from any authority external to the Organization. They shall refrain from any action which might reflect adversely on their positions as international officials ultimately responsible to the Council. Each member shall respect the exclusively international character of the responsibilities of the Executive Director and staff and shall not seek to influence them in the discharge of their responsibilities.

CHAPTER V. PRIVILEGES AND IMMUNITIES

Article 17
Privileges and immunities

1. The Organization shall have a legal personality. It shall in particular have the capacity to contract, to acquire and dispose of movable and immovable property, and to institute legal proceedings.

2. The status, privileges and immunities of the Organization, of its Executive Director, its staff and experts, and of representatives of members while in the territory of Japan shall continue to be governed by the Headquarters Agreement between the Government of Japan and the International Tropical Timber Organization signed at Tokyo on 27 February 1988, with such amendments as may be necessary for the proper functioning of this Agreement.

3. The Organization may conclude, with one or more countries, agreements to be approved by the Council relating to such capacity, privileges and immunities as may be necessary for the proper functioning of this Agreement.

4. If the headquarters of the Organization is moved to another country, the member in question shall, as soon as possible, conclude with the Organization a headquarters agreement to be approved by the Council. Pending the conclusion of such an Agreement, the Organization shall request the new host Government to grant, within the limits of its national legislation, exemption from taxation on remuneration paid by the Organization to its employees, and on the assets, income and other property of the Organization.

5. The Headquarters Agreement shall be independent of this Agreement. It shall, however, terminate:
   a) by agreement between the host Government and the Organization;

4. The eligibility of activities for assistance shall be governed by criteria to be developed by the Authority and adopted by the Conference at the first ordinary meeting of the Conference or as soon as possible thereafter.

5. When considering whether or not to approve a request for assistance the Conference or the Board shall consider, inter alia, the following:

a) the importance of a particular forest area in the world's heritage of nature and natural resources;

b) the resources available to the Party making the request;

c) the efforts and measures taken by the Party making the request in fulfilling its obligations under this Convention and its protocols;

d) the urgency of the request;

e) information and recommendations provided by the Authority;

f) the Long Term Action Plan prepared by the Authority; and

g) any other material consideration.

6. The Conference may approve requests for assistance from international intergovernmental agencies and bodies and international non-governmental organizations provided such requests meet the eligibility criteria referred to in paragraph 3 above. Joint applications from one or more Parties may be accepted.

7. The Conference or the Board shall define the procedure by which requests for assistance shall be made and considered.

8. The Conference or the Board shall develop and adopt procedures for the suspension of payments, or other sanctions it feels necessary if any conditions of the approved assistance are not fulfilled.

9. The Conference or Board shall give written reasons for all of its decisions, deliberations and guidelines.

Article 41
Liability and Compensation

1. Parties undertake to cooperate in the formulation and adoption of appropriate rules, regulations, and procedures for the determination of state and civil liability and compensation in respect of damage resulting to forests from acts or omissions in violation of the provisions of this Convention and protocols.

Article 42
Settlement of Disputes

1. In the case of a dispute between Parties as to the interpretation or application of this Convention and protocols Parties shall seek a solution to the dispute by negotiations or any other peaceful means of their own choice.

2. If the Parties concerned are unable to settle their dispute by negotiation, the dispute shall be submitted to an Ad Hoc organ set up at the instruction of the Conference by the Authority for this purpose. The Parties may choose mediation instead of arbitration before the Ad hoc organ. The conduct of the mediation or arbitration of disputes between Parties shall be in accordance with the provisions set forth in Annex III of this Convention.

3. The International Court of Justice shall have jurisdiction in any matter to which this Convention and protocols relate.

4. Parties shall use their best endeavours to secure the authorization by the General

Assembly of the United Nations of the Authority to seek where it considers appropriate, advisory opinions from the International Court of Justice on legal matters arising within the scope of its activities under Article 96(2) of the Charter of the United Nations.

PART IV. TECHNICAL ASPECTS

Article 43
Adoption of Additional Protocols

1. Parties may adopt at a diplomatic conference additional protocols to this Convention.
2. A diplomatic conference for the purposes of adopting additional protocols shall be convened by the Authority at the request of two-thirds of the Parties.

Article 44
Amendment of the Convention or Protocols

1. Any Party may propose amendments to this Convention or to any protocol.
2. Amendments to this Convention shall be adopted at a meeting of the Conference of the Parties. Amendments to any protocol shall be adopted at a meeting of the Parties to the protocol in question. The text of the proposed amendment to this Convention or to any protocol, except as may otherwise be provided in such protocol, shall be communicated to the Parties by the Authority at least six months before the meeting at which it is proposed for adoption. The Authority shall also communicate proposed amendments to the signatories to this Convention for information.
3. The Parties shall make every effort to reach agreement on any proposed amendment to this Convention by consensus. If all efforts at consensus have been exhausted and no agreement reached, the amendment shall as a last resort be adopted by a three-fourth majority vote of the Parties present and voting at the meeting, and shall be submitted by the Authority to all Parties for ratification, approval or acceptance.
4. The procedure mentioned in paragraph 3 above shall apply to amendments to any protocol, except that a two-thirds majority of the parties present and voting at the meeting shall suffice for their adoption.
5. Ratification, approval or acceptance of amendments shall be notified to the Authority in writing. Amendments adopted in accordance with paragraphs 3 and 4 above shall enter into force as between Parties having accepted them on the ninetieth day after the receipt by the Authority of their notification of their ratification, approval or acceptance by at least three-fourths of the Parties to this Convention or by at least two-thirds of the Parties to the protocols concerned, except as may otherwise be provided in the protocol. Thereafter amendments shall enter into force for any other Party on the ninetieth day after that Party deposits its instrument of ratification, acceptance or approval of the amendments.

Article 45
Adoption and Amendments of Annexes

1. The annexes to this Convention or to any protocol shall form an integral part of this Convention or such protocol, as the case may be, and, unless expressly provided otherwise, a reference to this Convention or its protocols constitutes at the same time a reference to any annexes thereto.
2. Except as may be otherwise provided in any protocol with respect to its annexes, the following procedure shall apply to the proposal, adoption and entry into force of additional annexes to this Convention or of annexes to a protocol:
a) Annexes to this Convention shall be proposed and adopted in accordance with the procedure laid down in paragraphs 2 and 3 of Article 44, while annexes to any protocol shall be proposed and adopted in accordance with the procedure laid down in paragraphs 2 and 4 of Article 44.
b) Any Party that is unable to approve an additional annex to this Convention or an annex to any protocol to which it is a Party shall so notify the Authority in writing within six months from the date of the communication of the adoption by the Authority. The Authority shall without delay notify all Parties of any notification received. A Party may at any time substitute an acceptance for a previous declaration of objection and the annexes shall thereupon enter into force for that Party.
c) On the expiry of six months from the date of circulation of the communication by the Authority, the annex shall become effective for all Parties to this Convention ...
3. If an additional annex or an amendment to an annex involves an amendment to this Convention or to any protocol, the additional annex or amended annex shall not enter into force until such time as the amendment to this Convention or to the protocol concerned enters into force.

Article 46
Relationship Between the Convention and its Protocols

1. A State or regional economic integration organization may not become a Party to a protocol unless it is, or becomes at the same time, a Party to the Convention.
2. Decisions concerning any protocol shall only be taken by the Parties to the protocol concerned.

Article 47
Relation to other Conventions and Coordination with Other Organizations

1. The provisions of this Convention shall in no way affect the rights or obligations of any Party deriving from existing conventions and agreements concerning conservation or utilization of forest resources.
2. The Authority shall make whatever arrangements are appropriate for consultation or cooperation with the United Nations and its organs, such as the Food and Agriculture Organization of the United Nations (FAO), the United Nations Environment Programme (UNEP), the United Nations Development Programme (UNDP), the United Nations Conference on Trade and Development (UNCTAD), the General Agreement on Tariffs and Trade (GATT) and such other specialized agencies of the United Nations and intergovernmental, govern-

mental and non-governmental organizations as may be appropriate including in particular, the International Tropical Timber Organization (ITTO).

Article 48
Signature

1. This Convention shall be open for signature by any State or regional economic integration organization at [...] from [...] until [...].

Article 49
Ratification, Acceptance or Approval

1. This Convention and any protocol shall be the subject of ratification, acceptance or approval by Parties who shall deposit instruments of ratification, acceptance or approval with the Depository.
2. Any regional economic integration organization which becomes a Party to this Convention or any protocol without any of its Member States being a Party shall be bound by all the obligations under the Convention or the protocol as the case may be. In the case of such organization, one or more of whose Member States is a Party to the Convention or the relevant protocol, the organization and its Member State shall decide on their respective responsibilities for the performance of their obligations under the Convention or protocol, as the case may be. In such cases the organization and the Member State shall not be entitled to exercise rights under the Convention or the protocol concurrently.
3. In their instruments of ratification, acceptance or approval, the organization referred to in paragraph 2 above shall declare the extent of their competence with respect to the matters governed by the Convention or the relevant protocol. These organizations shall also inform the Depository of any substantial modification in the extent of their competence.

Article 50
Accession

1. This Convention and any protocol shall be open for accession States and regional economic organizations from the date on which the Convention or Protocol concerned is closed for signature. The instruments of accession shall be deposited with the Depository.
2. In their instruments of accession the organizations referred to ... above shall declare the extent of their competence in respect of the matters governed by the Convention or the relevant protocol. These organizations shall also inform the Depository of any substantial modification in the extent of their competence.
3. The provisions of paragraph 2 of Article 31 shall apply equally to regional economic integration organizations which accede to this Convention or any protocol.

Article 51
Right to Vote

1. Each Party to this Convention or to any Protocol shall have one vote.
2. Except as provided for in paragraph 1 above, regional economic integration

organizations, in matters within their competence, shall exercise their right to vote with a number of votes equal to the number of the member States which are Parties to the Convention or the relevant protocol. Such organization shall not exercise their right to vote if their member State exercises theirs, and visa versa.

Article 52
Entry into Force

1. This Convention shall enter into force on the ninetieth day after the date of the deposit of the twentieth instrument of ratification, acceptance, approval or accession.
2. Any protocol, except as otherwise provided in such protocol, shall enter into force on the ninetieth day after the date of the deposit of the eleventh instrument of ratification, acceptance or approval of accession thereto.
3. For each Party which ratifies, accepts or approves this Convention or accedes thereto after the twentieth instrument of ratification, acceptance, approval or accession, it shall enter into force on the ninetieth day after the date of the deposit by such Party of its instrument of ratification, acceptance, approval or accession.
4. Any protocol, except as otherwise provided in such protocol, shall enter into force for a Party that ratifies, accepts or approves that protocol or accedes thereto after its entry into force pursuant to paragraph 2 above, on the ninetieth day after the date on which that Party deposits its instrument of ratification, acceptance, approval or accession, or on the date on which the Convention enters into force for that Party, whichever shall be the later.
5. For the purposes of paragraph 1 and 2 above, any instrument deposited by a regional economic integration organization shall not be counted as additional to those deposited by member States of such organizations.

Article 53
Reservations

1. No reservations shall be made to this Convention.

Article 54
Denunciation

1. At any time after five years from the date on which the Convention entered into effect for a Party, that Party may denounce this Convention by giving written notification to the Depository.
2. Except as may be provided for in any protocol, at any time after five years from the date on which such protocol has entered into force for a Party, that Party may denounce the protocol by giving written notification to the Depository.
3. Any such denunciation shall take effect one year after the date of its receipt by the Depository, or on such later date as may be specified in the notification to the Depository.
4. Any Party which denounces this Convention shall be considered as also having denounced any protocol to which it is a Party.

Article 55
Depository

1. Ratification, acceptance, approval or accession shall be effected by the deposit of a formal instrument with . . ., hereinafter designated as the Depository.
2. The Depository shall inform the Parties of:
a) the signature of this Convention and of any protocol, and the deposit of instruments of ratification, acceptance, approval or accession;
b) the date on which the Convention or any Protocol shall enter into effect;
c) notification of denunciations in respect of the Convention or any protocol;
d) all communication relating to the adoption and approval of annexes; and
e) notification by regional economic integration organizations of the extent of their competence with respect to matters governed by this Convention or any protocols.

Article 56
Authentic Texts and Certified Copies

1. The present Convention has been drawn up in a single copy in the English language. Official translations into Arabic, Chinese, French, Russian and Spanish shall be prepared. These shall be equally authentic and shall be deposited with the Depository.

IN WITNESS WHEREOF the undersigned, being duly authorized by their Governments, have signed the present Convention
DONE AT .... this .... day of .... one thousand nine hundred and ....

Annexes [not attached to this draft]

I. Statute of the Authority
II. Rules, Regulations and Procedures for the Calculation and Collection of Contributions to the Fund
III. Arbitration and Mediation

Recommended Further Reading

REFERENCE

Mather, Alexander S, *Global Forest Resources*, London: Belhaven, 1990

TROPICAL FOREST POLITICS

Banuri, Tariq and Marglin, Frédérique Apffel, *Who Will Save the Forests: Knowledge Power and Environmental Destruction*, London: Zed Books, 1993
Colchester, Marcus and Larry Lohmann (eds) *The Struggle for Land and the Fate of the Forests*, London: Zed Books, 1993
Hurst, Philip, *Rainforest Politics: Ecological Destruction in South-East Asia*, London: Zed Books, 1990
Utting, Peter, *Trees, people and power*, London: Earthscan, 1993
World Rainforest Movement, *Rainforest Destruction: Causes, Effects and False Solutions*, Penang: World Rainforest Movement, 1990

TROPICAL FORESTS

Brown, Katrina and David W Pearce (eds) *The Causes of Tropical Deforestation: The economic and statistical analysis of factors giving rise to the loss of tropical forests*, London: UCL Press, 1994
Grainger, Alan, *Controlling Tropical Deforestation*, London: Earthscan, 1992
Jepma, C J, *Tropical Deforestation: A Socio-Economic Approach*, London: Earthscan Publications Ltd, 1995
Park, Chris C, *Tropical Rainforests*, London: Routledge, 1992
Myers, Norman, *The Primary Source: Tropical Forests and Our Future*, London: W W Norton & Company, 1992
Poore, Duncan et al, *No Timber Without Trees: Sustainability in the Tropical Forest*, London: Earthscan Publications Ltd, 1989
Rietbergen, Simon (ed) *The Earthscan Reader in Tropical Forestry*, London: Earthscan Publications Ltd, 1994
Westoby, Jack, *Introduction to World Forestry: People and Their Trees*, Oxford: Basil Blackwell, 1989
Whitmore, T C, *An Introduction to Tropical Rainforests*, Oxford: Oxford University Press, 1990

TEMPERATE FORESTS

Dudley, Nigel, *Forests in Trouble: A Review of the Status of Temperate Forests Worldwide*, Gland, Switzerland: WWF International, 1992

Mather, Alexander (ed) *Afforestation: Policies, Planning and Progress*, London: Belhaven, 1993

COUNTRY AND REGIONAL STUDIES

Aiken, Robert S and Colin H Leigh, *Vanishing Rain Forests: The Ecological Transition in Malaysia*, Oxford: Clarendon Press, 1992

Barraclough, Solon L and Krishna B Ghimire, *Forests and Livelihoods: The Social Dynamics of Deforestation in Developing Countries*, London: Macmillan/ UNRISD, 1995

Cowell, Adrian, *The Decade of Destruction*, Sevenoaks: Hodder & Stoughton/ Channel 4, 1990

Cummings, Barbara J, *Dam the Rivers, Damn the People: Development and Resistance in Amazonia Brazil*, London: Earthscan Publications Ltd, 1990

Gradwohl, Judith and Russell Greenberg, *Saving the Tropical Forests*, London: Earthscan Publications Ltd, 1990

Hall, Anthony L, *Developing Amazonia: Deforestation and social conflict in Brazil's Carajás Programme*, Manchester: Manchester University Press, 1991

Hecht, Susanna and Alexander Cockburn, *The Fate of the Forest: Developers, Destroyers and Defenders of the Amazon*, London: Penguin, 1989

Leach, Gerald and Robin Mearns, *Beyond the Woodfuel Crisis: People, Land and Trees in Africa*, London: Earthscan Publications Ltd, 1988

Monbiot, George, *Amazon Watershed: The New Environmental Investigation*, London: Michael Joseph, 1991

Munslow, Barry et al, *The Fuelwood Trap: A Study of the SADCC Region*, London: Earthscan Publications Ltd, 1988

Peluso, Nancy Lee, *Rich Forests, Poor People: Resource Control and Resistance in Java*, Berkeley CA: University of California Press, 1992

Poffenburger, Mark (ed) *Keepers of the Forest: Land Management Alternatives in Southeast Asia*, West Harford: Kumarian Press, 1990

Price, David, *Before the Bulldozer: The Nambiquara Indians and the World Bank*, Washington DC: Seven Locks Press, 1989

Revkin, Andrew, *The Burning Season: The Murder of Chico Mendes and the Fight for the Amazon Rain Forest*, London: Collins, 1990

Vansina, Jan, *Paths in the Rainforests: Towards a History of Political Tradition in Equatorial Africa*, London: James Currey, 1990

TROPICAL FORESTRY ACTION PROGRAMME

Colchester, Marcus and Larry Lohmann, *The Tropical Forestry Action Plan: What Progress?*, Penang, Malaysia/Sturminster Newton, Dorset: World Rainforest Movement/The Ecologist, 1990

Cort, Cheryl, *Voices from the Margin: Non-Governmental Organization Participation in the Tropical Forestry Action Plan*, Washington DC: World Resources Institute, 1991

Halpin, Elizabeth A, *Indigenous Peoples and the Tropical Forestry Action Plan*, Washington DC: World Resources Institute, 1990

Lynch, Owen J, *Whither the People? Demographic, Tenurial and Agricultural Aspects of the Tropical Forestry Action Plan*, Washington DC: World Resources Institute, 1990

Sargent, Caroline, *Defining the Issues: Some thoughts and recommendations on recent critical comments of TFAP*, London: IIED, 1990

Winterbottom, Robert, *Taking Stock: The Tropical Forestry Action Plan After Five Years*, Washington DC: World Resources Institute, June 1990

INTERNATIONAL TROPICAL TIMBER ORGANIZATION

Friends of the Earth and World Rainforest Movement, *The International Tropical Timber Organization: Conserving the Forests or Chainsaw Charter? A critical review of the first five years of the International Tropical Timber Organization*, London: Friends of the Earth, 1992

Hpay, Terence, *The International Tropical Timber Agreement: Its prospects for tropical timber trade, development and sustainable management*, IUCN/IIED Tropical Forest Policy Paper No 3, London: IIED/Earthscan Publications Ltd, 1986

THE UNCED FORESTS NEGOTIATIONS

No book is devoted in its entirety to this subject. Readers are referred to Chapter 7 of:

Johnson, Stanley (ed) *The Earth Summit: The United Nations Conference on Environment and Development*, London: Graham and Trotman, 1993

THE TIMBER TRADE

Barbier Edward et al, *The Economics of the Tropical Timber Trade*, London: Earthscan Publications Ltd, 1995

Bari, Judi, *Timber Wars*, Monroe, Maine: Common Courage Press, 1994

Dudley, Nigel, Jean-Paul Jeanrenaud and Francis Sullivan, *Bad Harvest?: The Timber Trade and the Degradation of the World's Forests*, London: Earthscan Publications Ltd, 1995

Friends of the Earth, *Plunder in Ghana's Rainforest for Illegal Profit: An exposé of corruption, fraud and other malpractices in the international timber trade*, London: Friends of the Earth, 1992

Nectoux, Francois and Yoichi Kuroda, *Timber from the South Seas: An Analysis of Japan's Tropical Timber Trade and its Environmental Impact*, Gland, Switzerland: WWF International, 1989

Panayotou, Theodore and Peter S Ashton, *Not by Timber Alone: Economics and Ecology for Sustaining Tropical Forests*, Washington DC: Island Press, 1993

Raphael, Ray, *More Tree Talk: The People, Politics and Economics of Timber*, Washington DC: Island Press, 1994

FOREST PRODUCTS CERTIFICATION

Bass, Christopher and Stephen, *The Forest Certification Handbook*, London: Earthscan Publications Ltd, 1995

TEXTS ON THE INTERNATIONAL RELATIONS OF THE ENVIRONMENT CONTAINING SECTIONS ON FORESTS

Glasbergen, Pieter and Andrew Blowers (eds) *Environmental Policy in an International Context: Perspectives on Environmental Problems*, London: Arnold, 1995

Hurrell, Andrew and Benedict Kingsbury (eds) *The International Politics of the Environment: Actors, Interests, and Institutions*, Oxford: Clarendon Press, 1992

Lipschutz, Ronnie D and Ken Conca (eds) *The State and Social Power in Global Environmental Politics*, New York: Columbia University Press, 1993

Porter, Gareth and Janet Welsh Brown, *Global Environmental Politics*, Boulder, Colorado: Westview Press, 1991

Thomas, Caroline, *The Environment in International Relations*, London: Royal Institute of International Affairs, 1992

Vogler, John and Mark Imber (eds) *The Environment and International Relations*, London: Routledge, 1996

Acronyms

ACP	African, Caribbean and Pacific
AFOS	Agriculture, Forestry and Other Human Activities (workshop of the IPCC)
ASEAN	Association of Southeast Asian Nations
BCSD	Business Council for Sustainable Development
C&I	Criteria and Indicators for Sustainable Development
CAN	Canada, Australia, New Zealand group (UN caucus)
CARICOM	Caribbean Community
CCNFP	Country Capacity for National Forest Programmes
CFCs	Chlorofluorocarbons
CFDT	Committee on Forest Development in the Tropics (of the FAO)
CH4	Methane
CITES	Convention on International Trade in Endangered Species of Wild Fauna and Flora, 1973
CO2	Carbon Dioxide
COFO	Committee on Forestry (of the FAO)
COICA	Coordinating Body for the Indigenous Peoples' Organizations of the Amazon Basin
CSA	Core Support Agency (for NFAP)
CSCE	Conference on Security and Cooperation in Europe
CSD	Commission on Sustainable Development (of the UN)
CSE	Centre for Science and Environment (India)
CSA	Core Support Agency (for NFAP)
DTI	Department of Trade and Industry (UK)
EC	European Community
ECOSOC	Economic and Social Council (of the UN)
ELCI	Environment Liaison Centre International (Kenya)
EU	European Union
FAO	Food and Agriculture Organization of the UN
FAG	Forestry Advisers Group (formerly the TFAP Forestry Advisers Group)
FFDC	Forestry Forum for Developing Countries

FoE	Friends of the Earth
FSC	Forest Stewardship Council
G7	Group of Seven Industrialized Countries
G15	Group of Fifteen (also known as the Summit Level Group of Developing Countries)
G77	Group of Seventy-Seven Developing Countries
GATT	General Agreement on Tariffs and Trade
GCC	General Coordinating Committee (of the Helsinki process)
GEF	Global Environment Facility
GFC	global forests convention
GFI	global forests instrument
GLOBE	Global Legislators' Organisation for a Balanced Environment
ICSU	International Council of Scientific Unions
IGO	Intergovernmental Organization
IIED	International Institute for Environment and Development
ILO	International Labour Organization
IMF	International Monetary Fund
INCB	International Narcotics Control Board
IPC	Integrated Programme for Commodities (of UNCTAD)
IPCC	Intergovernmental Panel on Climate Change
IPF	Intergovernmental Panel on Forests
IPPF	International Planned Parenthood Federation
ITTA	International Tropical Timber Agreement (1983 or 1994)
ITTC	International Tropical Timber Council (of the ITTO)
ITTO	International Tropical Timber Organization
IUCN	World Conservation Union (formerly the International Union for the Conservation of Nature and Natural Resources)
IWGF	Intergovernmental Working Group on Forests
LEEC	London Environmental Economics Centre
NFAP	National Forestry Action Plan/Programme
NGO	Non-Governmental Organization
NIEO	New International Economic Order
N_2O	Nitrous Oxide
ODA	Overseas Development Administration (UK)
OECD	Organization for Economic Cooperation and Development
OFI	Oxford Forestry Institute (of Oxford University)
PC/65	UN document A/CONF.151/PC/65 "Guiding Principles for a Consensus on Forests" (UNCED process)
PrepCom	Preparatory Committee (of the UNCED process or of the negotiations for the ITTA 1994)
SA	Supporting Agency (for a NFAP)
SAP	Structural Adjustment Programme

SCOPE	Scientific Committee on Problems of the Environment (of ICSU)
SWCC	Second World Climate Conference (Geneva, 1990)
TFAP	Tropical Forestry Action Programme, sometimes referred to as the Tropical Forests Action Programme (until 1990 the Tropical Forestry Action Plan)
TNC	Transnational Corporation
TRADA	Timber Research and Development Association (UK)
TRAFFIC	Trades Record Analysis of Fauna and Flora in Commerce
UN	United Nations
UNCED	United Nations Conference on Environment and Development (Rio de Janeiro, 1992)
UNCTAD	United Nations Conference on Trade and Development
UNDP	United Nations Development Programme
UNEP	United Nations Environment Programme
UNESCO	United Nations Educational, Scientific and Cultural Organization
UNFPA	United Nations Population Fund (formerly the United Nations Fund for Population Activities)
UNSO	United Nations Sudano-Sahelian Office
USA	United States of America
USAID	United States Agency for International Development
WALHI	Wahana Lingkungan Hidup Indonesia (Indonesian Environmental Forum)
WCFSD	World Commission on Forests and Sustainable Development
WG	Working Group (of the UNCED Preparatory Committees)
WMO	World Meteorological Organization
WRI	World Resources Institute
WRM	World Rainforest Movement
WTO	World Trade Organization
WWF	World Wide Fund for Nature (formerly the World Wildlife Fund)

Index

acid rain, 10
Adams, Patricia, 4
African–Caribbean–Pacific–European
   Community (ACP–EC) Joint
   Assembly, 153
   support for a global forests
      convention, 153
African Convention for the Conserva-
   tion of Nature and Natural
   Resources (1968), 131
African Development Bank, 39, 40
agenda formation, 18
Agenda 21, 51, 53–4, 89, 97, 100, 138,
   143–4, 167
AIDEnvironment, 123, 153
Amazon, 1, 3–6, 8, 11, 13–14, 17–18,
   26, 28, 46, 72, 141, 143–4
Amazonian Cooperation, 2, 131, 141
   Workshop to Define Criteria and
      Indicators of Sustainability in the
      Amazon, 136, 138, 141–4, 150,
      152
Amha Buang, 106, 114
Anderson, Patrick, 11
Apkindo (Indonesian Association of
   Plywood Makers), 7
Argentina, 42
Armenia, 54
Asian Development Bank, 39–40
Association of South East Asian Nations
   (ASEAN), 73, 77, 131
   Agreement on the Conservation on
      Nature and Natural Resources
      (1985), 131
Australia, 40, 59, 66, 70, 78, 85–6, 88,
   97, 100, 119, 121, 141
   and IIED Report on sustainable forest
      management, 66

and negotiation of ITTA (1994), 119,
   121
and Target 2000, 70, 110, 121
and UNCED forests negotiations, 88,
   97, 100
Austria, 40, 59, 74, 77, 140
   timber labelling case (1992), 74, 77,
      125

'Bangkok Workshop' (to explore
   options for global forestry
   management, 1991), 94–7
Barba, Jorge, 111
bargaining, distributive and integrative,
   159, 163
bargaining issue-linkages, 96–7, 101–2,
   110, 157, 165, 167
Barnett Report (Papua New Guinea), 7
Bautista, Germelino, 8
Beijing Ministerial Declaration on
   Environment and Development
   (1991), 89, 94–6
   and proposal for Green Fund, 96
Bellagio Strategy Meeting on TFAP
   (1987), 34
Bellagio II meeting on TFAP (1988), 34
Berthoud, Gérald, 5
Bhopal chemical factory explosion, 214
biodiversity, see biological diversity
biological diversity, 2, 16–17, 26, 45,
   70, 80, 85–6, 96, 99, 101–2, 123,
   143, 154
Blowers, Andrew, 23
Bohlen, Curtis, 101, 162
Bolivia, 9, 19, 20, 41, 53, 59, 141
   cocaine cultivation in, 9
   ITTO Chimanes project, 71
Boulter, David, 114

Boutros-Ghali, Boutros, 146
BR–364 Amazonian highway, 3, 9
Bramble, Barbara, 21, 62, 75
Brazil, 3–6, 8–9, 11, 17–18, 20, 26, 28, 50, 58–9, 60, 80, 86, 92–3, 106, 114, 141, 164
  and negotiation of ITTA (1994), 106, 114
  and TFAP restructuring process, 50
  and UNCED forests negotiations, 92–3
  cattle ranching in, 6, 9
  external debt of, 4, 28
  gold prospecting in, 5
  Grande Carajàs programme, 3
Britain, *see* United Kingdom
British Gas, 6
  oil prospecting in Ecuador, 6
Brown, Janet Welsh, 25, 157
Brundtland Commission, *see* World Commission on Environment and Development
Bulgaria, 54, 122
burden-sharing, 87, 93, 103, 163–5; *see also* common responsibility; common but differentiated responsibilities
Burkino Faso, 39
Burma, *see* Myanmar
Bush, George, 84, 116, 162
Business Council for Sustainable Development, 167

Caldwell, Lynton Keith, 25
Cambodia, 10, 19–20, 41
Cameroon, 5, 20, 39, 41–2, 59, 119
  National Forestry Action Plan of, 39, 42
Canada, 25, 35, 40, 59, 85, 88, 90–1, 94, 114–15, 118–21, 136–7, 140–1, 149, 153
  and deforestation in British Columbia, 1
  and Intergovernmental Working Group on Forests, 136–7
  and negotiation of ITTA (1994), 114, 118–19, 121, 153
  and Target 2000, 119, 121
  and UNCED forests negotiations, 88, 90–1, 94, 136
cannabis, 8

Capito, Eugene, 114
Caribbean, 42
Cartagena Commitment (of UNCTAD VIII), 114, 127
Carter, Jimmy, 116–17
Centre for International Environmental Law, *see* Foundation for International Law and Development
certification, *see* timber labelling
Chamorro, Violeta, 38
Charter of the Indigenous-Tribal Peoples of the Tropical Forests (1992), 27
Chernobyl nuclear explosion, 21
Chile, 41, 122, 141
China, 41, 59, 88–9, 94, 96–7, 101, 121, 129, 141
  and negotiation of ITTA (1994), 121, 129
  and UNCED forests negotiations, 88–9, 94, 96–7, 101
  joint negotiating positions with G77 during UNCED process, 88–9, 94, 96–7, 101
Chipko movement (India), 18
Clinton, Bill, 116
cocaine, 8–9
Cockburn, Alexander, 4, 11
Colchester, Marcus, 14, 46, 48, 57, 60, 67
Colombia, 3, 8, 20, 41, 59, 86, 141
  poppy cultivation in, 8
Commission on Sustainable Development, *see* United Nations Commission on Sustainable Development
Common Fund for Commodities, 62
  relations with ITTO, 62
common but differentiated responsibilities, 96–8, 163–5; *see also* burden sharing; common responsibility
common responsibility, 93, 95, 98, 103, 163
Commonwealth, 107, 118, 136–7
  Fourteenth Forestry Conference of, 107, 118, 136–7
compensation for opportunity cost foregone, 91, 97–100
Conference on Security and Cooperation in Europe, 107, 118–19, 140
  forests seminar (1993), 107, 118–19, 140; *see also* Montreal Process
confidence-building, 135–8, 145, 162–3

Congo, 42
Conservation International, 16
Convention for the Protection of the
    Ozone Layer (1985), 131
Convention for the Protection of the
    World Cultural and National
    Heritage (1972), 131
Convention on Biological Diversity
    (1992), 85, 89, 96, 102, 131, 164
Convention on Climate Change (1992),
    84–6, 89, 96, 102, 131, 164
Convention on European Wildlife and
    Natural Resources (1979), 131
Convention on Long-Range
    Transboundary Air Pollution (1979),
    131
Convention on Nature Protection and
    Wildlife Preservation in the Western
    Hemisphere (1940), 131
Convention on the International Trade
    in Endangered Species of Wild
    Fauna and Flora (CITES) (1973), 75,
    80–1, 127, 131–3, 154, 162
  lists commercially-traded tree species,
    80–1, 154
  relations with ITTO, 80–1, 132–3
  Timber Working Group of, 81
Convention on Wetlands of Inter-
    national Importance Especially as
    Waterfowl Habitat (1971), 131
Coordinating Body for the Indigenous
    Peoples' Organizations of the
    Amazon Basin (COICA), 46, 72
  and ITTO, 72
  and TFAP, 46
Costa Rica, 17, 39, 41, 47, 79, 122
  government's agreement with Merck
    and Company, 17
  National Forestry Action Plan of, 39
Counsell, Simon, 119
country capacity projects, 46, 53
  in NFAPs, 46, 53
Cranbrook, Earl of, 72, 168–9
criteria and indicators for sustainable
    forest management, 67–8, 115, 118,
    136–45, 148–52, 154
  debate on harmonization/convergence,
    141–2, 149

dams, construction of, 3, 10

debt, 3–5, 9, 28, 91, 95–6, 98–100, 102,
    160, 161, 165, 171
debt-for-nature swaps, 4–5, 73
debt-for-timber swap (between
    Cameroon and France), 5
declining terms of trade, 98, 161
deforestation, 1–22, 28–9, 31, 42–5, 53,
    67, 77, 83–4, 90–1, 95–6, 98,
    102–3, 111, 115, 120, 133, 138,
    144, 163, 166–8, 170–1
  as an ethical issue, 21
  cause–agency distinction, 2, 9, 11,
    13–14
  causes of, 2–15, 22, 28–9, 42, 44,
    102–3, 166–7, 171
  definition of, 1
  effects of, 2, 15–18
  in the Amazon, 1, 4–6, 11, 13–14,
    17–18; Brazil, 4; British Colom-
    bia, 1, 120; Cambodia, 10;
    Colombia, 3, 8; El Salvador, 10;
    Ecuador, 6, 9; Gabon, 3; Guate-
    mala, 8; Indonesia, 9; Myanmar,
    10; Panama, 3; Peru, 3, 8; Siberia,
    1; Tanzania, 11; Vietnam, 10;
    Zaire, 11
  role of military élites in, 11
  surveys of, 19–21
Denmark, 61
Department of Trade and Industry (UK),
    61, 73
  and British delegation to ITTO, 61,
    73
Deutsche Gesellschaft für Technische
    Zusammenarbeit, 13, 35
Diegues, Antonio Carlos, 4
distributive justice, 26
Dorner, Peter, 14
Drucker, Milton, 106, 114–17, 120, 125
drug cultivation, 8–9, 10
  in Bolivia, 9; Colombia, 8;
    Guatemala, 8; Peru, 8

eco-colonialism, 166
ecologism, 12
Economic and Social Council of UN
    (ECOSOC), 89, 114
ECOROPA, 19
  and campaign for emergency session
    of UN General Assembly, 19

Ecuador, 6, 9, 20, 39, 41–2, 58, 59, 80, 111
    Land Colonization Laws of, 9
    National Forestry Action Plans of, 39, 42
    oil exploration in, 6
Egypt, 58–9, 66
Ehrlich, Paul and Anne, 12
El Salvador, 10, 41, 122
Elliott, Chris, 75, 86
embedded liberalism, 156
Environment Liaison Centre International (Kenya), 46
European Community/European Union, 40, 59–60, 71, 77, 85, 88, 90–1, 97, 99, 106, 110, 114–15, 119–21, 125, 128–9, 130, 136, 141, 153, 161–2
    and negotiation of ITTA (1994), 106, 114–15, 119–21, 125, 128–30
    and UNCED forests negotiations, 90–1, 97, 99, 161–2
European Council, 84–6, 90
    Dublin summit (1990), 84, 86, 90
    proposes tropical forest protocol, 84–5
European Parliament, 84–6, 153, 156
    proposes a worldwide convention on forests, 85–6
European Union, *see* European Community
Experts Meeting on Tropical Forestry (1982), 32

financial transfers, claims for by South, 11–12, 91, 95–101, 105–6, 110, 117–18, 120, 125, 137, 145, 158–9, 161, 163–6
Finland, 35, 40, 58, 59, 139–40
First Ministerial Conference on the Protection of Forests in Europe (1990), 139
First Ministerial Meeting of Developing Countries on Environment and Development, *see* Beijing Ministerial Declaration on Environment and Development (1991)
First World Climate Conference, 16
Food and Agriculture Organization (FAO) of the United Nations, 2, 16, 19–20, 32–54, 56–7, 67–8, 79, 83–8, 90, 93–4, 108–9, 123–4, 135, 141, 144–5, 148–9, 152–4, 162–3
    and creation of TFAP, 32, 34–5
    and negotiation of ITTA 1983, 56
    and negotiation of ITTA 1994, 109
    and TFAP restructuring process, 36, 38, 42, 44–5, 49, 88
    Committee on Forest Development in the Tropics (CFDT) of, 32, 35–6, 47, 49, 51–2, 79
    Committee on Forestry (COFO) of, 32, 35, 37, 46–7, 49–50, 52, 79, 84, 86–7, 149, 152, 154
    Council of, 35, 47–52, 152
    creation of, 32
    definition of sustainable development, 68
    deforestation surveys of 1981 and 1993, 19–20
    Director-General of, 35, 50, 88
    draft global forests convention of, 85–8, 93–4, 123–4, 163
    European Commission on Forestry of, 154
    Forestry Department of, 32, 35, 37, 44, 49, 53, 79, 88
    Ministerial Meeting on Forestry of 1995, 144, 152, 154
    offers to provide forum for negotiation of global forests instrument, 83, 86, 163
    relationship with UNCED, 88, 93, 163
    responds to erosion of authority, 152–3
forest conservation problematic, 21–9, 155, 166–71
    causal dimension, 22, 166–8, 170–1
    institutional dimension, 22–4, 168–71
    proprietorial dimension, 24–8, 170–1
forest degradation, definition of, 1
Forest Stewardship Council, 74–5, 108, 133, 136, 149–52, 158
    founding assembly of, 74, 149, 158
    goal of, 74, 150
    principles of forest management, 150–1
    relations with ITTO, 74–5, 133
Forestry Advisers Group (FAG), 34–6, 38, 42, 46–7, 49–53, 78–9, 158
    creation of, 34–5
    mandate of, 35–6, 52
    TFAP restructuring process, 49, 51–2
    relations with ITTO, 78–9, 133

relations with TFAP, 35–6, 52
steering group of, 35
support for World Commission on
Forests and Sustainable
Development, 52
Forestry Canada, 118
Forestry Forum for Developing
Countries, 49–51, 54, 158, 163
and TFAP restructuring process,
49–51, 54, 158
Delhi Declaration (1993) of, 163
Foundation for International Environmental Law and Development, 123
France, 5, 35, 40, 58, 59, 66–7, 73–5,
79, 116, 119, 138–40
Freezailah bin Che Yeom, 60
Friends of the Earth, 3, 5–6, 8, 14, 46,
48, 62, 66–7, 71–5, 79, 109–10,
116, 119, 149
and TFAP, 46, 48, 79
criticism of ITTO Mission to
Sarawak, 72
critique of Brazilian mahogany
industry, 8; Ghanaian timber
industry, 8; ITTO project
approval process, 66, 71; World
Bank project in Gabon, 3
observer status at the ITTO, 62
position during negotiation of ITTA
(1994), 109–10, 116
timber labelling proposal at the ITTO,
73–5, 149
withdraws from FSC, 74
Fujimori, Alberto, 8

Gabon, 3, 20, 41, 58, 59, 114, 119
Gaia Foundation, 123
General Agreement on Tariffs and Trade
(GATT), 74–8, 81, 133
discrimination between like products,
76
dolphin–tuna case (1991), 76
extraterritoriality, 76
further processing of timber, 77
relations with ITTO, 75–8, 81, 133
Uruguay Round, 76, 78
George, Susan, 4–5, 9
German–Japanese Expert Meeting on
Tropical Forests (1992), 80, 153
Germany, 35, 40, 58, 59, 70, 101, 138,
140, 147

and UNCED forests negotiations, 101
government finances ITTO's natural
tropical forests guidelines, 70
Ghai, Dharam, 27
Ghana, 8, 41, 58, 59, 80, 94, 96, 106,
119
Glasbergen, Pieter, 23
Global Biodiversity Strategy, 16
global commons, 24–5, 90, 95, 99, 103
Global Environment Facility (GEF), 52,
96–8, 100
Global Forests Conference (1993), 146
global forests convention, 21, 83–8,
90–1, 93, 98–100, 102, 109, 116,
121–4, 129, 135–6, 141, 146, 148,
153–65
blocked by G77 during UNCED
forests negotiations, 21, 99–100,
102, 135, 159, 161–2
FAO draft of, 85, 86–8, 93, 123, 163
GLOBE International draft of, 123–4
North's demands for, 84–6, 90–1,
98–100, 102, 109, 116, 122–4,
129, 153, 155–7
possible features of, 123–4, 153, 165
proposals for, 84–5, 124, 156
prospects for future negotiation of,
124, 153–4, 164–6
global forests instrument, proposals for,
83–5
Global Legislators Organisation for a
Balanced Environment (GLOBE),
110, 116, 123–4, 153
draft global forests convention of,
123–4
writes to ITTO on illegal timber trade,
110
global warming, 2, 15–16, 84, 98, 145,
164
and predicted net rises in sea levels, 15
deforestation and, 15–16
Goedkoop van Opijnen, Yolanda, 111
good governance, 97–8
Gore, Al, 116–17
and GLOBE International, 116
Goree, Langston James, 137
Grandé Carajas programme, 3
Green Fund proposal, 96, 98, 100
green parties, 45, 47
Brussels meeting of (1989), 45
greenhouse effect, *see* global warming

Greenpeace, 74
  withdraws from FSC, 74
Group of Fifteen, *see* Summit Level
  Group of Developing Countries
Group of Seven Industrialized Countries
  (G7), 21, 45, 47, 84–5, 88, 91,
  156
  calls for reform of TFAP, 45
  Houston summit, 1990, 21, 45, 47,
  84, 91
  proposes a global forests convention,
  84–5, 91, 156
Group of Seventy-Seven Developing
  Countries (G77), 49, 51, 88–9,
  94–7, 99–102, 122, 146, 154,
  158–62, 171
  and Forestry Forum for Developing
  Countries, 49
  and UNCED forests negotiations,
  94–5, 122, 158, 171
  joint negotiating positions with China
  during UNCED process, 88–9,
  94, 96–7, 101
Guatemala, 8, 39, 41
  cannabis cultivation in, 8
  National Forestry Action Plan of, 39
Guyana, 94

Haas, Peter, 12, 156
Hague Recommendations on
  International Environmental Law
  (1991), 25
'hamburger connection', 6
Harare Declaration on Family Planning
  for Life (1989), 13
Harbinson, Rod, 10
Hardin, Garrett, 26–7
Harrison, Paul, 6, 24, 25, 37
Hasan, Mohamad 'Bob', 7
Hecht, Susanna, 4, 6, 11
Helsinki process, 115–18, 136, 138–44,
  150, 153
Honduras, 7, 41, 58, 59
  government's agreement with Stone
  Container Corporation, 7
Hopkinson, Nicholas, 10
Hurrell, Andrew, 4, 26, 160

India, 18, 20, 21, 41, 59, 95–7, 102,
  136, 158

and UNCED forests negotiations,
  95–7, 102, 136, 158
Indian–UK Workshop 'Towards Sus-
  ainable Forestry: Preparing for CSD
  1995' (1994), 136, 138
indigenous peoples, 3, 9, 17, 22–4,
  27–8, 43, 46, 48, 51, 67, 71–2,
  109–10, 120, 123, 148–9, 151, 169
  activity at the ITTO, 71–2, 169
  Charter of the Indigenous-Tribal
  Peoples of the Tropical Forests
  (1992), 27
Indonesia, 7, 9, 11, 20, 41, 46, 50, 58,
  59, 60, 74, 77, 80, 91, 106, 111,
  114, 147, 164
  and negotiation of ITTA, 111, 114;
  (1994), 106
  and TFAP restructuring process, 50
  National Forestry Action Plan of, 164
  transmigration programme, 9–10
inequities in land tenure, 12–13, 14, 170
institutional overload, 132
Integrated Programme for Commodities,
  55
Inter-American Development Bank, 3,
  39, 40
InterAction Council of Former Heads of
  State and Government, 146–7
intergenerational equity, 21, 162
Intergovernmental Panel on Climate
  Change, 16, 84, 86, 145
  proposal for a global forest instrument
  by the AFOS workshop of, 84
  Responses Strategies Working Group
  of, 84
Intergovernmental Panel on Forests, 51,
  135, 144, 145, 147–9, 152–3,
  162–3, 169
  creation of, 51, 144, 162
  programme of work of, 144–5, 163
Intergovernmental Working Group on
  Forests, 135–7, 144–5, 149, 163
  role in defining agenda of Intergovernmental Panel on Forests, 144–5,
  163
international civil society, 1, 136, 147
International Conference on Trees and
  Forests (1986), 32
International Council of Scientific Union
  (ICSU), 15

Scientific Committee on Problems of
the Environment (SCOPE) of, 15
International Institute for Environment
and Development, 35, 38–40, 42,
57, 66–8, 79, 123
and Forestry Advisers Group, 35
and GLOBE International draft global
forests convention, 123
and National Forestry Action Plans,
35, 38–9, 42
hosts NGO meeting calling for
ratification of ITTA 1983, 57
report for ITTO on sustainable forest
management, 66–8
study on ITTO-TFAP relationship, 79
International Labour Organization, 27–8
Convention No 107 (1957), 27
Convention No 169 (1989), 27–8
International Monetary Fund, 5, 10
International Narcotics Control Board, 8
International Negotiating Committee on
Climate Change, 89
International Planned Parenthood
Federation, 13
International Tropical Timber Agreement (1983), 55–68, 70, 73, 76–8,
81, 88, 105–9, 114, 116, 118,
121–2, 132–3, 169
entry into legal effect of, 55, 57
IUCN intervention during the
negotiation of, 56–7
objectives of, 57, 60–2, 65–8, 70, 76–7
preparatory committee meetings for
negotiation of, 56, 64
UNCTAD conference for negotiation
of, 107
International Tropical Timber Agreement (1994), 55, 69, 77–8, 105,
121–6, 128–30, 133, 136, 153,
162–4, 166, 169
amended articles, 127–8
emphases on non-tropical timber, 128
monitoring, 123, 129, 130
new articles in, 126
references to Target/Objective 2000,
129
remit of, 125
reservations tabled, 121, 128–9
International Tropical Timber
Agreement (1994), negotiation of,
69, 105–33, 135, 166

and Bali Partnership Fund, 121, 125,
129–30
and global forests convention debate,
109, 153
Committee on Finance and Administration, 109, 113, 130
consumers' 'non-paper', 119, 121
consumers' statement on Target 2000,
121
consumers' positions, 106, 108–10,
117, 120–1, 123–4
consumers/timber traders alliance,
109–10, 125
European Community/European
Union's position, 106, 111,
114–15, 118–21, 128–9
FAO's position, 109
financial transfers issue, 105–6, 110,
117–18, 120, 125, 163
illegal timber trade, 110
incentives, 110, 133
indigenous peples, 109–10, 120, 123
'Issues and Options' paper, 106, 111
local communities, 109–10, 127, 130
national reporting, 109–11
NGO participation, 111, 114, 119;
NGOs' positions, 106–11, 115,
129–30, 125, 132
non-ITTO members partaking in,
121–2
preparatory meetings for, 105–11,
130
president's informal discussions,
117–18
producers/NGOs' alliance, 108–11,
125, 132
producers' positions, 106–11, 117
projects, 105, 108
scope, 105–6, 108–10, 114, 117–21
Target 2000, 109–11, 115–17,
119–21, 125–6, 128–9
technology transfers issue, 105, 110,
114, 118, 125–6, 163
timber traders' positions, 109–10
trade discrimination, 105, 107, 115,
117–20, 124
International Tropical Timber Organization, 4, 7, 27, 40, 51, 55–81, 83,
87–8, 102, 105–6, 108–11, 114–30,
132–3, 145, 149, 158–9, 162, 164,
166, 168–71

action plan of, 69, 127
administrative account of, 62
and indigenous forest peoples, 27, 67, 71–2
choice of headquarters site, 60, 63
executive director, 60, 63, 65, 70, 78–9, 110–11, 130
Expert Panel for Technical Appraisal of Project Proposals, 65–6
financing of, 62
Guidelines for the Conservation of Biodiversity in Tropical Production Forests, 70
Guidelines for the Establishment and Sustainable Management of Planted Tropical Forests, 70
Guidelines for the Sustainable Management of Natural Tropical Forests, 70–1, 83, 87–8
IIED report on sustainable forest management for, 66–8
incentives debate, 72–4, 81, 110, 130, 133, 158–9
institutional structure of, 60–2
ITTC, 55–6, 60–7, 69–75, 78–81, 106–7, 110–11, 114–16, 120, 122, 125, 129, 158, 164, 169
labelling debate, 72–5, 77, 81, 133, 149, 158–9
Mission to Sarawak, 72, 115, 168, 171
Permanent Committees, 60–1, 63–6, 69, 73, 75, 78, 107–9, 121, 130
project cycle, 63, 65, 71
project work of, 62, 63, 65, 105
relations with CITES, 80–1, 132–3; with Common Fund for Commodities, 62; with FAG, 78–9; with FSC, 74–5, 133; with GATT, 75–8, 81, 133; with TFAP, 75, 78–81
special account of, 62
'sunset provision', 65
Target 2000, *see* separate entry
voting system, 57–8, 60, 75
International Union for the Conservation of Nature and Natural Resources (IUCN), 6, 16, 25, 38, 40, 46, 56–7, 61, 67, 69–70, 77, 79–81
18th General Assembly of (1990), 70, 85–6

and International Environmental Law Conference (1991), 25
and negotiation of ITTA, 1983, 56–7
and negotiation of ITTA (1994), 120
and Netherlands delegation to ITTO, 61
and TFAP, 38, 40, 46
and WWF's 1995 Target, 69
guidelines for oil exploration in the tropics, 6
intervention during the negotiation of the ITTA 1983, 56
International Union of Biological Sciences, 15
International Union of Geodesy and Geophysics, 15
International Whaling Commission, 124
International Year of the World's Indigenous People (1993), 28
Ireland, 138

Japan, 7–8, 40, 58, 59–60, 80, 88, 97, 106, 110, 114–15, 121, 141, 153
and negotiation of ITTA (1994), 106, 114
and Target 2000, 121
and UNCED forests negotiations, 88, 97
government's financial contributions to the ITTO, 60, 62
proposes an international tropical timber agreement, 56
Japan Tropical Forest Action Network (JATAN), 62
observer status at ITTO, 62
Johnson, Brian, 158
Johnson, Stanley, 158, 162
Jönsson, Christer, 163

Kant, Immanuel, 26
Formula of Humanity of, 26
Keohane, Robert O, 155, 156, 160
Khmer Rouge, 10
Kismadi, M S, 91
Korea, South, 59, 141, 154
support for a global forests instrument, 154
Kuala Lumpur, 47, 64, 75, 86, 89, 101, 107, 114–18, 131, 136–7, 141
Declaration on Environment and Development (1992), 101

Kufuor, Edward, 94–5

Laos, 42
Leach, Gerald, 13
Levy, Marc A, 156
Liburd, Charles, 93
Litfin, Karen, 22, 156
Lithuania, 54
local communities, 17, 22–3, 26–7, 34,
    48, 71, 109–10, 127, 130, 148, 151,
    168–71
Loeis, Wisber, 111, 128, 164
Lohmann, Larry, 14
Lomé IV Convention, 130–1, 136, 153
    protocol on the Sustainable Manage-
    ment of Forest Resources to,
    130–1, 153
London Convention on the Dumping of
    Waste at Sea (1972), 162
London Environmental Economics
    Centre (LEEC), 73–5, 130
    incentives study for ITTO, 73
Lovejoy, Thomas, 4
Lutzenberger, José, 14

McLeary, Rachel, 26, 28
Mahathir Bin Mohamad, 101, 118
Mahony, Rhona, 4
Malaysia, 3, 7–8, 20, 40, 48, 50, 58–9,
    60–1, 67, 72–4, 80–1, 86, 91, 94–7,
    101–2, 106, 111, 114–15, 118, 132,
    136–7, 149, 158, 162–3, 169
    and Intergovernmental Working
    Group on Forests, 136–7
    and ITTO Mission to Sarawak, 72,
    115, 168
    and negotiation of ITTA (1994), 106,
    111
    and TFAP restructuring process, 48,
    50
    and UNCED forests negotiations, 91,
    94–7, 101, 136, 158
    proposes intergovernmental task force
    on forestry, 136–7
Malaysian–Indonesian Joint Working
    Group on Forestry, 111
Malthus, Thomas, 12–14
marine life, 15
marine pollution, 19
Marshall, George, 43
Mearns, Robert, 13

Mendes, 'Chico', 18
Merck and Company, 17
    agreement with government of Costa
    Rica, 17
Mexico, 3, 20, 32, 41, 76, 122, 141,
    150–1
Monbiot, George, 27
Montreal process, 118, 131, 136,
    138–44, 149–50
Myanmar, 10–11, 20, 41–2, 122
    Karen hill tribe, 10
    Moung Tai Army, 10
Myers, Norman, 1, 10, 13, 19–20
    deforestation surveys of 1980 and
    1989, 19–20

national accounting systems, 145
National Forestry Action Plans
    (NFAPs), 31–2, 34–46, 48, 52–4,
    78–80, 158, 164, 166
    donor participation in, 40, 52–3
    for Burkino Faso, 39; Cameroon, 39;
    Ecuador, 39; Guatemala, 39;
    Nicaragua, 38; Papua New
    Guinea, 78; Zaire, 39
    national lead institutions in, 36,
    38
    planning process, 221
    steering committees in, 37, 39, 53
National Wildlife Federation, 61, 116,
    118
    and negotiation of ITTA (1994), 116,
    118
    and US delegation to ITTO, 61
natural disasters, 14
Nepal, 42
Netherlands, 25, 34–5, 40, 58, 59, 61,
    65, 69, 80–1, 111, 115, 120
    government hosts meeting on TFAP
    (1985), 34–5
    government's support for WWF's
    1995 target, 69, 115
    position at Second Ministerial
    Conference on the Protection of
    Forests in Europe, 115
    proposals listing of tropical timber
    species at CITES, 81
net South-to-North financial flows, 4,
    11–12, 95, 99, 102, 160, 171
Netherlands Timber Trade Association,
    69

New International Economic Order, 96, 157–8, 160–1
New Zealand, 40, 59, 88, 141
NFAP Support Unit, 35–6, 42, 53–4
Nicaragua, 38
Nissho Iwai, 7
NOAA–9 satellite, 18
non-governmental organizations (NGOs), 2–4, 6–7, 14, 16, 18–19, 21–2, 24, 32, 34–5, 37–40, 42–3, 45–52, 54, 56–7, 60–3, 66–72, 74–5, 78–9, 84–6, 90, 106–11, 114–17, 119–20, 123, 125, 128–30, 132–3, 139, 148–9, 152–3, 167–9
   and negotiation of International Tropical Agreement, 1994, 106–11, 114–15, 119, 125, 129–30, 132
   calls for ITTO liaison officer, 63
   calls to World Bank to reform to TFAP, 45
   criticism on national reporting to the ITTO, 71
   draft resolution to ITTO on forest peoples, 71
   lobbying, at G7 summit (1990), 84; of USA on expansion of scope of ITTA, 116
   observer status at the ITTO, 61–2
   participation at TFAP Bellagio Strategy Meeting (1987), 34
   statement at the Second World Climate Conference, 85–6
non-wood products, 17
Nor, Salleh Mohammed, 43
Norway, 138
Nuu-Chah-Nulth Tribal Council, 120

Objective 2000, *see* Target 2000
Oil Industry Exploration and Production Forum, 6
   guidelines for oil companies operating in tropical forests, 6
oil production, 6
Oksanen, Tapani, 53
Osherenko, Gail, 159–60, 163, 165
Overseas Development Administration (UK), 61, 70, 73
   and British delegation to ITTO, 61
Oxford Forestry Institute, 73

ozone destruction, 19

Pan-American Highway, 3, 9
Papua New Guinea, 7, 20, 39, 41, 59, 78, 119
   Barnet Report on illegal practices in timber trade, *see* Barnett Report
   National Forestry Action Plan of, 39, 78
partnership in additionality, 97–8
Penan (Sarawak), 18
Perez, Carlos Andres, 99–100, 170
Pérez de Cuéllar, Javier, 19
Peru, 3, 8, 20, 41, 59, 141
   cocaine cultivation in, 8
   Sendero Luminoso (Shining Path), 8
Petro Ecuador, 6
Philippines, 8, 11, 20, 41, 58, 59, 74, 114
   and negotiation of ITTA (1994), 114
Plumwood, Val, 11
Poland, 140
Poore, Duncan, 66–7
poppy cultivation, 8
population, growth, 12–14; movements, 9
Porter, Gareth, 21, 25, 62, 75, 157
Portugal, 59, 140
postmodernism, 148, 150
Potter, David, 171
poverty, 11–14, 84, 91, 95, 98–9, 162

radiative forcing, 15–16
regime theory, 155–63
   cognitive factors, 160
   interest-based factors, 156–7
   metaphors and, 162–3
Renninger, John P, 160
Research Institute for Forestry and Forest Products (Hamburg), 70
Rittberger, Volker, 222
Rome Statement on Forestry (1995), 152, 154
Romm, J, 25–6
rosi periwinkle plant, 16
Routley, Richard, 11
Rubber Tappers' Association, Brazil, 18
Ruggie, John Gerard, 156
Russian Federation, 59, 110, 141, 153
   assumes obligations of USSR at the ITTO, 59

Siberia, 1, 107
Rwanda, 11, 41
    civil war between Hutu and Tutsi
        tribes, 11

Sabah, 8
Sachs, Wolfgang, 12
Sahabat Alam Malaysia, 67
Salim, Emil, 147
São Paulo Declaration of IPCC–AFOS
    workshop, 84
Sarawak, 18, 72, 115, 120, 168, 171
    ITTO Mission to, 72, 115, 168,
        171
Sargent, Caroline, 79
Sayer, Jeff, 86
Schmidt, Helmut, 147
Schücking, Hewfa, 11
scientific forest management, 138, 145,
    147–9
Second Account of the Common Fund
    for Commodities, 62
    relations with the ITTO, 62
Second Ministerial Conference on the
    Protection of Forests in Europe
    (1993), 107, 115–18, 136, 139
    follow-up process, *see* Helsinki
        process
    organization of, 139
Second Ministerial Meeting of Develop-
    ing Countries on Environment and
    Development, *see* Kuala Lumpur
    Declaration on Environment and
    Development (1992)
Second Regional African Ministerial
    Conference on Environment and
    Development (1991), 99–100, 170
Second World Climate Conference, 16,
    85–6, 88, 90
    conference statement of, 16, 85–6
    ministerial declaration of, 85–6
    NGO statement at, 85–6
Sendero Luminoso (Shining Path), 8
Senegal, 41, 154
    support for a global forests
        instrument, 154
Serra-Vega, José, 13
Shell, 6
    tree plantation guidelines (with
        WWF), 6
shifted cultivators, 9–10, 13

shifting cultivators, 9
Siberia, logging of, 1, 107
sic utere tuo ut alienum non laedas, 25
Sierra Club, 116
Singapore, 74
Slovakia, 54
South African Development
    Coordination Conference, 13
sovereignty over forest resources,
    24–6, 48, 50, 87, 93, 99–100,
    102–3, 132, 137, 141, 146, 158,
    161, 163, 166–71
Sri Lanka, 100
Stettin Bay Lumber Company, 7
stewardship, 87, 93–5, 97, 98, 103, 160,
    163–5, 170
    and the Svalbard Archipelago, 165
Stone Container Corporation, 7
    agreement with government of
        Honduras, 7
Strong, Maurice, 25, 158
Structural Adjustment Programmes
    (SAPs), 5, 9, 12
Sullivan, Francis, 61
Summit Level Group of Developing
    Countries (G15), 89, 99, 170
    Caracas meeting of (1991), 89, 99,
        170
Surinam, 41, 141
Survival International, 9, 62
    observer status at ITTO, 62
sustainable development, 21, 67–8, 83,
    89, 97, 107, 118, 126, 137–8, 140,
    143, 152–3
    Brundtland definition of, 21
    FAO definition of, 67–8
sustainable forest management, 21,
    66–9, 72–3, 115–18, 119, 120–1,
    136–45, 148–52, 154, 163–4
    criteria and indicators for, 67–8, 115,
        118, 136, 137, 138–45, 148–51,
        154
    Helsinki process definition of, 150
    ITTO definition of, 67–8, 72
    problems in defining, 67–9, 72, 148–9
sustainable timber production, 67
Sweden, 35, 40, 43, 58, 59, 114–15
    and Helsinki Ministerial Conference
        for the Protection of Forests in
        Europe, 115
    and negotiation of ITTA (1994), 114

Switzerland, 35, 40, 59, 153
  support for a global forests
    convention, 153

Tanzania, 11, 39
Target 2000, 68–72, 79, 109–11, 115–
  17, 119–21, 125–6, 128–9, 139,
  158–9, 164
  and negotiation of ITTA (1994), 69,
    109–11, 115–17, 119–21, 125–6,
    128–9
  and sustainable management of forest
    resources protocol to Lomé IV
    Convention, 154
  Australia, 119, 121
  Brazil, 164
  Canada, 119, 121
  change of name to Objective 2000,
    69, 119, 129
  controversy over definition of, 69, 79,
    109–10, 129
  European Community/European
    Union, 111, 115, 120–1
  Indonesia, 164
  ITTO consumers' statement on, 121
  Japan and, 121
  Netherlands, 115, 139
  origins of, 69
  producers' claims for consumers to
    meet 'full incremental costs' of,
    120, 159, 164
  references to in ITTA (1994), 129
  Sweden, 115
  USA, 115–17, 121, 139
Technical Workshop to Explore Options
  for Global Forestry Management
  (Bangkok, 1991), 94–7
technology transfers, claims for by
  South, 91, 96–102, 105, 110, 114,
  118, 125–6, 137, 144–5, 153,
  158–9, 161, 163, 165–6
terms of trade, 98, 161
TFAP Forestry Advisers Group, *see*
  Forestry Advisers Group
Thailand, 10, 20, 41, 59, 74, 94, 133
  logging ban, 133
Thiesenhusen, William, 14
timber labelling, 72–5, 77, 81, 125, 133,
  145, 149–50, 152, 158–9
Timber Research and Development
  Association (TRADA) (UK), 73

timber trade, 6–8, 10, 17, 61–3, 68–78,
  106, 109–11, 115, 132, 149–52
  illegal trade, 7–8, 10, 110–11
  in Brazil, 8, 60; Ghana, 8; Malaysia,
    7–8, 60, 73, 132; Philippines, 8,
    11
  malpractices in, 7–8
  representation at the FSC, 74, 149; at
    the ITTO, 61
Timber Trade Federation (UK), 69
  relationship with WWF, 69
Ting Wen Lian, 91, 94, 96, 162
Töpfer, Klaus, 101–2
TRAFFIC, 7, 77, 81
  lobbies ITTO on endangered species,
    81
tragedy of the commons, 26–7
Trail Smelter Arbitration on
  transboundary air pollution, 25
transnational corporations (TNCs), 2,
  4–8, 17, 22–3, 167–8
Trans-Sumatra Highway, 9
transfer pricing, 7–8
  in Malaysia, 8
  in Papua New Guinea, 7
  in Philippines, 8
  role of Japanese companies in, 8
Tropical Forestry Action Plan/
  Programme (TFAP), 31–54, 75,
  78–80, 83–5, 88, 102, 108, 124,
  146, 158, 166, 168–71
  and FAO Ad Hoc Group, 49–50, 54,
    170; FAO Council, 35, 47–52;
    FAO's Committee on Forest
    Development in the Tropics
    (CFDT), 35–6, 51, 79; FAO's
    Committee on Forestry, 32, 35,
    46
  and Forestry Advisers Group, 34–7,
    42, 51–2, 79
  'Bellagio II' meeting (1988), 34; Bella-
    gio Strategy Meeting (1987), 34
  consultative group debate, 46–51,
    108, 158, 168, 171
  creation of, 31–2
  five action programmes of, 31–4, 36,
    78
  Geneva meeting of experts (1991),
    46–8, 50, 168–9
  independent review of, 42–6, 84, 88,
    124, 146

institutional structure of, 34, 44
legitimacy crisis, 42–5, 168
New York meeting of cosponsors
(1991), 47–8
NFAP Support Unit, 35–6, 42, 53–4
Paris consultation, 47, 48–9
regional exercises, 52
relations with ITTO, 75, 78–81,
restructuring process of, 46–54, 83,
108, 166
World Rainforest Movement review
of, 43–5, 47
World Resources Institute review of,
43–4, 46–7, 84, 86

Ullsten, Ola, 43, 124, 146–7
and independent review of TFAP, 43,
124, 146
and World Committee on Forests and
Sustainable Development, 124,
146
UNCED Statement of Forest Principles,
89–103, 118, 125–7, 153–4, 163,
169–70
contents of, 102–3, 153, 163, 169
negotiation of, *see* next entry
UNCED Statement of Forest Principles,
negotiation of, 97–102, 106, 122–3,
125, 135–6, 138, 141, 144
Ad Hoc Sub Group on Forests, *see*
Contact Group on Forests
burden sharing, 87, 93, 103
common but differentiated
responsibility, 96–8
common responsibility, 93–5, 97–9,
103, 163
compensation for opportunity cost
foregone, 91, 97–100
Contact Group on Forests, 91–4, 97,
100–1
decision 1/14, 91; decision 2/13, 92–4
declining terms of trade, 98, 161
document A/CONF.151/PC/65, 93–5,
97, 99
'Draft Synoptic List', 91–3
external indebtedness, 91, 95–6,
98–100, 102, 171
FAO and, 88, 93, 163
financial transfers issue, 91, 95–101,
105, 110, 158–9

first PrepCom for (1990), 86, 90–1;
second PrepCom for (1991), 88,
91–3, 95–7, 99; third PrepCom
for (1991), 88, 93–7, 99–100;
fourth PrepCom for (1992),
99–101, 167
G77 and, 89, 94–102, 122
global commons, 25, 95
partnership in additionality, 97–8
stewardship, 93–4, 97–8, 103, 160,
163, 170
technology transfers issue, 91,
96–102, 105, 110, 158–9
Working Party on Forests of, 91–3
United Nations Centre on Transnational
Corporations, 4
United Nations Commission on Sustainable Development, 51–2, 89, 117,
127, 132, 135–8, 144–5, 147, 149,
152–5, 164–5
Ad Hoc Working Group on Sectoral
Issues, 144
Intergovernmental Panel on Forests
of, *see* separate entry
third session of (1995), 135, 144–5,
147, 155, 164–5
United Nations Conference on Environment and Development, Rio de
Janeiro, 1992 (UNCED), 11, 19, 21,
25, 27, 52–4, 49, 68, 83, 86–9,
91–102, 105–6, 109–10, 115–16,
118, 122–3, 125–7, 130, 135–6,
138, 141, 145–7, 152, 154, 156–63,
165, 167, 169–71
Agenda 21, 51, 53–4, 88, 97, 100,
138, 144, 167
negotiation of 'Non-legally binding
authoritative statement of principles for a global consensus on
the management, conservation
and sustainable development of
all types of forests', *see* UNCED
Statement of Forest Principles
NGO Global Forum, 169
organization of conference, 101
organization of preparatory
committee, 88
Rio Declaration on Environment and
Development of, 89, 164, 170
secretariat of, 88, 91–3, 97, 165

United Nations Conference on the
Human Environment, Stockholm,
1972, 24, 102, 170
Principle 21 of the Stockholm
Declaration of the, 24–5, 28, 98,
102, 170
United Nations Conference on Trade
and Development, 55–6, 63–4, 105,
111, 114, 126, 164
and negotiation of ITTA (1983),
55–7, 63, 105; (1994), 105, 111,
114, 164
fourth session of (1976), 55; eighth
session of (1992), 114, 127
Resolution 93(IV) of, 55
United Nations Development Programme (UNDP), 12, 32–5, 38, 40,
47–8, 51, 53, 78, 98
and TFAP, 32–5, 38, 40, 47, 48, 51,
53
Capacity 21 programme, 53
United Nations Educational, Scientific
and Cultural Organization
(UNESCO), 7, 16, 32, 35
United Nations Environment Programme, 2, 16, 32, 40, 51, 57, 77,
80, 89, 98, 145
exercises catalytic mandate with
respect to creation of TFAP, 32,
57; with respect to ratification of
ITTA (1983), 57
United Nations Fund for Population
Activities (UNFPA), 12, 40
United Nations General Assembly, 19,
22, 25, 28, 45, 83, 146
campaign for emergency session on
tropical deforestation of, 19, 45
Resolution 44/228 (1989), 83
United States of America, 10, 25, 40,
59, 61, 76, 85, 91, 94, 97, 106, 110,
114–18, 121, 140–1, 149, 153, 156,
162
and negotiation of ITTA (1994), 106,
114
and Target 2000, 115–17, 121
and UNCED forests negotiations, 91,
94, 97, 101
symposium, 'Forests and the Environment: A US Response to the Rio
Earth Summit', 107, 116–17

United States Senate Committee on
Foreign Relations, 45
criticism of TFAP, 45
unsustainable patterns of production
and consumption, 95, 102, 171
USAID, 10
Utting, Peter, 10, 67

Venezuela, 19, 20, 41, 99, 122, 141,
170
veto coalitions, 157–9, 161–2
Vietnam, 10, 20, 39, 41–2
Vogler, John, 25

WALHI (Indonesian Environmental
Forum), 46
and TFAP, 46
wars, 8, 10–11; Cambodia, 10; Central
America, 11; El Salvador, 10;
Myanmar, 10; Peru, 8; Rwanda, 11;
Vietnam, 10
Weinberg, Bill, 11
Westoby, Jack, 9
Witte, John,
woodfuel, 12–13, 17
Woods Hole Research Center, 146
Working Group on Criteria and Indi-
cators for the Conservation and
Sustainable Development of
Temperate and Boreal Forests, *see*
Montreal process
World Bank, 3, 5, 10, 16, 32, 34–5, 38,
40, 43, 45, 47, 51, 78–9, 98
and TFAP, 32, 34–5, 38, 40, 43, 45,
47, 51
Environment Department of, 3
Forest Sector Policy Paper, 3
World Commission on Environment and
Development, 4, 21, 23, 25
World Commission on Forests and
Sustainable Development, 52, 124,
136, 146–7, 152, 169
mandate of, 147
organization of, 147
receives support from Forestry
Advisers Group, 52, 146
World Conservation Monitoring Centre,
80
ITTO study on endangered species, 80
World Conservation Strategy, 21, 77

World Conservation Union, *see* International Union for the Conservation of Nature and Natural Resources (IUCN)
World Forestry Congress, 32, 47, 48, 92
  ninth congress (Mexico, 1985), 32
  tenth congress (Paris, 1991), 47, 48, 92
World Rainforest Movement, 14, 24, 27, 43–8, 71–2
  calls for TFAP funding moratorium, 44
  criticism of ITTO Mission to Sarawak, 72
  review of TFAP, 43–5, 47
  views on land reform, 14
  vision for a TFAP Consultative Group, 48
World Resources Institute, 16, 32–5, 38–9, 43–49, 51, 79, 84, 141
  and creation of TFAP, 32
  and Forestry Advisers Group, 35
  and National Forestry Action Plans, 38–9
  ceases funding TFAP activities, 45, 48
  proposal for a TFAP consultative group, 44, 46, 48–9
  review of TFAP, 43–4, 46–7, 84, 86
World Trade Organization, 77–8, 81, 133
World Wide Fund for Nature (WWF), 6, 16, 35, 38, 42, 45, 53, 61, 74–7, 79–80, 81, 133, 149, 167
  1995 Group, 68–9, 115
  1995 Plus Group, 69

  1995 target, 68–9
  and creation of Forest Stewardship Council, 74–5, 149, 158
  and debt-for-nature swaps, 4–5
  and Forestry Advisers Group, 35
  and negotiation of ITTA (1994), 111, 114–15, 118
  and TFAP, 35, 38, 42, 45, 47, 53, 80
  cooperation with UK Timber Trades Federation, 69
  criticism of ITTO Mission to Sarawak, 72, 115
  lobbies President Clinton, 116
  recommends ITTO to seek waiver from some GATT clauses, 77
  representation on national delegations to ITTO, 61
  research on temperate deforestation, 111
  timber labelling policy, 74–5
  tree plantation guidelines (with Shell), 6
  withdraws officers from national delegations to ITTO, 61, 106
World Wildlife Fund, *see* World Wide Fund for Nature

Young, Oran R, 159–60, 163, 165
Yudelman, Montague, 43

Zacher, Mark W, 161
Zaire, 11, 39, 42
Zentilli, Bernardo, 165

Other relevant publications from Earthscan

The Forest Certification Handbook
Christopher Upton and Stephen Bass

The first book to provide a thorough explanation and practical guide to what is involved in the certification of forests and timber products. Using case studies to illustrate all aspects of the process, the book systematically explains the components of a certification programme, the standards employed and the steps to achieving certification and maintaining quality management. It includes comprehensive directories of certification agencies, certified forests, consultancies and sources of further information.

£19.95 paperback ISBN 1 85383 222 7

Bad Harvest: The Timber Trade and the Degradation of the World's Forests
Nigel Dudley, Jean-Paul Jeanrenaud and Francis Sullivan

Based on 15 years of extensive research, the book presents an incisive account of the role that the timber trade had played in the loss and degradation of forests around the world. It examines the environmental consequences of the trade on boreal, temporal and tropical regions, and its impact for local people living and working in the forests. It also looks at the changing nature of the trade, and assesses current national and international initiatives to address the impact of deforestation. Finally, the authors describe some of the existing solutions, and show how things could be still further improved by presenting a new strategy for sustainable forests.

£12.95 paperback ISBN 1 85383 188 3

The Greening of Machiavelli: The Evolution of International Environmental Politics
Tony Brenton

A masterful survey of 25 years of international negotiations over the environment, by an author who was directly involved in much of it. The major conferences and agreements are analysed and the lessons for the future discussed.

£14.95 paperback ISBN 1 85383 211 1 £29.95 hardback ISBN 1 85383 214 6

For details of these and other publications, please contact:
Earthscan Publications Ltd
120 Pentonville Road, London N1 9JN
Tel: 0171 278 0433
Fax: 0171 278 1142
email: earthinfo@earthscan.co.uk
http://www.earthscan.co.uk